TABLE OF ATOMIC WEIGHTS—1961
(BASED ON CARBON-12)

Element	Symbol	Atomic Number	Atomic Weight	Element	Symbol	Atomic Number	Atomic Weight
Actinium	Ac	89		Mercury	Hg	80	200.59
Aluminum	Al	13	26.9815	Molybdenum	Mo	42	95.94
Americium	Am	95		Neodymium	Nd	60	144.24
Antimony	Sb	51	121.75	Neon	Ne	10	20.183
Argon	Ar	18	39.948	Neptunium	Np	93	
Arsenic	As	33	74.9216	Nickel	Ni	28	58.71
Astatine	At	85		Niobium	Nb	41	92.906
Barium	Ba	56	137.34	Nitrogen	N	7	14.0067
Berkelium	Bk	97		Nobelium	No	102	
Beryllium	Be	4	9.0122	Osmium	Os	76	190.2
Bismuth	Bi	83	208.980	Oxygen	O	8	15.9994a
Boron	B	5	10.811a	Palladium	Pd	46	106.4
Bromine	Br	35	79.909b	Phosphorus	P	15	30.9738
Cadmium	Cd	48	112.40	Platinum	Pt	78	195.09
Calcium	Ca	20	40.08	Plutonium	Pu	94	
Californium	Cf	98		Polonium	Po	84	
Carbon	C	6	12.01115a	Potassium	K	19	39.102
Cerium	Ce	58	140.12	Praseodymium	Pr	59	140.907
Cesium	Cs	55	132.905	Promethium	Pm	61	
Chlorine	Cl	17	35.453b	Protactinium	Pa	91	
Chromium	Cr	24	51.996b	Radium	Ra	88	
Cobalt	Co	27	58.9332	Radon	Rn	86	
Copper	Cu	29	63.54	Rhenium	Re	75	186.2
Curium	Cm	96		Rhodium	Rh	45	102.905
Dysprosium	Dy	66	162.50	Rubidium	Rb	37	85.47
Einsteinium	Es	99		Ruthenium	Ru	44	101.07
Erbium	Er	68	167.26	Samarium	Sm	62	150.35
Europium	Eu	63	151.96	Scandium	Sc	21	44.956
Fermium	Fm	100		Selenium	Se	34	78.96
Fluorine	F	9	18.9984	Silicon	Si	14	28.086a
Francium	Fr	87		Silver	Ag	47	107.870b
Gadolinium	Gd	64	157.25	Sodium	Na	11	22.9898
Gallium	Ga	31	69.72	Strontium	Sr	38	87.62
Germanium	Ge	32	72.59	Sulfur	S	16	32.064a
Gold	Au	79	196.967	Tantalum	Ta	73	180.948
Hafnium	Hf	72	178.49	Technetium	Tc	43	
Helium	He	2	4.0026	Tellurium	Te	52	127.60
Holmium	Ho	67	164.930	Terbium	Tb	65	158.924
Hydrogen	H	1	1.00797a	Thallium	Tl	81	204.37
Indium	In	49	114.82	Thorium	Th	90	232.038
Iodine	I	53	126.9044	Thulium	Tm	69	168.934
Iridium	Ir	77	192.2	Tin	Sn	50	118.69
Iron	Fe	26	55.847b	Titanium	Ti	22	47.90
Krypton	Kr	36	83.80	Tungsten	W	74	183.85
Lanthanum	La	57	138.91	Uranium	U	92	238.03
Lead	Pb	82	207.19	Vanadium	V	23	50.942
Lithium	Li	3	6.939	Xenon	Xe	54	131.30
Lutetium	Lu	71	174.97	Ytterbium	Yb	70	173.04
Magnesium	Mg	12	24.312	Yttrium	Y	39	88.905
Manganese	Mn	25	54.9381	Zinc	Zn	30	65.37
Mendelevium	Md	101		Zirconium	Zr	40	91.22

aThe atomic weight varies because of natural variations in the isotopic composition of the element. The observed ranges are boron, ±0.003; carbon, ±0.00005; hydrogen, ±0.00001; oxygen, ±0.0001; silicon, ±0.001; sulfur, ±0.003.

bThe atomic weight is believed to have an experimental uncertainty of the following magnitude: bromine, ±0.002; chlorine, ±0.001; chromium, ±0.001; iron, ±0.003; silver, ±0.003. For other elements the last digit given is believed to be reliable to ±0.5.

Printed by permission of the International Union of Pure and Applied Chemistry and Butterworths Scientific Publications.

CHEMICAL CALCULATIONS

an introduction to the use of
mathematics in chemistry

CHEMICAL
CALCULATIONS

———

an introduction to the use of mathematics in chemistry

———

SIDNEY W. BENSON

*Chairman, Department of Thermochemistry
and Chemical Kinetics
Stanford Research Institute*

———

Third Edition

John Wiley & Sons, Inc., New York·London·Sydney·Toronto

PREFACE

The continued upgrading of the elementary chemistry curriculum by the incorporation of more physical chemistry has prompted me to extend the sections of this book on thermochemistry (new chapter), rate constants, and titration. Other changes are comprised of newer units (torr instead of mm Hg) and a more consistent use of significant figures. The student who is weak on mathematical operations is urged to review sections 1 to 4 of Appendix I, before proceeding. Appendices II and III provide a quick summary of descriptive chemistry for students desiring it.

The present revision is a result of the suggestions of many colleagues in different parts of the United States for changes to fit their varying needs. Some requests were contradictory, and I have tried to strike the best compromise. I am indebted to all of my colleagues for their kind assistance.

SIDNEY W. BENSON

PREFACE TO THE SECOND EDITION

It is now twelve years since the preparation of the first edition of this book. In that time, the conversion factor approach to chemical calculations has had an extremely gratifying success. For us at the University of Southern California, the result has been a liberation of teaching and discussion time from the once tedious but elementary problems of stoichiometry to the more sophisticated problems in chemical equilibria and the more detailed consideration of molecular chemistry. This has in turn generated a demand for the revisions incorporated in the present book. They include a larger number and variety of problems, with more weight on "difficult" problems. In addition, my experience in developing the subject of competing equilibria has led to a complete rewriting of the sections dealing with acid-base phenomena and complex ionic equilibria, particularly buffers. I believe that the present approach makes for an easier understanding of the subtleties of hydrolysis and related systems of competing equilibria.

In addition, I have added some sections on descriptive material in the Appendix. These have turned out to be quite useful in my own classes, and I hope they will help to make somewhat more palatable the large doses of descriptive material which are still after all the subject matter of chemistry.

Finally, I should like to extend my thanks to the numerous colleagues and readers who have been interested enough to write to me about the various errors and obscurities in the first printings. Their observations have been duly recorded in the present version and have been greatly appreciated.

SIDNEY W. BENSON

November 1962

PREFACE TO THE FIRST EDITION

The student beginning his studies in chemistry today undertakes a prodigious labor. He must learn rapidly an almost entirely new language, replete with names, symbols, and rules of grammar. Speed is essential since he is expected to use this language in thinking about and memorizing a huge amount of loosely related and descriptive material. Finally, he must absorb the theoretical structure, expressed in a mathematical language, which provides a basis for unifying a great deal of the subject material.

Although I shall not pretend to say here which part of this gargantuan body of knowledge is important and which is not, one thing is certain. For the student who plans to continue the study of chemistry, it is essential that he obtain a thorough understanding of the way in which mathematical thinking is incorporated into chemistry. It is for these students and for those others who, though taking chemistry for only one year, may be expected to master chemical calculations that this book is written.

The neophyte finds his greatest difficulty usually with chemical problems. I trust that he will not be too surprised when I say that this is not due to his inadequate understanding of mathematics. The amount of mathematical understanding required for 80 per cent of first-year chemistry problems is microscopically small. The student's difficulty in doing problems arises not from a lack of mathematical background but rather from a lack of familiarity with the way in which chemists use mathematics. Most college students have, or soon acquire, quite enough mathematical understanding to pay for a "coke," borrow money from a friend, and place bets in a card game. If they had the same familiarity with chemical terms and units that they do with these more eclectic enterprises, they would show an equal agility with their homework and examinations.

In the present book an attempt is made to cope with this very real difficulty by presenting all terms, units, and definitions that will be employed mathematically, in the form of equations. This approach provides a unified method for attacking chemical problems. The student using this book is urged first to read Chapters I and II on measurement and chemical units. Although a good part of these chapters may seem self-evident and oversimplified, they nevertheless contain precisely that material which in my own experience usually has been needed to clarify the thinking of the student and deepen his understanding of chemical calculations. If the student is to learn to do chemical arithmetic, then he should do it logically and well.

This book represents the cumulative experience of many years of experimenting with the presentation of chemical arithmetic. To the many hundreds of my own students who have borne patiently, if not ecstatically, the various phases through which these pedagogical researches have passed, this book is dedicated.

I wish to express my appreciation to Miss Diane Frost, Secretary of the Chemistry Department, for the very generous assistance given in typing this manuscript. I also wish to express my thanks to Victoria Von Hagen, without whom this book would have been ready six months earlier.

Sidney W. Benson

Los Angeles
February, 1952

CONTENTS

VII. The Concept of Combining Power—Valence   **85**

VIII. Measurement of Solutions   **95**

CHAPTER I

Measurement

1. The Language of Chemistry

In this book we are going to be concerned with that part of chemistry which can be expressed in terms of numbers. To clarify our understanding of this area, something must first be said of its relation to the whole structure of the science.

One of the goals of chemistry is to make simple, compact statements about the properties of matter. These statements must be clear and, as far as possible, devoid of any ambiguity. For this purpose, the language of everyday conversation is inadequate. The statements of chemistry must be free from the kind of double meanings which might confound, for example, the conversation of the cannibal and the missionary's wife, each of whom can say, "I am fond of missionaries."

To achieve such clarity, chemists have had to construct a vocabulary in which each word is precisely defined. Thus one definition of an acid is: "A substance whose water solution will turn blue litmus red." Such a definition permits of a clear and simple test for an acid.

The beginning student of chemistry will find that a great deal of his difficulty arises from a lack of appreciation of the fact that he is using a precise language, and that the usual liberties taken with our everyday speech are not permitted. A vague understanding of a word, which may suffice for the general intent of an ordinary statement, may lead to a total block in understanding a chemical statement. To obtain a genuine mastery, the student must constantly subject his study to the self-criticism of definition: "What does this word mean?"

Such caution is doubly necessary when the statements are quantitative rather than qualitative. Quantitative statements are preferred in science since they are the most compact and the least subject to ambiguity. To say that lead has a density of 11.34 grams per cubic centimeter is much more precise and informative than to say that lead is a dense metal. Such compactness, however, is apt to be misleading. Though we can read it and say it more quickly, until we are quite familiar with the terms used, it actually calls for far more thought and reflection than the almost equally brief but qualitative statement. The difference will be found in the loose, qualitative idea conveyed by the word "dense," as opposed to the precise meaning contained in the word "density."

2. Measurement

Chemistry, like most natural science, starts by making qualitative or descriptive statements about nature. Thus, "copper sulfate is a blue, crystalline material that is soluble in water." How does it proceed to the quantitative? What is required in order that we shall be able to make statements in the language of mathematics? The answer lies in what we mean by measurement.

We may observe that a certain stick is long. We may measure its length with a ruler and find it to be 14.5 inches.

A property such as length may be expressed quantitatively, that is, it may be measured, if:

1. We can define the property precisely (e.g., length is the shortest distances between two points).
2. We have a standard object with which we can compare our object to be measured (e.g., a ruler).
3. We have a means of making the numerical comparison between the property of the standard object (the ruler) and the object whose property we wish to measure (the stick). This will consist in placing the two objects side by side and reading, from the divisions on the ruler, the length of the stick.

To summarize, then, a property can be expressed or measured quantitatively if three conditions are satisfied: 1. *precise definition;* 2. *a standard;* 3. *a means of comparison.*

3. Standards

Not all properties are capable of being expressed in numerical fashion. Taste and smell, which are certainly distinctive properties of a great many substances, have so far defied attempts to reduce them to exact measurement.

Of the large number of other properties which have been reduced to quantitative measurement, however, a few have come to be considered as basic. Among these are mass, length, time, and temperature. It would be most agreeable if a single set of standards were adopted for the measurement of these quantities. Unfortunately, such is not the case, and there are at present two different sets of standards in common usage. One of them is known as the English system of units, and the other as the metric system. Since the metric system is the simpler, being based on a decimal system of subdivision, it is the one which is universally employed by pure scientists. The English system finds its adherents principally among engineers (in 1971, American only).

Table I indicates the definitions, units, and means of measurement employed.

TABLE I

BASIC STANDARDS FOR MEASUREMENT

| Quantity | Definition | Standard | | Common Device for Measurement |
		English	Metric	
Mass	Quantity of matter. Resistance offered by a body to a change in its velocity.	Pound (lb.)	Kilogram (kg.)	Balance
Length	Shortest distance between two points.	Foot (ft.)	Meter (m.)	Straightedge ruler
Time	Interval between two events.	Second (sec.)	Second (sec.)	Clock
Temperature	Degree of hotness or coldness which determines the flow of heat from one body to another.	Degree Fahrenheit (°F.)	Degree centigrade (°C.)	Thermometer

The actual systems which have been adopted are quite arbitrary. Thus there is a lump of platinum in the Bureau of Standards at Washington, D.C., which has a mass of 1 pound. This has become the primary standard of mass against which all other standards have been compared by means of a balance. A larger or smaller lump of platinum could equally well have been chosen, and the same applies for the standards of length, time, and temperature.

We can observe that each basic quantity has associated with it a standard of measurement and a unit designating the standard itself. Our principle of measurement is based on the assumption that we can so design our measuring apparatus (e.g., the ruler) that we can express a given property such as length in multiples or fractions of the standard unit. Thus one object may be 20 meters long; another object may be only $\frac{1}{25}$ (0.040) meter long. Without the ability to make these numerical comparisons, exact definitions and standards are of no avail for quantitative work. Thus we may take a sample of perfume to represent the standard scent of, perhaps, orange blossom. Until we can then devise an instrument or method for comparing some other sample with our standard perfume in such a way that the observer can say, "This is twice as 'odoriferous' as the standard," we will not be able to measure orange blossom quantitatively.

4. Units

It is frequently very clumsy or inconvenient to use the standard units directly. Thus a biologist measuring the diameter of a blood cell under the microscope will find it awkward to have to express his results as 0.0000075 meter. Similarly, a surveyor will not want to write his distances on maps as, perhaps, 1,432,000 meters. To obviate such inconveniences, subsidiary sets of units have been adopted which are exactly defined by means of algebraic equations in terms of the major units. In the metric system, these auxiliary units are defined by attaching appropriate prefixes to the standard unit. Some common prefixes are shown in Table II.

TABLE II
SOME AUXILIARY METRIC UNITS (PREFIXES)

Prefix	Definition	Prefix	Definition
Tera-	1×10^{12}	Centi-	1×10^{-2}
Giga-	1×10^{9}	Milli-	1×10^{-3}
Mega-	1×10^{6}	Micro-	1×10^{-6}
Kilo-	1000	Nano-	1×10^{-9}
Deka-	10	Pico-	1×10^{-12}

In terms of these units the diameter of the blood cell can now be written as 7.5 micrometers (generally shortened to microns) and our surveyor's distance as 1432 kilometers, or 1.432 megameters. (The reader can ponder on the savings in printers' ink and paper, if the Treasury Department, newspapers, magazines, etc., would refer to items such as our national debt in terms of megabucks, kilobucks, etc.)

It is unfortunate that the list of metric prefixes is so limited, since science is now measuring distances as small as 0.000000000001 meter and as large as 10^{20} meters. Because of this large range we will frequently find other auxiliary units defined. The angstrom unit Å is now in quite common usage in the discussion of sizes of atoms and molecules. It is defined as one-ten-billionth of a meter:

$$1 \text{ angstrom} = 0.0000000001 \text{ meter} = 1 \times 10^{-10} \text{ meter} \quad \text{(definition)}$$

With the development of supersensitive balances it is also quite common to hear talk of the gamma, defined as

$$1 \text{ gamma} = 0.000001 \text{ gram} = 1 \times 10^{-6} \text{ gram} \quad \text{(definition)}$$

In the Table of Common Units (Appendix II) the student will find a more complete list of the metric units in common usage.

Do problems 1 and 2 at the end of the chapter.

5. Relations between Different Systems of Measurement

The existence of two such different systems as the English and the metric for measuring the same property requires some means for translation from one system to the other. This can be provided only by direct experiment. That is, we must take, for example, the pound and measure its mass in kilograms, or vice versa. Careful measurements have been made for all the different systems of units. A few of the results obtained are indicated by the following equations.

$$1 \text{ meter} = 3.2808 \text{ feet} \quad \text{(experiment)}$$

$$1 \text{ kilogram} = 2.2046 \text{ pounds} \quad \text{(experiment)}$$

As will be seen shortly, one such relation enables us to relate any auxiliary unit in one system to any auxiliary unit in the other. By means of the above equation for the relation between lengths in the metric and English systems, we can show that 1 inch = 2.540 centimeters, 1 millimeter = 0.03937 inch, etc.

It is important to observe the distinction between the equations given above for relating similar properties in two different systems, which must be determined experimentally, and the equations given in section 4, which relate auxiliary units to standard units and which are mathematically exact.

6. Intensive and Extensive Properties

Properties that depend on the quantity of matter being measured are known as *extensive* properties. Mass, length, and volume are examples. If we take twice as much matter of a given substance, it will contain twice as much volume and twice as much mass.

Properties that do not depend on the quantity of matter being measured are called *intensive* properties. They are important properties because they often depend only on the nature of the material being measured and not on the quantity involved. Density and temperature are examples of *intensive* properties. At a given temperature the density of water is the same whether we measure the density of one drop of water or one quart, namely, 1 gram per cubic centimeter. This is then a property of water itself and can be used to identify water, since in general all pure liquids will have different densities. Studies of intensive properties may throw great light on the fundamental nature of substances. We shall observe this in many of the quantitative laws.

Do problem 3 at the end of the chapter.

7. Conversion of Units—Conversion Factors

When properties have been measured in one set of units, it frequently is necessary to know the property in a different set of units. This translation from one set of units into a different set of units may be performed mathematically and is known as a conversion.

Conversions can be performed only if we have an equation relating the two units in question.

Example: Mass: What is the mass in pounds of an object whose mass is 900 g.?
Answer: The two units in question are *pounds* and *grams*, and the fundamental equation relating them is

$$1 \text{ lb.} = 454 \text{ g.} \qquad \text{(see Table of Units)}$$

The answer is then

$$\frac{900}{454} = 1.98 \text{ lb.} \qquad \text{(see significant figures in Appendix)}$$

From the preceding example we have seen that equations relating different units can be used to convert quantities expressed in one of these units to the other unit. However, it is not always obvious when to multiply and when to divide in such conversions. This difficulty is completely avoided if we put units in our mathematical operations and cancel and multiply them as if they were numbers. Conversions of units are most conveniently performed by means of what are known as conversion factors.

Definition: A conversion factor is a numerical ratio of units which is equal to the pure number 1.

Example: The ratio $\left(\dfrac{454 \text{ g.}}{1 \text{ lb.}}\right) = 1$; also: $\left(\dfrac{1 \text{ lb.}}{454 \text{ g.}}\right) = 1$. These relations followed directly from our fundamental equation:

$$454 \text{ g.} = 1 \text{ lb.}$$

If we divide both sides of this equation by the quantity 1 lb., we obtain the first conversion factor. If we divide both sides by 454 g., we obtain the second conversion factor. These operations are valid since equals divided by equals remain equal.

$$\frac{454 \text{ g.}}{1 \text{ lb.}} = \frac{1 \text{ lb.}}{1 \text{ lb.}} = 1; \quad \text{or} \quad \frac{1 \text{ lb.}}{454 \text{ g.}} = \frac{454 \text{ g.}}{454 \text{ g.}} = 1$$

By similar treatment every equation involving units can be made to give conversion factors. The usefulness of these conversion factors lies in the fact that they are equal to 1. We can multiply or divide any quantity by a conversion factor and the result will still be equal to the original quantity. This follows since multiplication or division by the number 1 does not change the value of the original quantity. When this is done, so that the units of the original quantity are canceled by some units in the conversion factors, we have performed a conversion of units.

Example:

$$2 \text{ lb.} = 2 \text{ lb.} \times \left(\frac{454 \text{ g.}}{1 \text{ lb.}}\right) = 908 \text{ g.} \rightarrow 900 \text{ g.}$$
$$\text{significant figures}$$

We see in the above example that by multiplying the quantity 2 lb. by the conversion factor relating pounds and grams we were able to convert from the units of pounds to grams. If we had multiplied by the second conversion factor (1 lb./454 g.) the units of pounds would not have been canceled and we would not have obtained an answer in grams. (The student should verify this.)

The following simple example is easily remembered and should serve as a guide in converting units.

Example: How many nickels are there in 6 quarters?
Answer: The fundamental equation is

$$1 \text{ quarter} = 5 \text{ nickels}$$

From this equation we want a conversion factor to cancel the units of quarters when we multiply it by 6 quarters. The conversion factor is

$$\left(\frac{5 \text{ nickels}}{1 \text{ quarter}}\right) = 1$$

Then

$$6 \text{ quarters} = 6 \text{ quarters} \times \left(\frac{5 \text{ nickels}}{1 \text{ quarter}}\right) = 30 \text{ nickels}$$

Summary

1. *In order to convert from one set of units to another we must first have an equation which gives a relation between these two units.** *(See Table of Units.)*

2. *By dividing one side of the equation by the other we obtain a conversion factor (equal to 1) which we can use to convert from one set of units to the other.*

3. *Since we can divide either side of an equation by the other, every equation will give two conversion factors. The one to use is the one that will cancel the units we wish to eliminate.*

Example: How many centimeters are there in 2 ft.?
Answer: The fundamental relations are

$$1 \text{ ft.} = 12 \text{ in.}$$

$$1 \text{ in.} = 2.54 \text{ cm.}$$

The conversion factors are

$$\left(\frac{12 \text{ in.}}{1 \text{ ft.}}\right) \quad \text{and} \quad \left(\frac{2.54 \text{ cm.}}{1 \text{ in.}}\right)$$

Then

$$2 \text{ ft.} = 2 \text{ ft.} \times \left(\frac{12 \text{ in.}}{1 \text{ ft.}}\right) \times \left(\frac{2.54 \text{ cm.}}{1 \text{ in.}}\right) = 61 \text{ cm.} \rightarrow 61 \text{ cm.}$$

Example: A sign says, "Rowboats for rent, two bits per hour." What will it cost in dollars to rent a rowboat for two weeks?
Answer: (To the student: write down all the fundamental equations involved.)

$$2 \text{ wk rental} = 2 \text{ wk rental} \times \left(\frac{7 \text{ days}}{1 \text{ wk}}\right) \times \left(\frac{24 \text{ hr.}}{1 \text{ day}}\right) \times \left(\frac{2 \text{ bits}}{1 \text{ hr.}}\right) \times \left(\frac{1 \text{ quarter}}{2 \text{ bits}}\right)$$
$$\times \left(\frac{1 \text{ dollar}}{4 \text{ quarters}}\right) = \frac{2 \times 7 \times 24 \times 2}{2 \times 4} \text{ dollars} = 84 \text{ dollars}$$

Note: The only way to achieve proficiency in problems is by practice. Most problems can be solved by using conversion factors, and with a little practice the answers can be written down on inspection. From the unit equations in the Table of Units, practice writing conversion factors and then obtain relations between units which are not given.

* Whenever a problem involving conversion of units is given, the first step is always to seek for the fundamental relation *or relations* which equate the units concerned. (More than one relation may be involved.)

The student will observe that every one of the quantities in parentheses is a conversion factor and numerically equal to unity. Multiplying a great many of these still preserves the value of the original quantity.

Example: A car is moving at a rate of 30 miles/hr. What is this in centimeters per second?

Answer: (To the student: Write down all the fundamental equations involved.)

$$\frac{30 \text{ miles}}{1 \text{ hr.}} = \frac{30 \text{ miles}}{1 \text{ hr.}} \times \left(\frac{5280 \text{ ft.}}{1 \text{ mile}}\right) \times \left(\frac{12 \text{ in.}}{1 \text{ ft.}}\right) \times \left(\frac{2.54 \text{ cm.}}{1 \text{ in.}}\right) \times \left(\frac{1 \text{ hr.}}{60 \text{ min.}}\right)$$

$$\times \left(\frac{1 \text{ min.}}{60 \text{ sec.}}\right) = \left(\frac{30 \times 5280 \times 12 \times 2.54}{60 \times 60}\right) \frac{\text{cm.}}{\text{sec.}} = 1340 \text{ cm./sec.}$$

The method outlined above for the conversion of one set of units to another may at first seem unorthodox or too obvious to be worth the effort of going through. However, it represents a fundamental approach to such problems and, as we shall see, to most of the problems in chemistry. The experience of the author has been that, once students overcome their conservatism and unfamiliarity and master the method, they will gain an extraordinary facility in doing problems. The method permits a rapid solution of the problem with an instant check on basic errors. Answers to problems can be written down very quickly, units canceled, and finally all numbers gathered so that the arithmetic may be left for last.

For the students who find the arithmetical steps or notation troublesome, a section on simple algebraic definitions and manipulations is included in Appendix I.

The Table of Common Units in Appendix IV gives fundamental relations between some of the common units. However, from these it is possible to derive other fundamental relations by the previous method. Such relations can then be applied in future problems and so save work.

Example: Derive a relation between ounces (oz.) and milligrams (mg.), given 16 oz. = 1 lb.

Answer: Write the other relations needed:

$$1 \text{ oz.} = 1 \text{ oz.} \times \left(\frac{1 \text{ lb.}}{16 \text{ oz.}}\right) \times \left(\frac{454 \text{ g.}}{1 \text{ lb.}}\right) \times \left(\frac{1000 \text{ mg.}}{1 \text{ g.}}\right)$$

$$= 28,400 \text{ mg.}$$

Example: How many milligrams are there in 0.74 oz.?

Answer: This can be solved by usual methods.   However, if we take the relation from the previous example,

$$1 \text{ oz. } = 28{,}400 \text{ mg.}$$

we can use it to give a direct conversion:

$$0.74 \text{ oz. } = 0.74 \ \cancel{oz.} \ \times \left(\frac{28{,}400 \text{ mg.}}{1 \ \cancel{oz.}} \right) = 21{,}000 \text{ mg.}$$

The student may find it useful to make up tables of such additional relations. Handbooks such as the *Handbook of Chemistry and Physics* and Lange's *Chemical Handbook* will be found to contain many such tables.

Do problems 4–7 at the end of the chapter.

8. Properties Expressed in Complex Units

In the previous sections we have placed principal emphasis on those properties which can be expressed in terms of single, basic units (e.g., length, mass).   It is possible to define more complex properties having important physical meaning which are related to these basic properties.

The property of "area" is one example.   The area of a rectangle is defined as the product of its width and its height.   Expressed as an algebraic equation,

$$\text{Area} = \text{Width} \times \text{Height} \qquad \text{(definition)}$$

We can now inquire about the units in which area will be expressed. The width of a rectangle is a property expressed in the units of length. Similarly, the height of a rectangle is expressed in the units of length. If we multiply a unit of length by itself, as indicated in the above equation, we will obtain, following the notation of algebra, the units of length squared.

Example: What is the area of a rectangle whose width is 6 cm. and whose height is 3 cm.?

Answer:

$$\text{Area} = 6 \text{ cm. } \times 3 \text{ cm. (from above definition)}$$

$$= 18 \text{ cm.}^2 \text{ (or 18 sq. cm.)}$$

Proceeding from this definition, geometry has developed theorems which permit us to compute the areas of much more complex figures, such as triangles, circles, etc., in terms of their dimensions.   In each one the final result, the area, will be expressed as the square of the unit of length. How can we convert the units of a complex property, such as area, from one system to another? The answer is given by the procedure already outlined.

Example: Express the results of the preceding example (Area = 18 cm.^{2}) in square inches.

Answer: From our Table of Units we find

$$1 \text{ in.} = 2.54 \text{ cm.}$$

If we square both sides of this equation we do not disturb the equality:

$$(1 \text{ in.})^2 = (2.54 \text{ cm.})^2$$

$$= (2.54)^2 \text{ cm.}^2$$

$$1 \text{ in.}^2 = 6.45 \text{ cm.}^2$$

This last equation now provides us with a fundamental relation between square inches and square centimeters, and thus a conversion factor between the two sets of complex units. Hence

$$18 \text{ cm.}^2 = 18 \text{ cm.}^2 \times \left(\frac{1 \text{ in.}^2}{6.45 \text{ cm.}^2} \right)$$

$$= 2.79 \text{ in.}^2 \rightarrow 2.8 \text{ in.}^2$$

In a similar fashion we find that the property of volume is expressed as the cube of the unit of length (i.e., cm.^{3} or in.^{3} or ft.^{3}, etc.). The same methods can be applied to units of volume.

Example: The volume of a rectangular solid is given by the product of its length times its width times its height. What is the volume of a rectangular solid whose dimensions are 60 cm. by 8 cm. by 10 cm.? Express this in cubic feet.

Answer:

$$\text{Volume} = 60 \text{ cm.} \times 8 \text{ cm.} \times 10 \text{ cm.}$$

$$= 4800 \text{ cm.}^3 \text{ (i.e., cubic centimeters)}$$

$$= 4800 \text{ cm.}^3 \times \left(\frac{1 \text{ in.}}{2.54 \text{ cm.}} \right)^3 \times \left(\frac{1 \text{ ft.}}{12 \text{ in.}} \right)^3$$

$$= 4800 \text{ cm.}^3 \times \frac{\text{in.}^3}{(2.54)^3 \text{ cm.}^3} \times \frac{\text{ft.}^3}{1728 \text{ in.}^3} = \frac{4800 \text{ ft.}^3}{16.4 \times 1728}$$

$$= 0.169 \text{ ft.}^3 \rightarrow 0.17 \text{ ft.}^3$$

Some other important physical properties, together with their definition and units, are given in Table III. (A more complete list will be found in the Table of Units in Appendix IV.)

TABLE III
DEFINITION AND UNITS OF SOME COMPLEX PROPERTIES

Property	Verbal Definition	Algebraic Definition	Usual Units
Velocity (v)	Distance traveled divided by the time of travel.	$Velocity = \dfrac{Distance}{Time}$	cm./sec. ft./sec. miles/hr.
Kinetic energy $(K.E.)$	One-half of the mass of a body times the square of its velocity.	$K.E. = \dfrac{mv^2}{2}$	g.-cm.^{2}/sec.^{2} (1 g.-cm.^{2}/ sec.^{2} = 1 erg) lb.-ft.^{2}/sec.^{2}
Density (d)	The mass of a body divided by its volume.	$d = \dfrac{m}{v}$	g./cm.^{3} lb./ft.^{3}
Acceleration (a)	The amount by which the velocity of a body changes divided by the time during which it changes.	$a = \dfrac{Change\ in\ velocity}{Time}$	cm./sec.^{2} ft./sec.^{2}
Force (f)	The push or pull that is capable of changing the velocity of (i.e., accelerating) a body. More precisely, force equals the mass of a body times the acceleration imparted to it (Newton's Law).	$f = ma$	g.-cm./sec.^{2} $\left(1\ dyne = \dfrac{1\ g.\text{-}cm.}{sec.^2}\right)$ lb.-ft./sec.^{2} $\left(1\ poundal = \dfrac{1\ lb.\text{-}ft.}{sec.^2}\right)$
Pressure (P)	The force acting on an object divided by the area over which the force is exerted.	$P = \dfrac{f}{A}$	dynes/cm.^{2} $\left(\dfrac{1\ dyne}{cm.^2} = \dfrac{1\ g.}{cm.\text{-}sec.^2}\right)$ lb. force/in.^{2}

Note: There may be some confusion in the English system since units of pounds have sometimes been used for the measurement of mass and sometimes for the measurement of force. Proper usage is units of pounds for mass and units of *poundals* for force. When pounds are used to express force, it should be observed that 1 lb. force = 32.2 poundals force = 32.2 lb.-ft./sec.^{2}. Similarly, when grams are used to express force, 1 g. force = 980 dynes. In this book both pounds and grams will be used as units of mass unless explicitly designated otherwise.

As a final example of the conversion of units for these more complex properties we shall take the following:

Example: The density of metallic mercury is 13.55 g./cm.^{3} at 0°C. Convert this to pounds per cubic foot.

Answer: The student should write out the various fundamental equations and compute the conversion factors given as follows.

$$\frac{13.55 \text{ g.}}{\text{cm.}^3} = \left(\frac{13.55 \text{ g.}}{\text{cm.}^3}\right) \times \left(\frac{1 \text{ lb.}}{454 \text{ g.}}\right) \times \left(\frac{2.54 \text{ cm.}}{1 \text{ in.}}\right)^3 \times \left(\frac{12 \text{ in.}}{1 \text{ ft.}}\right)^3$$

$$= \left(\frac{13.55 \text{ g.}}{\text{cm.}^3}\right) \times \left(\frac{\text{lb.}}{454 \text{ g.}}\right) \times \left(\frac{16.4 \text{ cm.}^3}{\text{in.}^3}\right) \times \left(\frac{1728 \text{ in.}^3}{\text{ft.}^3}\right)$$

$$= \left(\frac{13.55 \times 16.4 \times 1728}{454}\right) \frac{\text{lb.}}{\text{ft.}^3}$$

$$= 846 \text{ lb./ft.}^3$$

Do problems 8–15 at the end of the chapter.

9. Relation between Different Properties

In the preceding section we have shown how complex properties may be defined in terms of simpler properties. We have been able to apply the laws of algebra to such defined properties because the definitions could be expressed in algebraic form. These definitions, given in the form of equations involving properties, can be treated as algebraic equations and "juggled" mathematically to give new equations.

Thus density is defined as the mass of an object divided by its volume. Expressed algebraically as an equation:

$$\text{Density} = \frac{\text{Mass}}{\text{Volume}}$$

If we multiply both sides of this equation by "Volume" we have:

$$\text{Density} \times \text{Volume} = \frac{\text{Mass}}{\text{Volume}} \times \text{Volume} = \text{Mass}$$

This is a new form of the original equation which tells us that multiplying the density of an object by its volume will give us the mass of the object. The student should verify the following relation which can be derived from the original equation:

$$\frac{\text{Mass}}{\text{Density}} = \text{Volume}$$

We observe that the original definition represents an algebraic relation between three properties: the mass, the volume, and the density of an object. The equations given above, which were derived from it, all represent the same relation but in different forms.

The original equation contained three properties. If we have numbers given for any two of the properties, it is always possible to calculate the numerical value of the third property from the equation. The derived equations merely represent convenient methods for doing this.

Example: The density of mercury (Hg) at 0°C. is 13.55 g./cm.^{3}.　What volume will be occupied by 20 g. of mercury?

Answer: From our definition of density:

$$\text{Density} = \frac{\text{Mass}}{\text{Volume}}$$

we can write the derived equation:

$$\text{Volume} = \frac{\text{Mass}}{\text{Density}}$$

On substitution of the above numbers,

$$\text{Volume} = \frac{20 \text{ g.}}{13.55 \text{ g./cm.}^3} = \frac{20 \text{ g.-cm.}^3}{13.55 \text{ g.}} \quad \text{(inversion of fractions)}$$

$$= 1.48 \text{ cm.}^3 \ \rightarrow \ 1.5 \text{ cm.}^3$$

Observe that in this example, through consistent and correct use of units, the volume is calculated not only as a number but in the proper units of cubic centimeters (cm.^{3}).　Had an error been made in the equation, it would have led to a result expressed in the wrong units. Thus the use of units provides an additional check on our method of calculation.

If it had been desired to express the answer in other units, conversion factors could now be applied to the present answer to change it.

10. An Interpretation and a General Method for Solving Problems

We can infer from the preceding discussion a general principle involved in calculations.　If most problems are analyzed we will see that they can be interpreted as follows:

1. A property or set of properties is given, and we are requested to calculate from these some other property or set of properties.
2. The problem may have the added complication that the data given may be expressed in one set of units, whereas the answer requested may be in a different set of units.

From this analysis of problems we can outline the following set of rules for solving the problem.

1. Write down the given data and the data requested in the answer. These will generally be properties of one kind or another.

2. The task of the problem is then to find a law, definition, or laboratory observation which relates the properties given to the properties requested. *This is the heart of the problem.*
3. The law, definition, or datum, once discovered, is expressed in algebraic form and rearranged to yield an equation which expresses the property sought for in terms of the properties given.
4. The numbers and *units* for the given data are now substituted in this derived equation and the appropriate arithmetic performed. Units are canceled and multiplied just as if they were numbers.
5. The answer obtained must have the proper units; otherwise a mistake has been made. If a different set of units is requested, appropriate conversion factors may be employed directly in the original equation or else in the final answer.

Example: The density of grain alcohol at 20°C. is 0.79 g./cm.^{3}. What is the mass of 250 cm.^{3} of alcohol?

Answer: The data (or properties) given are: density (0.79 g./cm.^{3}) and volume (250 cm.^{3}). We are requested to find *mass*. The relation between these properties is provided by the definition of density:

$$\text{Density} = \frac{\text{Mass}}{\text{Volume}}$$

Solving this for mass we find:

$$\text{Mass} = \text{Density} \times \text{Volume}$$

Substituting the proper numbers we find:

$$\text{Mass} = \frac{0.79 \text{ g.}}{\text{cm.}^3} \times 250 \text{ cm.}^3 = 198 \text{ g.}$$

Example: The density of water is 62.4 lb./ft.^{3} at 20°C. What is the volume in liters occupied by 800 g. of water?

Answer: The data (or properties) given are: density (62.4 lb./ft.^{3}) and mass (800 g.). We are requested to find volume (liters). Proceeding as formerly, we write the fundamental equation connecting these properties and rearrange to solve for volume.

$$\text{Volume} = \frac{\text{Mass}}{\text{Density}}$$

Substituting we have:

$$\text{Volume} = \frac{800 \text{ g.}}{62.4 \text{ lb./ft.}^3} = \frac{800 \text{ g.-ft.}^3}{62.4 \text{ lb.}} \quad \text{(by inversion of fraction containing units)}$$

We now apply the proper conversion factors to obtain the desired units.

$$\text{Volume} = \frac{800 \text{ g.-ft.}^3}{62.4 \text{ lb.}} \times \left(\frac{1 \text{ lb.}}{454 \text{ g.}}\right) \times \left(\frac{12 \text{ in.}}{1 \text{ ft.}}\right)^3 \times \left(\frac{2.54 \text{ cm.}}{1 \text{ in.}}\right)^3 \times \left(\frac{1 \text{ l.}}{1000 \text{ cm.}^3}\right)$$

$$= \frac{800 \times 1728 \times 16.4 \text{ ft.}^3}{62.4 \times 454 \times 1000} \times \frac{\text{in.}^3}{\text{ft.}^3} \times \frac{\text{cm.}^3}{\text{in.}^3} \times \frac{\text{l.}}{\text{cm.}^3}$$

$$= 0.800 \text{ l.}$$

Once facility is gained in doing such problems, the student will find that he can combine the substitution in the fundamental equation together with the conversion factors and write the answer as one single equation combining both.

Do problems 16 and 17 at the end of the chapter.

11. Property Conversion Factors

A definition or a law or a direct observation may be thought of as expressing a relation between two different kinds of properties. Thus the definition of density expresses a relation between the property of mass and the property of volume. The complex property of density can be thought of as a conversion factor which can be used to convert from the property of mass to the property of volume or vice versa.

When we say that the density of mercury is 13.55 g./cm.^{3} at 0°C., we are in effect saying that 13.55 g. of mercury are the same as 1 cm.^{3} of mercury. Algebraically:

$$1 \text{ cm.}^3 \text{ Hg} = 13.55 \text{ g. Hg}$$

That is, the same quantity of mercury (Hg) has the volume of 1 cm.^{3} and the mass of 13.55 g. Either property thus provides an equally valid way of measuring out a quantity of mercury. The above equation expresses an equivalence between these two different properties of mercury and can be used to obtain conversion factors between the volume occupied by mercury and the mass of this same quantity of mercury.

In a similar fashion other complex properties (like density) can be looked upon as expressing an equivalence between two different sets of properties of a substance, and the algebraic relation stating this equivalence can be used to provide conversion factors for going from one of these sets of properties to the other. Such conversion factors are not the same as the conversion factors connecting sets of units, and we shall designate them as *property conversion factors*. However, the use of these property conversion factors, mathematically, is exactly the same as ordinary conversion factors.

The above equation for the density of mercury provides us with two alternative property conversion factors: $\left(\dfrac{1 \text{ cm.}^3 \text{ Hg}}{13.55 \text{ g. Hg}}\right)$ and $\left(\dfrac{13.55 \text{ g. Hg}}{1 \text{ cm.}^3 \text{ Hg}}\right)$. (Note that we write the symbol for mercury as part of the unit, since it is not true or meaningful to say that 1 cm.^{3} is equivalent to 13.55 g.)

This way of regarding complex properties provides us with a short method of doing problems. We can now analyze a problem as consisting of a combination of property conversion steps and unit conversion steps.

Example: If 80 lb. of apples cost 63 cents, what is the cost in quarters of 200 kg. of apples?

Answer: We are given the mass of apples (80 lb.) and total cost (63 cents) and asked to find the total cost (quarters) of another mass (200 kg.) of apples. The given data provide an equivalence between two quantities, mass of apples and cents. Expressing this algebraically we find:

$$80 \text{ lb. apples} = 63 \text{ cents}$$

This will be used to get a conversion factor between pounds of apples and cents. To complete the problem we need regular conversion factors. Thus, to convert 200 kg. of apples to the equivalent number of quarters:

$$200 \text{ kg. apples} = 200 \text{ kg. apples} \times \left(\frac{1000 \text{ g.}}{1 \text{ kg.}}\right) \times \left(\frac{1 \text{ lb.}}{454 \text{ g.}}\right) \times \left(\left(\frac{63 \text{ cents}}{80 \text{ lb. apples}}\right)\right)$$
$$\times \left(\frac{1 \text{ quarter}}{25 \text{ cents}}\right)$$

$$= \frac{200 \times 1000 \times 63}{454 \times 80 \times 25} \text{ quarters}$$

$$= 13.9 \text{ quarters}$$

Note: The property conversion factor in the above equation has been placed in double parentheses for purposes of emphasis and to distinguish it from the other unit conversion factors.

---

12. Summary

1. *A unit conversion factor (or, simply, conversion factor) is a ratio of two different units which measure the same property. The ratio is numerically equal to unity.*

2. *Unit conversion factors are obtained from definitions or experiments and can be used for expressing a given property in one set of units if it has been expressed in a different set of units.*

3. *Scientific laws or direct observations may express an equivalence between two different properties of the same substance. When expressed as equations, such laws or observations may be used to obtain ratios of two different sets of properties. These are called property conversion factors and may be used to calculate one property if the related property is given*

4. *Most calculations may be considered as involving sequences of unit conversion factors and property conversion factors. If all the laws, observations, and definitions involved in the problem are known, then the answer may be written out as a product of such conversion factors.*

13. A General Approach to Problem Solving

In the illustrative examples used in this chapter, we have deliberately presented a variety of methods for working out problems using conversion factors. In the author's experience, however, there is one method which appears to be uniformly applicable to such problems. It involves first analyzing a problem into one of two categories:

A. **Conversions of properties from one set of units to another.** This includes simple or complex properties; e.g., converting pounds to grams; converting cm.^{2} to ft.^{2}; converting g./cm.^{3} to lb./ft^3.

B. **Deducing a complex property from a set of observations on simpler properties.** For example, computing the density of a substance from measurements of its mass (or related property) and volume (or related property).

In solving Case A, we always start with an identity involving the given property and proceed to convert to the desired property by the stepwise use of conversion factors. If the reader scans back over most of the examples solved, he will note that they usually start in this fashion. Thus the preceding example in which we found the cost of 200 kg. of apples started with the identity:

$$200 \text{ kg. apples} = 200 \text{ kg. apples}$$

The examples on page 15 can be solved by this method as well. For the alcohol problem we write the identity:

$$250 \text{ cm.}^3 \text{ alc.} = 250 \text{ cm.}^3 \text{ alc.}$$

Using density as a property conversion factor we then write:

$$250 \text{ cm.}^3 \text{ alc.} = 250 \text{ cm.}^3 \text{ alc.} \times \left(\left(\frac{0.79 \text{ g. alc.}}{1 \text{ cm.}^3 \text{ alc.}} \right) \right) = 198 \text{ g. alc.}$$

For the water problem we write:

$$800 \text{ g. water} = 800 \text{ g. water} \times \left(\frac{1 \text{ lb.}}{454 \text{ g.}}\right) \times \left(\left(\frac{1 \text{ ft.}^3 \text{ water}}{62.4 \text{ lb. water}}\right)\right) \times$$

$$\left(\frac{12 \text{ in.}}{1 \text{ ft}}\right)^3 \times \left(\frac{2.54 \text{ cm.}}{1 \text{ in.}}\right)^3 \times \left(\frac{1 \text{ l.}}{1000 \text{ cm.}^3}\right)$$

$$= 0.800 \text{ l. water}$$

Note that in each example we have two important checks. First the units must cancel properly. Secondly, each individual conversion factor must equal unity, i.e., its denominator and numerator must be equal. Note also that property conversion factors such as density always include the substance as part of the units.

In solving Case B our starting point is the definition or equation for the complex property. We then substitute into this starting equation the given data or their equivalent and then apply conversion factors to obtain the final units as needed.

Example: A falling object is found to move a distance of 148 m. in 5.50 sec. What is its average velocity in miles per hour?

Answer: From Table III (p. 12) we write the definition of velocity.

$$\text{Velocity} = \frac{\text{Distance}}{\text{Time}}$$

Substituting the specific information:

$$\text{Velocity} = \frac{148 \text{ m.}}{5.50 \text{ sec.}}$$

We now apply suitable conversion factors:

$$\text{Velocity} = \frac{148 \text{ m.}}{5.50 \text{ sec.}} \times \left(\frac{60 \text{ sec.}}{1 \text{ min.}}\right) \times \left(\frac{60 \text{ min.}}{1 \text{ hr.}}\right) \times \left(\frac{100 \text{ cm.}}{1 \text{ m.}}\right) \times \left(\frac{1 \text{ in.}}{2.54 \text{ cm.}}\right) \times$$

$$\left(\frac{1 \text{ ft.}}{12 \text{ in.}}\right) \times \left(\frac{1 \text{ mile}}{5280 \text{ ft.}}\right) \quad \frac{148 \times 60 \times 60 \times 100}{5.50 \times 2.54 \times 12 \times 5280} \text{ miles/hr.}$$

$$= 60.3 \text{ miles/hr.}$$

Example: Specific rate of production is defined as the number of units produced per man per day. Its units are units/man-day.

In a given factory it is found that 14 men produce 119 trays in 6 days. Calculate the specific rate of production (S.R.P.).

Answer: From the definition of S.R.P.

$$\text{S.R.P.} = \frac{\text{units}}{\text{man} \times \text{days}}$$

$$= \frac{119 \text{ trays}}{14 \text{ men} \times 6 \text{ days}} = \frac{119}{14 \times 6} \text{ trays/man-day}$$

$$= 1.42 \text{ trays/man-day}$$

14. Problems

1. Express the following quantities with the metric prefixes:

(a) 0.0002 of an inch.   (d) 100,000,000 years.
(b) 0.01 mile.   (e) 7,000,000,000 watts.
(c) 12,000 dresses.   (f) 0.0000035 ton.

2. Write as ordinary numbers in basic units:

(a) 3.5 kilobucks.   (e) 95 gammas.
(b) 75 microns.   (f) 2.7 milligallons.
(c) 42 kilowatts.   (g) 1.6 kilosheep.
(d) 0.16 megapound.

3. Which of the following are extensive and which are intensive properties?

(a) Density.   (d) Cost.
(b) Weight.   (e) Color of copper.
(c) Temperature.   (f) Color of the sky.

4. Convert 25 g. to pounds.

5. Convert 16 mg. to kilograms.

6. What is the weight in tons of 120 mg. of radium?

7. The selling price of radium is now $25,000 per gram.   How much will 2 lb. of radium cost?

8. Convert 20 sq. yd. to square centimeters.

9. A coal bin has the dimensions: 8 ft. by 12 ft. by 15 ft.   How many cubic meters capacity does it have?

10. The wavelength of the yellow light from a sodium lamp is roughly 5900 A. What is this in inches? in centimeter? in microns?

11. The density of carbon tetrachloride is 1.60 g./cm.^{3}.   What is it in tons per cubic yards?

12. A sign in a Mexican town gives the speed limit as 40 km./hr.   What is this in miles per hour? in centimeters per second?

13. The force of the earth's gravitational field will produce an acceleration of 32 ft./sec.^{2}.   What is this in centimeters per square second? in meters per square minute?

14. Convert a force of 980 g.-cm./sec.^{2} (980 dynes) to kilogram-meters per square minute.

15. The volume of a solid is stated by an eccentric person to be 70 in.^{2}-cm.   Is this a possible set of units for a volume?   Explain why and then proceed to express it in more conventional units.

16. 400 lb. of iron metal occupy a volume of 0.0234 m.^{3}.   Calculate the density of iron in grams per cubic centimeter.

17. The density of benzene is 0.88 g./cm.^{3} at 20°C.   How many milligrams of benzene are there in 25 cm.^{3}?

18. If 24 hens can lay 150 eggs in a week, how many days will it take 5 hens to lay 250 eggs?   (*Note:* Calculate the rate at which one hen lays eggs.   What are its units?)

19. An atom of oxygen has a mass of 2.68×10^{-23} g.   How many atoms of oxygen are there in 10.0 lb. of oxygen?

20. A molecule of water has a cross-sectional area of 10 A.². (*a*) How many molecules of water will be required to cover a square centimeter of surface? (*b*) How many molecules will be required to cover the total surface of a cube whose edge is 1.0 mm.? (*Note:* Assume that the molecules have the shape of squares.)

21. What is the density in g./cm.³ of a rectangular block of plastic 30 in. high × 16 in. wide × 1.0 ft. long weighing 12 lb.

22. A sample of water is found to contain some dissolved salt (NaCl). What measurements would be needed to express this information in quantitative form? Devise a complex unit which would represent the salt content of the water as an intensive property.

23. The formula for the volume of a sphere is given by: $V = \frac{4}{3}\pi r^3$, where r is its radius. What is the weight in tons of a sphere of water 1.0 mile in radius, the density of water being 1 g./cm.³?

24. A block of iron weighing 3600 lb. rests on a base which is a rectangle 2.4 in. × 1.6 in. Calculate the pressure on this base in lb./sq. in.

25. The blade of an ice skate makes contact with the ice over a length of about 6 in. and a width of 0.02 in. Calculate the pressure on the ice produced by a 150 lb. ice-skater skating on one blade.

26. A phonograph needle makes contact with a record surface over a circular area whose radius is about 0.2 mm. Calculate the pressure on the record in lb./sq. in. produced by a record arm weighing 4 oz.

27. A molecule having a mass of 2.0×10^{-22} g. has a velocity of 4.0×10^4 cm./sec. What is its kinetic energy in ergs?

28. When the molecule in problem 27 strikes a wall, it is estimated that it comes to rest in a time of the order of 2.0×10^{-13} sec. Calculate its negative acceleration (deceleration, see Table III) in cm./sec.². What is this in ft./sec.²? Compare this with the acceleration produced by the earth's gravity, $g = 980$ cm./sec.² at 41° latitude.

CHAPTER II

Methods of Measuring
Quantities of Matter

In the first chapter we have outlined the basis for the measurement of properties of matter and a method for dealing with calculations involving these properties. The remainder of this book will be devoted to an application of these concepts to the solution of chemical problems. In the present chapter we shall turn our attention to the atomic theory of matter and the way in which it affords us a method of measuring quantity of matter in chemical units.

1. Practical Units

A straightforward method for measuring a quantity of matter would be in terms of mass. This is readily accomplished in the laboratory by means of a balance. This is, however, not always the most convenient method available. Thus it is more convenient to measure liquids by volume, with a graduated cylinder or buret. If we know the temperature and the density of the liquid at that temperature, we can easily convert the measured volume to its equivalent in terms

TABLE IV

PRACTICAL METHODS FOR MEASURING QUANTITIES OF MATTER

Physical State	Method Employed	Property Measured	Additional Data Needed To Convert to Mass
Solid	Balance	Mass	None
Liquid	Balance	Mass	None
	Pipet, buret, graduated cylinder	Volume	Temperature and density
Gas	Gas buret, graduated cylinder, tank	Volume	Temperature, pressure, and gas density

of mass units. Similarly, when dealing with gases it is very difficult and awkward to measure a quantity of gas by means of a balance. In the laboratory we measure out quantities of gas by measuring the volume occupied by the gas, the temperature of the gas, and the pressure used to confine the gas to that particular volume. From a knowledge of the temperature, volume, and pressure of the particular gas we can then compute from the gas laws the quantity of matter in the gas. Table IV summarizes the customary methods employed in the laboratory for measuring quantities of matter.

2. Chemical Units

The practical units for measuring out quantities of matter, though the most direct, are not of the greatest significance. The atomic theory of matter tells us that matter is composed of units which we call atoms and molecules. The changes which occur in chemical transformations always involve whole numbers of atoms or molecules. From this theoretical point of view it is more significant to talk directly about the numbers of atoms or molecules which are reacting rather than some auxiliary property such as their mass or volume. In analogous fashion we might say that there are 185 g. of nickels in a dollar. This would hardly be as informative as saying that there are 20 nickels in a dollar. Similarly, the Census Bureau does not report the population of San Francisco as 50,000 tons of people but preferably as 700,000 individuals.*

How, then, can we go about expressing quantities of matter in terms of the number of molecules present? It would seem at first an impossible or at the very least a burdensome task, since even the smallest quantities of matter dealt with in the laboratory contain billions upon billions of molecules. And, further, how can we go about counting molecules, which, as everyone knows, are invisible? The answer to both of these difficulties is given by the atomic theory.

3. Chemical Formulae—Atomic Weights

Chemists have been able to develop methods (which we shall not discuss here) for measuring the weights of single atoms and molecules. They have also been able to determine through methods of chemical analysis the exact composition of compounds, that is, the relative number of atoms of each element that is present in a compound. This type of information is summarized in what we call chemical formulae.

---

* This assumes, of course, that the Census Bureau also has a conversion factor, namely, the average weight of a person.

Thus when we see the formula, NaCl, the symbols tell us that in any quantity of pure NaCl (sodium chloride) there will always be precisely 1 atom of sodium (Na) for every atom of chlorine (Cl). Similarly, the formula H_2SO_4 tells us that in pure sulfuric acid (H_2SO_4) there are precisely 4 atoms of oxygen (O) and 2 atoms of hydrogen (H) for every atom of sulfur (S). These formulae are determined by analysis and are brief, precise means of expressing the information gained by such analysis. If now we know the weights of the respective atoms in these compounds, we can calculate the weights of the molecule corresponding to the written formula. This information is provided by a Table of Atomic Weights.

Thus the weight of 1 molecule of sulfuric acid (H_2SO_4) is obtained by adding the weights of 1 atom of sulfur plus 2 times the weight of 1 atom of hydrogen plus 4 times the weight of 1 atom of oxygen. Performing these additions, we should find that the weight of 1 molecule of $H_2SO_4 = 32 + (2 \times 1) + (4 \times 16) = 98$.

But we are immediately struck by a difficulty. What are the units for this weight?* On further inspection of the table we find that the units are arbitrary. That is, the Table of Atomic Weights does not give the absolute weight of an atom in mass units such as grams but gives the relative weight of the atoms with respect to the weight of an isotopic atom of carbon of mass 12. Thus the atomic weight of sulfur (S), which is recorded as 32, simply means that an atom of sulfur is heavier than an atom of carbon in the ratio of 32 to 12. If we knew the absolute weight of an atom of Carbon 12 in grams, we could then calculate the weight of an atom of sulfur in grams. To designate the relative nature of the masses recorded in the Table of Atomic Weights we shall use the term, atomic mass units, and so for consistency we should write the atomic weight of sulfur as 32 a.m.u. (32 atomic mass units). Similarly the mass of 1 molecule of sulfuric acid computed above would then be properly written as 98 a.m.u. If we are now given the relation between atomic mass units and grams, we can quickly convert these relative weight units into grams. However, such information is seldom needed, as we shall soon see, and it is quite rarely that we are ever concerned with the absolute mass of atoms and molecules.

Do problem 1 at the end of the chapter.

---

* The Table of Atomic Weights should more properly be designated the Table of Atomic Masses, since we are really discussing the mass, not the weight. However, since common usage has been to call these quantities weights, we shall follow this designation with the understanding that masses are really meant.

4. Chemical Units—The Mole

Let us define a unit for the measurement of quantities of atoms and molecules as follows:

Definition:

1 mole of a compound = The formula weight of the element or compound expressed in units of grams*

Examples:

1 mole NaCl	=	58.5 g.
1 mole O_2	=	32 g.
1 mole H_2SO_4	=	98 g.
1 mole $Al_2(SO_4)_3$	=	342 g.
1 mole $Al_2(SO_4)_3 \cdot 18H_2O$	=	666 g.
1 mole Fe	=	56 g.

Note: 1 mole of a monatomic element = atomic weight in grams.

The importance of this unit, the mole, which we have defined above is the following:

One mole of every compound contains the same number of molecules.

Thus 1 mole of H_2SO_4 will contain the same number of molecules (or formula units) of H_2SO_4 as 1 mole of O_2 or 1 mole of H_2O or 1 mole of Fe, etc. The truth of this statement may be demonstrated as follows:

If we take 1 molecule of a compound, let us call it A, and 1 molecule of a second compound, B, then the ratio of the weights of these 2 molecules will be in the ratio of the molecular weight of A to the molecular weight of B.

If we take not 1 but 10 molecules of A and 10 molecules of B, then the ratio of the total weight of the 10 molecules of A to the total weight of the 10 molecules of B will still be in the ratio of the molecular weight of A to the molecular weight of B.

In fact, it doesn't matter how many total molecules of each we have;

* Since there is considerable confusion about whether a certain molecule as expressed by its formula really exists, we shall avoid any ambiguity by referring always to the formula. Thus NaCl is the formula for sodium chloride and as such gives its composition. However, solid sodium chloride is made of ions, not molecules. By defining 1 mole of sodium chloride as equal to the formula weight of NaCl expressed in grams, such questions are avoided.

as long as we have *equal numbers of molecules of A and B*, the ratio of the total weight of A to B will always be in the ratio of their molecular weights.

Conversely, if we have a certain weight of A and a certain weight of B and these weights are in the ratio of the molecular weights of A and B, then the number of molecules of A and B must be equal. This is illustrated in the following chart, with O_2 molecules and SO_2 molecules as examples.

Total Number of Molecules of O_2	Total Number of Molecules of SO_2	Ratio of Total Weights SO_2/O_2
1	1	$\dfrac{1 \times 64}{1 \times 32} = 2$
10	10	$\dfrac{10 \times 64}{10 \times 32} = 2$
500	500	$\dfrac{500 \times 64}{500 \times 32} = 2$
1×10^{20}	1×10^{20}	$\dfrac{1 \times 10^{20} \times 64}{1 \times 10^{20} \times 32} = 2$
X	X	$\dfrac{X \times 64}{X \times 32} = 2$

Thus we have succeeded in defining a unit, *the mole*, which is a direct measure of the number of molecules present in a compound and which also is capable of translation into practical units of grams. The student should observe, however, that this is true only if we know the formula for the compound.

The definition of the mole gives us an algebraic relation between the units of moles and the units of grams and so allows us to obtain a conversion factor between the two sets of units.

Example: How many moles of H_2SO_4 are there in 25 g.?
Answer:
$$1 \text{ mole } H_2SO_4 = 98 \text{ g. } H_2SO_4$$
Then
$$25 \text{ g. } H_2SO_4 = 25 \text{ g. } \cancel{H_2SO_4} \times \left(\frac{1 \text{ mole } H_2SO_4}{98 \text{ g. } \cancel{H_2SO_4}} \right)$$
$$= 0.255 \text{ mole } H_2SO_4 \rightarrow 0.26 \text{ mole } H_2SO_4$$

Example: How many grams of $CaCl_2$ are there in 2.5 moles?
Answer:

$$1 \text{ mole } CaCl_2 = 111 \text{ g. } CaCl_2$$

Then

$$2.5 \text{ moles } CaCl_2 = 2.5 \text{ moles } CaCl_2 \times \left(\frac{111 \text{ g. } CaCl_2}{1 \text{ mole } CaCl_2} \right)$$
$$= 278 \text{ g. } CaCl_2 \rightarrow 280 \text{ g. } CaCl_2$$

Example: Which will contain more molecules, 40 g. of H_2O or 60 g. of NaCl?
Answer:

$$40 \text{ g. } H_2O = 40 \text{ g. } H_2O \times \left(\frac{1 \text{ mole } H_2O}{18 \text{ g. } H_2O} \right) = 2.22 \text{ moles } H_2O$$

$$60 \text{ g. NaCl} = 60 \text{ g. NaCl} \times \left(\frac{1 \text{ mole NaCl}}{58.5 \text{ g. NaCl}} \right) = 1.03 \text{ moles NaCl}$$

Thus there are more moles and hence molecules of H_2O.

Do problems 2 and 3 at the end of the chapter.

5. The Number of Molecules in a Mole—Avogadro's Number

In the preceding section we have seen how it is possible to define a unit, the mole, which is proportional to the number of molecules in a quantity of matter and at the same time can be translated into mass units of grams. This would be a perfectly satisfactory system of units even if we were never to know exactly how many molecules were present in a mole.

Scientists, however, are never long satisfied by such a challenge, and a number of ingenious laboratory experiments have been devised which make it possible to measure numerically the number of molecules in a mole of substance. This number, known as Avogadro's number, is equal to 6.0221×10^{23} molecules. Thus we can write the very important algebraic relation (rounded off):

$$1 \text{ mole} = 6.02 \times 10^{23} \text{ molecules} \quad \text{(from experiment)}$$

This equation now gives us a means of translating from moles to number of molecules, and, since moles are related to units of grams, all three sets of units can be interconverted.

Example: How many molecules are there in 2 g. of water (H_2O)?
Answer: The two fundamental relations are:

$$1 \text{ mole } H_2O = 18 \text{ g. } H_2O$$

$$1 \text{ mole } H_2O = 6.02 \times 10^{23} \text{ molecules } H_2O$$

Then

$$2.0 \text{ g. } H_2O = 2.0 \text{ g. } H_2O \times \left(\frac{1 \text{ mole } H_2O}{18 \text{ g. } H_2O}\right) \times \left(\frac{6.02 \times 10^{23} \text{ molecules } H_2O}{1 \text{ mole } H_2O}\right)$$

$$= 6.69 \times 10^{22} \text{ molecules } H_2O \rightarrow 6.7 \times 10^{22} \text{ molecules } H_2O$$

Example: What is the mass of 1 atom of oxygen?
Answer: (Student should supply the fundamental relations.)

$$1 \text{ atom } O = 1 \text{ atom } O \times \left(\frac{1 \text{ mole } O}{6.02 \times 10^{23} \text{ atoms } O}\right) \times \left(\frac{16 \text{ g. } O}{1 \text{ mole } O}\right)$$

$$= 2.66 \times 10^{-23} \text{ g. } O$$

The following example shows how these methods can be applied to obtain interesting information.

Example: What is the average volume occupied by 1 molecule of water (H_2O)?
Answer: We can calculate the mass of 1 mole of H_2O (from its formula) to be 18 g. We also know how many molecules there are in 1 mole (6×10^{23}). The problem asks us, however, to calculate the property of volume. To do this we need a relation between mass of water and volume. This relation is provided by the density of water. So we can look through a table of densities and find that the density of water at 0°C. is 1.00 g./cm.^{3}. We now have all the necessary information and can proceed to write out the fundamental relations. The student should do this and check the following:

$$1 \text{ molecule } H_2O = 1 \text{ molecule } H_2O \times \left(\frac{1 \text{ mole } H_2O}{6.02 \times 10^{23} \text{ molecules } H_2O}\right)$$

$$\times \left(\frac{18 \text{ g. } H_2O}{1 \text{ mole } H_2O}\right) \times \left(\frac{1 \text{ cm.}^3 H_2O}{1 \text{ g. } H_2O}\right)$$

$$= 2.99 \times 10^{-23} \text{ cm.}^3$$

Note: If we imagine the molecule as fitting into a cube, the length of one edge of this cube is obtained by taking the cube root of the above volume or 3.1×10^{-8} cm. = 3.1 A. This method represents a crude way of obtaining diameters of molecules.

Do problems 4–7 at the end of the chapter.

6. Volumes of Gases—The Molar Volume

A study of the properties of gases shows us that equal volumes of gases at the same conditions of temperature and pressure will contain equal numbers of molecules. This statement is known as Avogadro's hypothesis and can be derived from the kinetic-molecular hypothesis. The converse of this statement, which is also true, is that equal numbers of molecules of different gases will occupy equal volumes when measured at the same conditions of temperature and pressure. But, since 1 mole of any substance has in it the same number of molecules as 1 mole of any other substance, if the two substances are both gases, then they will occupy equal volumes if kept at the same temperature and pressure.

As a consequence of this extraordinary property of gases, we can define the molar volume of a gas:

Definition: Molar volume of gas = Volume occupied by 1 mole of a gas (at 0°C. and a pressure of 1 atm.)

The temperature of 0°C. and the pressure of 1 atm. have been arbitrarily selected as the standard conditions for gas measurement and are referred to as STP (i.e., standard temperature and pressure).

By experiment the molar volume of a gas has been determined to be 22.4 l. STP. We can thus write the following relation *valid for ideal gases:*

$$1 \text{ mole gas} = 22.4 \text{ l. gas STP} \qquad \text{(experiment)}$$

This last relation is quite extraordinary in that it permits us to determine the molecular weight of a gas merely by weighing it, *without any prior knowledge of its formula or composition.*

Example: A sample of gas, occupying 150 cm.^{3} STP is found to weigh 0.624 g. What is its molecular weight?

Answer: The problem asks us to find the molecular weight, that is, the weight in grams, of 1 mole of the gas. We are told that 150 cm.^{3} STP = 0.624 g.

$$1 \text{ mole gas} = 1 \text{ mole gas} \times \left(\frac{22.4 \text{ l. STP}}{1 \text{ mole gas}} \right) \times \left(\frac{1000 \text{ cm.}^3}{1 \text{ l.}} \right) \times \left(\frac{0.624 \text{ g.}}{150 \text{ cm.}^3 \text{ STP}} \right)$$

$$= \frac{22.4 \times 1000 \times 0.624}{150} \text{ g.}$$

$$= 93.2 \text{ g.}$$

The molecular weight is thus 93.2 g.

Alternate Answer: We can also work out the same problem by the uniform method described in Chapter I, section 13. We are asked in this problem to obtain a complex quantity, molecular weight, and so it fits Case B. We thus start with the definition:

$$\text{Molecular weight} = \frac{\text{Grams of substance}}{\text{Moles of substance}}$$

We then substitute for the numerator and denominator in the above formula. However, we note that although we are given the mass of the substance (0.624 g.) we are not given its moles. We are given its volume so this must be converted into moles. We can do this independently:

$$150 \text{ cm.}^3 \text{ STP} = 150 \text{ cm.}^3 \text{ STP} \times \left(\frac{1 \text{ liter}}{1000 \text{ cm.}^3} \right) \times \left(\left(\frac{1 \text{ mole}}{22.4 \text{ l. STP}} \right) \right)$$

$$= \frac{150}{1000 \times 22.4} \text{ moles} = 6.70 \times 10^{-3} \text{ moles}$$

Substituting this into the above equation we find:

$$\text{Molecular weight} = \frac{0.624 \text{ g.}}{6.70 \times 10^{-3} \text{ moles}} = 93.2 \text{ g./mole}$$

7. Summary

From the point of view of the atomic theory, the fundamental chemical units are atoms and molecules. A unit is defined, the mole, which is related to the number of molecules in a quantity of matter. Since laboratory practice requires that we measure quantities of matter in mass units or volume units, we need a set of relations which will allow us to relate these practical, laboratory units to the more significant theoretical units. These relations are the following:

(1)   1 mole = Formula weight of a substance in grams   (definition)

(2)   1 mole = 6.02×10^{23} molecules   (experiment)

(3)   1 mole = 22.4 l. STP (*for gases only*)   (experiment)

Auxiliary units:

Since in the laboratory it is much more convenient to deal with milliliters (ml.) of gases or liquids and milligrams (mg.) of materials rather than liters and grams it is convenient to define the auxiliary unit, millimole.

1 millimole (mmole) = One-thousandth mole
$$= 1 \times 10^{-3} \text{ mole}$$ (definition)

In terms of this unit we can now reexpress the preceding quantities as:

(1)   1 mmole = Formula weight of a substance expressed in milligrams

(2)   1 mmole = 6.02×10^{20} molecules

(3)   1 mmole = 22.4 cm.^{3} STP (*for gases only*)

Thus we can equally well define molecular weight as milligrams of substance/millimoles of substance.

From these relations we can derive conversion factors for converting from any one of these units to any other. Since all these units are in frequent use, the student should gain facility in these conversions.

Example: How many grams of NH_3 (ammonia) are there in 800 cm.^{3} STP of NH_3 gas?

Answer:

$$800 \text{ cm.}^3 \text{ STP } NH_3 = 800 \text{ cm.}^3 \text{ STP } NH_3 \times \left(\frac{1 \text{ l.}}{1000 \text{ cm.}^3} \right)$$

$$\times \left(\frac{1 \text{ mole } NH_3}{22.4 \text{ l. STP } NH_3} \right) \times \left(\frac{17 \text{ g. } NH_3}{1 \text{ mole } NH_3} \right)$$

$$= \frac{800 \times 17}{1000 \times 22.4} \text{ g. } NH_3$$

$$= 0.607 \text{ g. } NH_3$$

Example: How many millimoles of H_2SO_4 are represented by 2×10^{18} molecules of H_2SO_4?

Answer:

$$2 \times 10^{18} \text{ molecules of } H_2SO_4 = 2 \times 10^{18} \text{ molecules } H_2SO_4$$

$$\times \left(\frac{1 \text{ mole } H_2SO_4}{6 \times 10^{23} \text{ molecules } H_2SO_4} \right) \times \left(\frac{1000 \text{ mmoles}}{1 \text{ mole}} \right) = 3.3 \times 10^{-3} \text{ mmole } H_2SO_4$$

Note on Significant Figures: For the purposes of this text, one per cent or slide rule accuracy is quite sufficient. We shall thus express our answers to at most three significant figures (see Appendix I.8).

8. Problems

1. Compute the molecular weights of the following substances:

(a) ZnS.
(b) Na_2O_2.
(c) $CuSO_4$.
(d) $Na_2S_2O_3$.

(e) $CuSO_4 \cdot 5H_2O$.
(f) C_6H_6.
(g) $Ba_3(AlO_3)_2$.
(h) H_4SiO_4.

2. Make the following conversions:

(a) 6.0 g. of NH_3 to moles.
(b) 4.7 mmoles of HCl to grams.
(c) 2.7 moles of $Ca_3(PO_4)_2$ to grams.
(d) 0.026 g. of $Cu(NO_3)_2$ to millimoles.
(e) 0.18 kg. of $Ba(ClO_3)_2$ to moles.
(f) 1.8 lb. of $(NH_4)_2SO_4$ to millimoles.
(g) 24 g. of $CuSO_4 \cdot 5H_2O$ to moles.
(h) 15 micrograms (μg.) of NH_3 to ml. STP of NH_3
(i) 8.0 gigamoles of H_2 to tons of H_2.
(j) 8.5 picoliters STP of Cl_2 to molecules of Cl_2.

3. Which has more molecules?

(a) 4 mmoles of HNO_3 or 80 mg. of HNO_3.
(b) 16 g. of NaCl or 20 g. of KCl.
(c) 4 moles of $Ca(NO_3)_2$ or 1 kg. of $PbCl_2$.

4. Make the following conversions:

(a) 0.24 mole of H_2O to molecules.
(b) 1.7 mmoles of HCl to molecules.
(c) 6.9×10^{16} molecules of CO_2 to moles.
(d) 2 lb. of Cl_2 to molecules.
(e) 6000 molecules of $Ca_3(PO_4)_2$ to tons.

5. How many millimoles of S are there in 20 mg. of H_2S?

6. If 1.8×10^{19} molecules of NO_2 are removed from 10 mg. of NO_2, how many moles of NO_2 are left?

7. The density of carbon tetrachloride (CCl_4) is 1.65 g./cc. at $-5°C$. What is the average volume occupied by 1 molecule of CCl_4 at this temperature?

8. Make the following conversions:

(a) 26 cc. STP of Cl_2 gas to millimoles of Cl_2.
(b) 2.1×10^{26} molecules of NO gas to liters STP of NO gas.
(c) 16 mg. H_2S to cubic centimeters STP of H_2S gas.
(d) 33 l. STP of SF_6 gas to kilograms of SF_6.

9. 620 mg. of an unknown gas occupy a volume of 175 cc. STP. What is the molecular weight of the gas?

10. One molecule of an unknown chemical compound is found to have a mass of 2.33×10^{-22} g. What is the molecular weight of the compound?

11. A 1.00-l. vessel is found to weigh 500.763 g. when evacuated. What will its weight be when it is filled with CO_2 gas STP? (Ignore buoyancy effects.)

12. It is found that 2.71×10^{19} molecules of an unknown compound have a mass of 3.76 mg. What is its molecular weight?

13. A 250-cc. flask has a weight of 261.023 g. when filled at STP with a gas. When evacuated its weight is 260.242 g. What is the molecular weight of the gas?

14. The density of CCl_4 (liquid) at $0°C$. is 1.600 g./cc. What is the volume occupied by 800 molecules of CCl_4? What is the volume occupied by a single molecule of CCl_4? From the latter result, what is the diameter in angstroms of a molecule of CCl_4?

15. Chemists find that *micro* techniques are much more convenient for chemical analysis than older techniques using large quantities of materials. Express the relations (1), (2), and (3) of the last section (section 7) and the definition of molecular weight in micro-units of matter.

16. An atom of Hg is supposed to have a radius of 1.45 A. Assuming that it is spherical in shape, what volume will it have? Assuming that in liquid Hg, the atoms are packed with 6 nearest neighbors (i.e., in a cubic, checkerboard type of array), what volume will be occupied by 1 mole of liquid Hg?

17. Spreading oil-like molecules in a unimolecular film on water is found to reduce the rate of evaporation of the water so covered. If the cross section of an oil-like molecule having the formula $C_{12}H_{26}O$ is 36 square angstroms, what weight of it will be required to cover a reservoir whose area is 1 square mile?

18. What is the mass in grams of one molecule of H_2SO_4?

19. How many molecules are there in 1 cc. STP of air?

20. On the average 100 molecules of our dry atmosphere consist of approximately 78 molecules of N_2, 21 molecules of O_2 and 1 molecule of Ar. What is the average molecular weight of the atmosphere?

21. Moist air contains molecules of H_2O in addition to the normal components N_2, O_2, and Ar.   At STP, which is heavier, 1 cc. of dry air or 1 cc. of moist air? Explain your answer as briefly as possible.

22. On a hot, humid day the atmosphere may contain as much as 6 molecules of H_2O out of every 100 molecules.   Assuming that the proportions of N_2, O_2, and Ar in this remaining 94 molecules are the same as in problem 20, calculate the average molecular weight of moist air.

CHAPTER III

Chemical Formulae

1. Interpretation of Chemical Formulae

In Chapter II we stated that a chemical formula represents a concise method of giving the chemical composition of an element or compound. Let us now be more specific about the information conveyed by a chemical formula.

The symbol H_2O is interpreted as representing 1 molecule of water. It further means that in 1 molecule of water there are 2 atoms of hydrogen (H) and 1 atom of oxygen (O). We shall now extend this interpretation to include the following: H_2O may also mean 1 mole as well as 1 molecule of water. The full meaning conveyed is that 1 mole of water (H_2O) contains 2 moles of H and 1 mole of O. In similar fashion we shall interpret the formula H_2SO_4 as representing 1 mole of sulfuric acid which contains 2 moles of H, 1 mole of S, and 4 moles of O.

This may appear to the reader like a step backwards. Are we not adding confusion by giving a chemical symbol two different meanings? The answer is, of course, that we are. However, the confusion will be small if not negligible since it will generally be quite apparent from the way in which these symbols are used just which interpretation is meant. On the other hand, by making this additional interpretation we are making it possible to discuss formulae in terms of theoretical units (moles) which represent a certain large number of molecules and which at the same time may be related to practical units such as grams.

The student should convince himself that the two interpretations are consistent. This can be seen as follows:

If we have 6.02×10^{23} molecules of H_2O, we have $2 \times 6.02 \times 10^{23}$ atoms of H and 6.02×10^{23} atoms of O. But, by the relation, 1 mole = 6.02×10^{23} molecules, these numbers represent in turn 1 mole of H_2O; 2 moles of H, and 1 mole of O. Thus the two interpretations are consistent.

2. Molar Composition of Chemical Compounds

We have seen from the discussion above that a chemical formula summarizes our knowledge of the chemical composition of a compound. It presents this information concerning composition not in terms of practical or laboratory units, such as grams, but in theoretical units. That is, it tells us the relative number of atoms of each element present in the compound or, alternatively, the relative number of moles of each element present in the compound. Chemical formulae thus give us the *molar composition* of a compound.

How does a chemist obtain the information which enables him to write a chemical formula? That is, how does he obtain the *molar composition* from which a formula can be written? In answering this very important question, let us proceed backwards and first show that if we know the *molar composition* we can write the chemical formula.

Suppose we find that 0.100 mole of calcium (Ca) metal react precisely with 0.100 mole of oxygen (O) to produce the single, pure substance calcium oxide. We can infer from this that the formula for calcium oxide must be CaO, since our experimental observation has informed us that there is 1 mole of oxygen for every mole of calcium in the compound calcium oxide.

Example: It is found experimentally that a certain quantity of the compound, zinc sulfate, contains 0.241 mole of zinc (Zn), 0.241 mole of sulfur (S), and 0.964 mole of oxygen (O). What is its chemical formula?

Answer: The above analytical evidence shows us that for every 1 mole of Zn there are 1 mole of S and 4 moles of O present in zinc sulfate. The formula must be $ZnSO_4$.

Example: It is found that a certain quantity of the compound $Al_2(CO_3)_3$ on analysis will yield 1.32 moles of aluminum (Al). How many moles of carbon (C) and oxygen (O) were present in this quantity of the compound?

Answer: From the formula, the molar composition is 2 moles of Al, 3 moles of C, and 9 moles of O. Thus there will be $\frac{3}{2} \times 1.32$ moles of C = 1.98 moles of C and $\frac{9}{2} \times 1.32$ moles of O = 5.94 moles of O.

Do problems 1–7 at the end of the chapter.

3. Composition Conversion Factors

It is possible to set the information conveyed by a chemical formula in terms of algebraic relations such that our methods of conversion factors may be employed.

The formula for calcium chloride, $CaCl_2$, tells us that 1 molecule of $CaCl_2$ contains 1 atom of Ca and 2 atoms of Cl or, alternatively,

1 mole of $CaCl_2$ contains 1 mole of Ca and 2 moles of Cl. From this information, then, for the compound $CaCl_2$ we can write all of the following relations:

$$1 \text{ mole } CaCl_2 = 1 \text{ mole } Ca; \quad 1 \text{ mole } CaCl_2 = 2 \text{ moles } Cl;$$

$$1 \text{ mole } Ca = 2 \text{ moles } Cl$$

But surely, you will protest, 1 mole of $CaCl_2$ is not equal to 1 mole of Ca. This objection is certainly true, and we must be careful, as we were in the discussion of complex properties (e.g., density), to place a restrictive interpretation on these equations. Strictly speaking, they represent properties of the same amount (1 mole) of a particular substance ($CaCl_2$), and as such they represent alternative methods of measuring this particular amount (1 mole) of this particular substance. Thus we can measure out 1 mole of $CaCl_2$ by taking that quantity of $CaCl_2$ which contains 1 mole of Ca or that which contains 2 moles of Cl.

From these relations between the composition properties of $CaCl_2$ we can make composition conversion factors that will be used like our regular conversion factors to convert from one aspect of the composition to another.

Example: How many moles of oxygen (O) are there in 0.265 mole of copper sulfate, $CuSO_4$?

$$0.265 \text{ mole } CuSO_4 = 0.265 \text{ mole } CuSO_4 \times \left(\frac{4 \text{ moles O}}{1 \text{ mole } CuSO_4} \right)$$

$$= 1.06 \text{ moles O}$$

Example: How many moles of Na_2CO_3 (sodium carbonate) will contain 0.124 mole of Na?

Answer:

$$0.124 \text{ mole Na} = 0.124 \text{ mole Na} \times \left(\frac{1 \text{ mole } Na_2CO_3}{2 \text{ moles Na}} \right)$$

$$= 0.062 \text{ mole } Na_2CO_3$$

Example: A certain quantity of barium phosphate, $Ba_3(PO_4)_2$, is found to contain 0.64 mole of O. How many moles of Ba were present also?

Answer:

$$0.64 \text{ mole O} = 0.64 \text{ mole O} \times \left(\frac{3 \text{ moles Ba}}{8 \text{ moles O}} \right)$$

$$= 0.24 \text{ mole Ba}$$

To summarize, we can say that the subscripts in a chemical formula represent the moles of each element present in 1 mole of the compound represented by that formula.

Composition conversion factors can be obtained by writing the ratio of the numbers of moles of each element present in the formula, and these can be used to convert from moles of one element present to moles of any other element or to moles of the parent compound.

4. Chemical Analysis—Weight Composition—Summary

We now come to the important question: How can we determine the formula of a compound? In order to determine the formula (that is, the molar composition) we must first know what elements are present, and, second, we must know the weight of each element present. This latter information is provided by the experimental techniques of quantitative analysis. The results of a direct experimental analysis generally give us the weight composition, that is, the weight of each element present in a compound. The formula can be obtained from this weight composition by conversion to molar composition with the atomic weights as conversion factors.

Example: It is possible by chemical analysis to show that in, let us say, 2.000 g. of copper chloride there is 0.945 g. of copper (Cu) and 1.055 g. of chlorine (Cl). From this information determine the formula for copper chloride.

Answer: We are given the weight composition and desire the molar composition.

$$0.945 \text{ g. Cu} = 0.945 \text{ g. Cu} \times \left(\frac{1 \text{ mole Cu}}{63.6 \text{ g. Cu}}\right) = 0.0148 \text{ mole Cu}$$

$$1.055 \text{ g. Cl} = 1.055 \text{ g. Cl} \times \left(\frac{1 \text{ mole Cl}}{35.5 \text{ g. Cl}}\right) = 0.0297 \text{ mole Cl}$$

Thus for the compound $CuCl_2$ there are 0.0297 mole of Cl for 0.0148 mole of Cu. Written algebraically:

$$0.0297 \text{ mole Cl} = 0.0148 \text{ mole Cu} \quad \text{(for copper chloride)} \quad \text{(experimental)}$$

Dividing both sides by 0.0148 (the smaller of the two quantities), we have

$$2.006 \text{ moles Cl} = 1 \text{ mole Cu}$$

Allowing then for the error of 6 parts in 2000, we see that the formula is $CuCl_2$.

Example: In 4.50 g. of an organic compound, acrylic acid, there are found to be 2.25 g. of carbon (C), 2.00 g. of oxygen (O), and 0.25 g. of hydrogen (H). What is its formula?

Answer: Using atomic weights we convert this weight composition to molar composition:

$$2.25 \text{ g. C} = 2.25 \text{ g. C} \times \left(\frac{1 \text{ mole C}}{12 \text{ g. C}} \right) = 0.187 \text{ mole C}$$

$$2.00 \text{ g. O} = 2.00 \text{ g. O} \times \left(\frac{1 \text{ mole O}}{16 \text{ g. O}} \right) = 0.125 \text{ mole O}$$

$$0.25 \text{ g. H} = 0.25 \text{ g. H} \times \left(\frac{1 \text{ mole H}}{1 \text{ g. H}} \right) = 0.25 \text{ mole H}$$

Dividing each of these by the smallest, 0.125 mole O, we find the molar composition conversion factors:

$$\frac{0.187 \text{ mole C}}{0.125 \text{ mole O}} = \frac{1.5 \text{ moles C}}{1 \text{ mole O}} ; \quad \frac{0.25 \text{ mole H}}{0.125 \text{ mole O}} = \frac{2 \text{ moles H}}{1 \text{ mole O}}$$

From this we can write the formula $C_{1.5}H_2O_1$. However, we want whole numbers in our formula, so we multiply all the subscripts by 2 and obtain $C_3H_4O_2$.

Do problems 8–14 at the end of the chapter.

Summary: *To obtain a chemical formula from the experimentally determined weight composition:*

1. *Convert the weights to mole composition, using the atomic weights as conversion factors.*

2. *Obtain composition conversion factors by dividing the number of moles of each element by the smallest number of moles of an element present.*

3. *Write the formula from the mole conversion factors using as subscripts the smallest whole numbers.*

5. Weight Per Cent

It is usual for analysts to express their results as weight per cent rather than in terms of grams. An analysis of the composition of water will be reported as 11.1% hydrogen (H) and 88.9% oxygen (O). This means that, if we take 100 weight units of water (100 lb., 100 g., etc.), 11.1 weight units will be hydrogen and 88.9 weight units will be oxygen. Thus, 100 tons of water contain 11.1 tons of hydrogen and 88.9 tons of oxygen. Percentage is an *intensive* property. It is independent of the quantity of material chosen. The percentage of hydrogen in a sample of water is always 11.1% even though the sample may weigh 1 g., 29.6 g., or 2 tons.

If a substance contains individual components A, B, C, D, etc., then the weight per cent of component A is given by the following definition.

Definition:

Wt. % A = Weight of A contained in 100 weight units of substance

or

$$\text{Wt. \% A} = \frac{\text{Weight of A}}{\text{Weight of substance}} \times 100$$

Example: What is the percentage of sulfur (S) in a mixture containing 80 lb. of sulfur and 70 lb. of iron?
Answer:

$$\text{Wt. \% S} = \frac{80 \text{ lb. S}}{150 \text{ lb. mixture}} \times 100$$

$$= 53.3$$

It is to be noted that the property of weight per cent can be used as a property conversion factor relating the number of weight units of a component to 100 weight units of the total sample.

To calculate formulae from composition reported as weight per cent we proceed by writing weight units, grams, in place of the units, per cent, and continue as usual.

Example: The composition of calcium pyrophosphate is: calcium (Ca) = 25.3%; phosphorus (P) = 39.2%; oxygen (O) = 35.5%.   What is its formula?
Answer:

Weight Composition for 100-g. Sample	Molar Composition	Dividing by the Smallest Quantity
25.3 % Ca = 25.3 g. Ca = $25.3 \text{ g. Ca} \times \left(\dfrac{1 \text{ mole Ca}}{40 \text{ g. Ca}}\right)$ = 0.633 mole Ca =		1
39.2 % P = 39.2 g. P = $39.2 \text{ g. P} \times \left(\dfrac{1 \text{ mole P}}{31 \text{ g. P}}\right)$ = 1.266 moles P =		$\dfrac{2 \text{ moles P}}{1 \text{ mole Ca}}$
35.5 % O = 35.5 g. O = $35.5 \text{ g. O} \times \left(\dfrac{1 \text{ mole O}}{16 \text{ g. O}}\right)$ = 2.22 moles O =		$\dfrac{3.5 \text{ moles O}}{1 \text{ mole Ca}}$

The formula is thus: $Ca_1P_2O_{3.5}$ or, eliminating fractions, $Ca_2P_4O_7$.

Do problems 15 and 17–20 at the end of the chapter.

6. Empirical Formulae and True Formulae

The weight composition or molar composition of a compound tells us only the ratio of atoms of each element present in a compound; it tells us nothing about molecules.   Indeed, if sodium chloride is used as an example, there are no molecules which exist as such in a crystal of table salt.

The formula which we write on the basis of a chemical analysis is the simplest formula or the *empirical formula*. The analysis of water tells us only that for every atom of oxygen there are 2 atoms of hydrogen. We could write for the formula of water: H_2O; H_4O_2; H_6O_3; H_8O_4; etc. Chemical analysis alone cannot tell us which is the proper formula for a molecule of water or indeed whether there is any such thing as a molecule of water.

In order to write the correct molecular formula, rather than the empirical or simplest formula, we must know:

       1. That a molecule really exists.
       2. What the molecular weight is.

Water, for example, can exist as a gas which must be, by the molecular theory, made up of molecules. From the density of the gas (water vapor) the molecular weight can be calculated; it is 18. Comparing this with the molecular weights of the possible formulae shown above, we see that H_2O is the true molecular formula for water vapor.

For salts made up of individual ions, there are no molecular units, and so empirical formulae are always written (e.g., $NaCl$, $CaSO_4$, $BaCl_2$, etc.).

Example: Benzene has the empirical formula CH. If its molecular weight (from gas density) is 78, what is its true molecular formula?

Answer: The true formula must be some multiple of CH, such as C_2H_2, C_3H_3, etc. One CH unit has a molecular weight of 13. Since the total molecular weight is 78, there are $78/13 = 6$ CH units in the molecule. Thus the true molecular formula is C_6H_6.

Do problem 16 at the end of the chapter.

7. Summary

1. The subscripts in a chemical formula give the number of atoms of each element present in the molecular unit designated by the formula.

2. The above molecular interpretation can be extended to include: The subscripts in a formula give the number of moles of each element present in 1 mole of the given formula.

3. From the ratio of any two subscripts we can obtain a composition conversion factor which enables us, for that particular compound, to convert from moles of one element to moles of another element.

4. Chemical analysis gives us the weight composition of a compound. To calculate the formula we must convert this to mole composition (by dividing by the atomic weights). From the mole composition we can obtain molar ratios and thus write the empirical formula for the compound.

5. *True molecular formula can be written only if we have additional information such as the molecular weight. Not all compounds are composed of molecules, notably salts!*

8. Problems

1. Interpret the following formulae in terms of molecules and atoms and in terms of moles. Which seem to be true formulae? Explain why.

(a) H_2O_2.
(b) Na_2SO_4.
(c) $Na_2SO_4 \cdot 10H_2O$.
(d) $Al_2(SO_4)_3$.

(e) CaC_2.
(f) N_2H_4.
(g) $2CaSO_4 \cdot H_2O$.
(h) $C_{10}H_{18}$.

2. How many moles of oxygen (O) are there in 0.280 mole of $Al_2(SO_4)_3$?

3. How many moles of sulfur (S) are there in 1.78 moles of $Na_4S_4O_6$?

4. How many moles of oxygen (O) are there in 0.120 mole of $Na_2CO_3 \cdot 10H_2O$?

5. How many moles of $CuSO_4$ will contain 6.3 mmoles of oxygen (O)?

6. How many moles of H_2O_2 will contain 24 mmoles of H?

7. A certain quantity of $Na_2B_4O_7$ contains 0.33 mole of O. How many millimoles of B does it contain?

8. How many moles of B are there in 20 g. of $Na_2B_4O_7$?

9. How many millimoles of S are there in 18 g. of $Al_2(S_2O_3)_3$?

10. How many grams of O are there in 64 mmoles of $CaSO_4$?

11. How many milligrams of Na are there in 240 mg. of Na_2CO_3?

12. How many grams of Ca are there in 1.0 ton of $Ca_3(PO_4)_2$?

13. How many atoms of O are there in 20 mg. of H_2O_2?

14. What is the weight per cent composition of the following:

(a) NaBr.
(b) $Ca(CN)_2$.
(c) $K_2S_2O_3$.

(d) $Na_2CO_3 \cdot 10H_2O$.
(e) $C_2H_4O_2$.
(f) $C_6H_5NO_2$.

15. Write the empirical formula for each of the following compounds whose weight per cent composition is given:

(a) Calcium (Ca) = 20.0%; bromine (Br) = 80.0%.
(b) Carbon (C) = 53.0%; oxygen (O) = 47.0%.
(c) Aluminum (Al) = 23.1%; carbon (C) = 15.4%; oxygen (O) = 61.5%.
(d) Strontium (Sr) = 65.7%; silicon (Si) = 10.4%; oxygen (O) = 23.9%.
(e) Manganese (Mn) = 56.4%; sulfur (S) = 43.6%.
(f) Calcium (Ca) = 18.3%; chlorine (Cl) = 32.4%; hydrogen (H) = 5.5%; oxygen (O) = 43.8%.

16. Given the following empirical formulae and molecular weights, compute the true molecular formulae:

Empirical Formula	Molecular Weight	True Formula
CH_2	84	C_6H_{12}
HO	34	
CH_2O	150	$C_5H_{10}O_5$
HgCl	472	
HF	80	H_4F_4

17. 9.7 g. of a hydrate of $CuSO_4$ loses 3.5 g. of water on heating. What is the empirical formula for the hydrate?

18. 2.50 g. of a compound containing chromium (Cr) and sulfur (S) contains a total of 1.20 g. of sulfur (S). What is its formula?

19. When 3.75 g. of platinum chloride is heated, chlorine escapes and 2.17 g. of platinum is left. What is its formula?

20. 2.80 g. of $CuCl_2$ combines with 2.11 g. of NH_3 to form a compound. What is its formula?

21. When sugar is heated to 600°C. in the absence of air, it turns to a black mass of carbon (C) and water (H_2O) which volatilizes as a gas. If 465 mg. of sugar yield 196 mg. of C, what is its empirical formula?

22. When very concentrated solutions of nitric acid (HNO_3) in water are cooled, a white crystalline solid separates which is found to contain 22.2% of water and 77.8% HNO_3 by weight. A number of chemists have argued that this solid is not a mixture but a new, true compound. Can you suggest any reasons from the analysis to support their conclusions?

23. An unknown compound has the composition 3.2% H, 37.5% C, and 59.3% F. What is its empirical formula? If 150 cc. STP of the vapor of this material weighs 1.07 g., what is its true formula?

24. A compound contains 52% N, 40% B, and the rest H. 60 cc. STP of its vapor weighs 210 mg. Calculate its molecular weight and true formula. Be careful of significant figures!

25. The composition of a material can be expressed in terms of mole per cent. Thus the mole per cent of any component in a mixture or compound indicates the total number of moles of that component present in a quantity of the material comprising 100 moles. The mole per cent of Na in NaCl is 50%. In each of the following cases give the mole per cent of the indicated material.

    (a) H_2O mole % H = _____

    (b) NH_3 mole % N = _____

    (c) H_3PO_4 mole % O = _____

    (d) $C_6H_{12}O_6$ mole % C = _____

    (e) $Na_2B_4O_7 \cdot 10H_2O$ mole % O = _____   mole % H = ____

    (f) Dry air (see problem 20, Chapter II); mole % O_2 = _____

26. 15 g. of water are mixed with 11 g. of alcohol (C_2H_5OH). What is the mole per cent of water in the mixture? (See problem 25 for definition.)

27. An alloy of Cu and Zn contains 74.5 wt. % of Cu. Calculate the mole per cent of Cu in the alloy.

CHAPTER IV

Chemical Reactions

1. Interpretation of Equations: Molecular

Just as a chemical formula is a concise way of representing the results of the analysis of a chemical compound in theoretical units (molecules, moles), so a chemical equation is an equally precise way of representing our *experimentally determined* information about a chemical reaction. Let us see how this information is acquired and how it is presented.

It is possible to determine in the laboratory that, when ferrous sulfide is heated in oxygen gas, the products of the reaction are ferric oxide and sulfur dioxide gas. We can represent this information as follows:

Ferrous sulfide + Oxygen (gas) → Ferric oxide + Sulfur dioxide (gas)

By chemical analysis we can determine the formulae for all the starting materials (reactants) and for the products. Substituting this information in our written equation we have:

$$FeS + O_2 \rightarrow Fe_2O_3 + SO_2 \quad \text{(skeleton equation)}$$

This skeleton, or unbalanced, equation, as it is called, now tells us what the reactants are, what the products of the reaction are, and also the composition of each of these. This equation may be "balanced"; that is, it can be written in such fashion as to tell us precisely what amounts of each substance takes part in the reaction. The justification for such balancing is provided by the atomic theory, which states that atoms can be neither created nor destroyed in chemical reactions.

This means that there must be the same number of atoms of each element present at the start and at the end of the reaction. (That is, there must be present the same number of atoms of each element distributed among the products of the reaction as there were in the reactants.)

The method of balancing is to put the *proper coefficients* before the formula for each substance in such a way that a final equation will result in which the equality of atoms is achieved. For simple equations this can be done by trial and error. For more complicated equations we shall develop special methods in a later chapter on oxidation and reduction.

The skeleton equation with which we started is balanced in the following form:

$$4FeS + 7O_2 \rightarrow 2Fe_2O_3 + 4SO_2$$

Atoms Present in Reactants	Atoms Present in Products
Fe: 4 (in FeS)	Fe: 4 (in Fe_2O_3)
S: 4 (in FeS)	S: 4 (in SO_2)
O: 14 (in O_2)	O: 14 (in SO_2 and Fe_2O_3)

The equation can be interpreted in molecular terms:

4 molecules FeS + 7 molecules O_2 → 2 molecules Fe_2O_3
$$+ \; 4 \text{ molecules } SO_2$$

This is now a precise statement of the quantities of matter that take part in the chemical change.

Do problem 1 at the end of the chapter.

2. Chemical Conversion Factors: Molecular

The ratio of any two quantities in the balanced equation gives us a "chemical" conversion factor which allows us to convert from molecules of one substance to the equivalent number of molecules of the other substance involved in the particular reaction.

From the balanced equation:

$$4FeS + 7O_2 \rightarrow 2Fe_2O_3 + 4SO_2$$

we can write the following *chemical conversion factors:*

$$\left(\frac{4 \text{ molecules FeS}}{7 \text{ molecules } O_2}\right); \quad \left(\frac{4 \text{ molecules FeS}}{2 \text{ molecules } Fe_2O_3}\right); \quad \left(\frac{7 \text{ molecules } O_2}{4 \text{ molecules } SO_2}\right); \text{ etc.}$$

These conversion factors can now be used in performing calculations.* See caution on p. 32.

* We will continue to use the word molecule to refer to the formula, even if the particular substance indicated by the formula is not composed of molecules.

Example: How many molecules of SO_2 can be made, using the above reaction, from 20 molecules of FeS?

Answer: Our conversion factor is

$$\left(\frac{4 \text{ molecules } SO_2}{4 \text{ molecules FeS}}\right)$$

Then

$$20 \text{ molecules FeS } = 20 \text{ molecules FeS} \times \left(\frac{4 \text{ molecules } SO_2}{4 \text{ molecules FeS}}\right)$$

$$= 20 \text{ molecules } SO_2$$

Example: How many molecules of O_2 are needed to react with 36 molecules of FeS?

Answer:

$$36 \text{ molecules FeS } = 36 \text{ molecules FeS} \times \left(\frac{7 \text{ molecules } O_2}{4 \text{ molecules FeS}}\right)$$

$$= 63 \text{ molecules } O_2$$

3. Chemical Conversion Factors: Molar

Molecules are not practical units for laboratory work. Thus the above interpretation of equations in terms of molecules is not very useful for actual experiments. We need another interpretation of equations in terms of some unit which will have the theoretical significance of molecules and which can be related to practical, laboratory units such as grams. This unit, as we have already seen, is the *mole.*

Just as formulae can be interpreted directly in terms of moles or of molecules, so equations can be also interpreted directly in terms of moles as well as molecules. To show this, let us multiply every term on both sides of our equation by the number 6.02×10^{23}. This does not alter the equality. The resulting equation is

$$4(6.02 \times 10^{23})\text{FeS} + 7(6.02 \times 10^{23})O_2 \rightarrow 2(6.02 \times 10^{23})\text{Fe}_2O_3$$
$$+ 4(6.02 \times 10^{23})SO_2$$

But observe that 6.02×10^{23} molecules of a substance is precisely 1 mole of that substance. We can thus replace this number by its equivalent in moles, and our equation becomes

$$4 \text{ moles FeS } + 7 \text{ moles } O_2 \rightarrow 2 \text{ moles Fe}_2O_3 + 4 \text{ moles } SO_2$$

Thus we have shown that balanced equations can be interpreted directly in terms of moles as well as molecules. Since moles can be related directly to practical units of mass, this is the interpretation we shall most frequently make. Just as the ratio of any two quantities in an equation can be used to obtain chemical conversion factors in terms of molecules, our added interpretation allows us to obtain chemical conversion factors in terms of moles.

Thus we have the following conversion factors:

$$\left(\frac{4 \text{ moles FeS}}{7 \text{ moles O}_2}\right); \quad \left(\frac{7 \text{ moles O}_2}{4 \text{ moles SO}_2}\right); \quad \left(\frac{4 \text{ moles FeS}}{2 \text{ moles Fe}_2\text{O}_3}\right);$$

$$\left(\frac{2 \text{ moles Fe}_2\text{O}_3}{4 \text{ moles SO}_2}\right); \text{ etc.}$$

This can be used in solving problems in units of moles.

Example: How many moles of SO_2 can be made from 0.24 mole of FeS?
Answer:

$$0.24 \text{ mole FeS} = 0.24 \text{ mole FeS} \times \left(\frac{4 \text{ moles SO}_2}{4 \text{ moles FeS}}\right)$$

$$= 0.24 \text{ mole SO}_2$$

Example: How many moles of oxygen are needed to produce 2.50 moles of Fe_2O_3?
Answer:

$$2.50 \text{ moles Fe}_2\text{O}_3 = 2.50 \text{ moles Fe}_2\text{O}_3 \times \left(\frac{7 \text{ moles O}_2}{2 \text{ moles Fe}_2\text{O}_3}\right)$$

$$= 8.75 \text{ moles O}_2$$

Do problems 2–4 at the end of the chapter.

4. Summary

1. *Balanced chemical equations can be interpreted in terms of molecules or moles directly. The coefficients in the balanced equation indicate the number of moles (or molecules) of each substance taking part in the chemical reaction.*

2. *The ratio of any two coefficients in a chemical equation gives us a chemical conversion factor for converting from moles (or molecules) of one substance to moles (or molecules) of the other substance.*

5. Calculations in Mixed Units

From the foregoing discussion we have seen that problems concerning chemical reactions can be solved in one step by means of chemical conversion factors.   This is true, however, only if the units are moles.   But most practical questions involving chemical reactions are based not on molar units but rather on practical units such as grams, or, if gases are concerned, the volume of the gas at STP.

We have already seen, in Chapter II, how to convert these practical units directly to molar units and vice versa.   Every calculation involving chemical reactions can be solved by the following scheme:

1. Convert given units (e.g., grams, volumes, etc.) to moles.
2. Convert from moles of given substance to moles of desired substance by means of the appropriate chemical conversion factor.
3. Convert from moles of desired substance to whatever units are requested in the answer.

Note that these three steps can be combined into a single product of conversion factors.

Example:
$$Na_2CO_3 + 2HNO_3 \rightarrow 2NaNO_3 + H_2O + CO_2$$

How many grams of $NaNO_3$ can be made from 10 g. of $NaCO_3$?

Answer: (1) Convert 10 g. of Na_2CO_3 to moles.   Fundamental equation is

$$1 \text{ mole } Na_2CO_3 = 106 \text{ g. } Na_2CO_3$$

Conversion factor is

$$\left(\frac{1 \text{ mole } Na_2CO_3}{106 \text{ g. } Na_2CO_3} \right)$$

Then

$$10 \text{ g. } Na_2CO_3 = 10 \text{ g. } \cancel{Na_2CO_3} \times \left(\frac{1 \text{ mole } Na_2CO_3}{106 \text{ g. } \cancel{Na_2CO_3}} \right) = 0.094 \text{ mole } Na_2CO_3$$

(2) Convert moles of Na_2CO_3 to moles of $NaNO_3$.   Fundamental equation is the balanced chemical equation.   Chemical conversion factor is

$$\left(\frac{2 \text{ moles } NaNO_3}{1 \text{ mole } Na_2CO_3} \right)$$

Then

$$0.094 \text{ mole } Na_2CO_3 = 0.094 \text{ } \cancel{\text{mole } Na_2CO_3} \times \left(\frac{2 \text{ moles } NaNO_3}{1 \text{ } \cancel{\text{mole } Na_2CO_3}} \right)$$

$$= 0.188 \text{ mole } NaNO_3$$

(3) Finally, convert moles of $NaNO_3$ to grams of $NaNO_3$.   Fundamental equation is

$$1 \text{ mole } NaNO_3 = 85 \text{ g. } NaNO_3$$

Conversion factor is

$$\left(\frac{85 \text{ g. NaNO}_3}{1 \text{ mole NaNO}_3} \right)$$

Then

$$0.188 \text{ mole NaNO}_3 = 0.188 \text{ mole NaNO}_3 \times \left(\frac{85 \text{ g. NaNO}_3}{1 \text{ mole NaNO}_3} \right) = 15.98 \text{ g. NaNO}_3$$

Notice that the three steps could have been combined:

$$
\overset{}{10 \text{ g. Na}_2\text{CO}_3} \times \overset{(1)}{\left(\frac{1 \text{ mole Na}_2\text{CO}_3}{106 \text{ g. Na}_2\text{CO}_3} \right)} \times \overset{(2)}{\left(\frac{2 \text{ moles NaNO}_3}{1 \text{ mole Na}_2\text{CO}_3} \right)} \times \overset{(3)}{\left(\frac{85 \text{ g. NaNO}_3}{1 \text{ mole NaNO}_3} \right)}
$$
$$= 16 \text{ g. NaNO}_3$$

Example:
$$\text{MnO}_2 + 4\text{HCl} \rightarrow \text{MnCl}_2 + \text{Cl}_2 + 2\text{H}_2\text{O}$$

How many liters of Cl_2 gas STP can be made from 20 g. of HCl?
 Answer:

$$
20 \text{ g. HCl} = 20 \text{ g. HCl} \times \overset{(1)}{\left(\frac{1 \text{ mole HCl}}{36.5 \text{ g. HCl}} \right)} \times \overset{(2)}{\left(\frac{1 \text{ mole Cl}_2}{4 \text{ moles HCl}} \right)} \times \overset{(3)}{\left(\frac{22.4 \text{ l. STP}}{1 \text{ mole}} \right)}
$$
$$= 3.06 \text{ l. STP Cl}_2 \rightarrow 3.1 \text{ l. STP Cl}_2$$

Example:
$$\text{Al}_4\text{C}_3 + 12\text{H}_2\text{O} \rightarrow 4\text{Al(OH)}_3 + 3\text{CH}_4$$

How many molecules of CH_4 will be produced when 0.2 g. of Al(OH)_3 are prepared?
 Answer:

$$
0.2 \text{ g. Al(OH)}_3 = 0.2 \text{ g. Al(OH)}_3 \times \overset{(1)}{\left(\frac{1 \text{ mole Al(OH)}_3}{78 \text{ g. Al(OH)}_3} \right)} \times \overset{(2)}{\left(\frac{3 \text{ moles CH}_4}{4 \text{ moles Al(OH)}_3} \right)}
$$
$$
\times \overset{(3)}{\left(\frac{6.02 \times 10^{23} \text{ mol.}}{1 \text{ mole}} \right)} = 1.16 \times 10^{21} \text{ molecules CH}_4
$$
$$\rightarrow 1.2 \times 10^{21} \text{ molecules CH}_4$$

Example:
$$4\text{FeS} + 7\text{O}_2 \rightarrow 2\text{Fe}_2\text{O}_3 + 4\text{SO}_2$$

How many grams of Fe_2O_3 can be made, using 600 cc. STP of O_2?
 Answer:

$$
600 \text{ cc. O}_2 \text{ STP} = 600 \text{ cc. O}_2 \text{ STP} \times \left(\frac{1 \text{ l.}}{1000 \text{ cc.}} \right) \times \left(\frac{1 \text{ mole}}{22.4 \text{ l. STP}} \right)
$$
$$
\times \left(\frac{2 \text{ moles Fe}_2\text{O}_3}{7 \text{ moles O}_2} \right) \times \left(\frac{160 \text{ g. Fe}_2\text{O}_3}{1 \text{ mole Fe}_2\text{O}_3} \right) = 1.22 \text{ g. Fe}_2\text{O}_3
$$
$$\rightarrow 1.2 \text{ g. Fe}_2\text{O}_3$$

Do problems 5–18 at the end of the chapter.

Note: For the same reasons as already described we can also interpret our equations directly in terms of millimoles or micromoles. The former in particular is a very useful unit, especially in mixed unit problems when milligrams and milliliters are used.

Example:

$$4FeS + 7O_2 \rightarrow 2Fe_2O_3 + 4SO_2$$

How many cc STP of O_2 are needed to burn 160 mg. of FeS?

$$160 \text{ mg. FeS} = 160 \text{ mg. FeS} \times \left(\frac{1 \text{ mmole FeS}}{88 \text{ mg. FeS}} \right) \times \left(\frac{7 \text{ mmoles } O_2}{4 \text{ mmoles FeS}} \right)$$

$$\times \left(\frac{22.4 \text{ ml. STP } O_2}{1 \text{ mmole } O_2} \right)$$

$$= \frac{160 \times 7 \times 22.4}{88 \times 4} \text{ ml. STP } O_2 \text{ gas}$$

$$= 71 \text{ ml. STP } O_2 \text{ gas} \qquad (1 \text{ ml.} = 1 \text{ cc.})$$

---

6. Problems

1. Translate the following statements into chemical equations and then balance the equations (see text for formulae):

(a) Chlorine gas burns in hydrogen gas to give hydrogen chloride.

(b) Barium chloride reacts with zinc sulfate to give zinc chloride and a precipitate of barium sulfate.

(c) Calcium nitrate reacts with sodium phosphate to give sodium nitrate and a precipitate of calcium phosphate.

(d) Potassium chlorate, when heated, gives a mixture of potassium chloride and potassium perchlorate.

(e) Hydrogen sulfide gas burns in air to give water and sulfur dioxide.

(f) Aluminum metal replaces iron from ferric oxide, giving aluminum oxide and iron.

(g) Hydrogen gas combines with nitrogen to give ammonia.

(h) Copper dissolves in dilute nitric acid to give copper nitrate, water, and nitric oxide.

(i) Phosphorus burns in oxygen to give phosphorus pentoxide.

(j) Sodium metal reacts with water to give sodium hydroxide and hydrogen gas.

(k) Carbon disulfide burns in air to give carbon dioxide and sulfur dioxide.

(l) Sodium hypochlorite when heated gives a mixture of sodium chloride and sodium chlorate.

2. $4HCl + O_2 \rightarrow 2H_2O + 2Cl_2$. How many moles of HCl are needed to form 0.35 mole of Cl_2?

3. $16HCl + 2KMnO_4 \rightarrow 2MnCl_2 + 2KCl + 4H_2O + 5Cl_2$. How many moles of chlorine gas will be produced from 3.20 moles of hydrochloric acid?

4. $3CaCl_2 + 2K_3PO_4 \rightarrow Ca_3(PO_4)_2 + 6KCl$. How many moles of potassium phosphate are needed to produce 0.076 mole of potassium chloride?

5. $CS_2 + 3O_2 \rightarrow CO_2 + 2SO_2$. How many cubic centimeters STP of SO_2 will be produced by burning 3.0 g. of CS_2?

6. $FeS + 2HCl \rightarrow H_2S + FeCl_2$. How many grams of FeS are needed to produce 100 g. of H_2S? What volume does this H_2S occupy at STP?

7. $2KClO_3 \rightarrow 2KCl + 3O_2$. How many liters STP of oxygen will be produced by decomposing 25 g. of $KClO_3$?

8. $2NaOH + H_2SO_4 \rightarrow Na_2SO_4 + 2H_2O$. How many moles of sodium hydroxide are needed to neutralize 100 g. of sulfuric acid?

9. $2Na_2O_2 + 2H_2O \rightarrow O_2\uparrow + 4NaOH$. How many cubic centimeters STP of O_2 gas can be made from 224 mg. of Na_2O_2?

10. In the contact process for the production of sulfuric acid, the sulfur present in FeS_2 is eventually converted into H_2SO_4. Assuming that the conversion is complete, how many kilograms of H_2SO_4 may be made from 2.0 tons of FeS_2?

11. If the H_2SO_4 produced in the preceding problem has a density of 1.86 g./cc., what volume will it occupy?

12. $2C_4H_{10} + 13O_2 \rightarrow 8CO_2 + 10H_2O$. How many cubic centimeters STP of CO_2 can be produced from 15 cc. of liquid butane (C_4H_{10})? The density of liquid butane is 0.60 g./cc. at 0°C.

13. From the reaction indicated in problem 3, how many grams of $KMnO_4$ are needed to produce 200 cc. STP of Cl_2 gas?

14. $Al_4C_3 + 12H_2O \rightarrow 4Al(OH)_3 + 3CH_4$. How many grams of Al_4C_3 are required to produce 250 l. STP of CH_4? How many grams of water are needed for this reaction? What is the liquid volume occupied by the water? (Density $H_2O = 1.00$ g./cc.)

15. In the Fischer-Tropsch process, coal is made to react with water over a catalyst to produce hydrocarbons. In this step 34% of the coal is converted to hydrocarbons. Of the total hydrocarbons produced, 6% of the carbon present in them is in the form of hexane, C_6H_{14}. How many kilograms of hexane can be produced from 1 ton of coal? (Consider coal as carbon, C.)

16. A metallurgical process for the extraction of uranium starts with carnotite ore which is 3.5% U_3O_8. This is treated in a number of steps and finally converted to uranyl sulfate hydrate, $(UO_2)SO_4 \cdot 3H_2O$. What is the maximum number of kilograms of the hydrate which can be made from 1.00 ton of carnotite ore?

17. $Zn + 2HCl \rightarrow ZnCl_2 + H_2$.
   (a) If 2.0 moles of Zn is mixed with 1.6 moles of HCl, what substances will be present when the reaction is over? What is their quantity?
   (b) If 14 g. of Zn is mixed with 18 g. of HCl, what substances will be present when the reaction is over? What is the weight of each?

18. $3Cu + 8HNO_3 \rightarrow 3Cu(NO_3)_2 + 2NO + 4H_2O$.
   (a) If 4 moles of Cu is added to 16 moles of HNO_3, what substances will be present when the reaction is over? How many moles of each?
   (b) If 24 g. of Cu is added to 12 g. of HNO_3, what substances will be present when the reaction is over? How many grams of each?

19. The density of liquid benzene, C_6H_6, is 0.88 g./cc. at 20°C. It burns in O_2 as follows:

$$2C_6H_6 + 15O_2 \rightarrow 12CO_2 + 6H_2O$$

How many cubic centimeters STP of O_2 gas are needed to burn 3.5 cc. of liquid benzene?

20. $NaBH_4 + 4HOH \rightarrow NaOH + 4H_2 + H_3BO_3$. How many milliliters STP of H_2 gas can be produced from 25.0 mg. of $NaBH_4$?

CHAPTER V

Energy and Chemical Changes

1. Energy and Work—Conservation of Energy

Next to the direct analysis of the products of chemical reactions, the greatest insight into the structure of matter has been obtained through the consideration of the energy changes which accompany the transformation of matter.

All chemical changes are accompanied by the absorption or liberation of energy by the reacting system. These energy changes may be manifested as thermal energy (heat), mechanical energy (pressure of gases from an explosion, sound, etc.), radiant energy (light, infrared or visible, ultraviolet, X-rays, etc.), or electric energy (storage batteries, nerve cell stimulation, electric eels, etc.).

Before discussing these energy changes in detail, let us first define the terms we shall use.

Definition: By *energy* we shall mean that which is capable of doing work.

Definition: By *work* we shall mean mechanical work, that is, the result achieved when a force moves through a distance. (See Table III.)

Since, as we shall see, all types of energy can be transformed by various laboratory devices, one into the other, all types of energy may be transformed into mechanical work, and vice versa. These devices for transforming different types of energy into each other and into mechanical work also provide a means whereby the units for measuring these various types of energy can be related to the units for mechanical energy or work.

If a force is applied to an object, its application will be manifested as a change in the velocity of the object (acceleration). The product of the force times the distance along which the force moves is defined as mechanical work.

Definition:

Mechanical work = Force × Distance traversed by force

The product of the mass of an object times one-half its velocity squared is defined as its kinetic energy (energy of motion).

Definition:

$$\text{Kinetic energy} = \frac{1}{2} \times \text{Mass} \times (\text{Velocity})^2$$

The change in the velocity of an object when a force acts on it produces a change in its kinetic energy. If there is no production of other types of energy, then the mechanical work done by the force is transformed entirely into kinetic energy. This represents an illustration of a very important law of energy, the *law of conservation of energy*.

Law of conservation of energy: Energy can neither be created nor destroyed. It can only be transformed from one form into another.

2. Units of Energy and Work

In Table V are listed the more important types of energy and the units in which they are measured.

TABLE V

TYPES OF ENERGY AND THEIR UNITS

Types of Energy	Definition	Units
Mechanical	Force × Distance	dyne-cm; erg (1 erg = 1 dyne-cm.)
Kinetic	$\dfrac{\text{Mass} \times (\text{Velocity})^2}{2}$	$\text{g.-}\dfrac{\text{cm.}^2}{\text{sec.}^2}$; erg $\left(1 \text{ g.-}\dfrac{\text{cm.}^2}{\text{sec.}^2} = 1 \text{ erg}\right)$
Thermal	Quantity of heat. 1 cal. is the amount of heat necessary to raise the temperature of 1 g. of water by 1°C. at 15°C.	calorie; kilocalorie (1 cal. = 4.185×10^7 ergs)
Electrical	Work done by electrical forces. 1 joule = work done by electrical forces in moving 1 *coulomb* of charge through a potential drop of 1 volt.	joule (1 joule = 0.24 cal. = 1×10^7 ergs) (1 watt-sec. = 1 joule)

Do problems 1, 2, and 3 at the end of the chapter.

3. Heat—Temperature Scale

Temperature is a measure of the heat energy contained in a substance. We define a temperature scale as follows:

TABLE VI
TEMPERATURE SCALES

Standard	Centigrade Scale	Fahrenheit Scale	Absolute Scale (Kelvin)	Rankine Scale
The temperature of a mixture of pure ice and water.	0°C.	32°F.	273°K.	491.6°R.
The temperature of boiling water at a pressure of 76 cm. of Hg. (1 atm.)	100°C.	212°F.	373°K.	671.6°R.

We can see from this scale that the temperature interval between melting water and freezing water is divided into 100 parts in the centigrade scale and in the absolute scale, but into 180 parts in the Rankine and Fahrenheit scales. From the above definitions we can derive the following equations relating the various scales:

$$°C. = \frac{5}{9}(°F. - 32); \quad °F. = \frac{9}{5}°C. + 32; \quad °F. = °R. - 459.6$$

$$°K. = °C. + 273; \quad °C. = °K. - 273; \quad °C. = \frac{5}{9}(°R. - 491.6)$$

Observe, however, that a *change* of 1°C. is equivalent to a *change* of $\frac{9}{5}$°F., whereas a change of 1°C. is equal to a change of 1°K. That is, if the temperature of an object is 10°C. and changes to 11°C., in the Fahrenheit scale it will be 50°F. and change to 51.8°F. (a change of 1.8°F. = $\frac{9}{5}$°F.), whereas in the absolute or Kelvin scale it will go from 283°K. to 284°K. (change of 1°K.). Most of the time we are concerned with temperature changes rather than actual temperatures, and the relations between temperature changes then become important.

Note that the apparatus for measuring temperature is the mercury thermometer. The property involved is the change in volume of the liquid mercury with temperature. Thus, by measuring the volume occupied by a drop of mercury in the thermometer bulb, we can relate this volume to the temperature of the mercury.

Do problems 4 and 5 at the end of the chapter.

4. Heat Energy—Specific Heat

Since the energy involved in chemical changes is usually in the form of thermal energy, this type of energy will be emphasized. We have seen that temperature is a measure of the intensity of heat energy of a body. When two bodies have the same intensity of heat energy they will be at the same temperature and if brought together no heat energy will pass from one to the other. When two bodies are at different temperatures, heat energy will flow from the hotter one to the colder one until they both come to the same temperature. The absolute temperature scale is theoretically important since 0°K. is the lowest attainable temperature (0°R. on the Rankine scale).

It is found experimentally that 1 cal. of heat energy is required to raise the temperature of 1 g. of water by 1°C. (starting at 15°C.). While defining the calorie of heat energy, this observation also defines an *intensive property* of water known as the *specific heat*. In general, 1 g. of each different substance will require a different number of calories to raise its temperature by 1°C. This number of calories is called the *specific heat* of the substance.

Definition:

$$\text{Specific heat of a substance} = \frac{\text{Change in heat energy}}{\text{Mass} \times \text{Temperature change}}$$

The units of specific heat are calories per gram per degree Centigrade (cal./g.-°C.). The following Table VII gives the specific heats of a few common substances.

TABLE VII
Specific Heats of Some Common Substances at 15°C.

Substance	Specific Heat (cal./g.-°C.)	Substance	Specific Heat (cal./g.-°C.)
Water	1.000	Iron (α)	0.107
Mercury	0.0333	Zinc	0.092
Graphite (C)	0.17	Sodium chloride	0.206
Diamond (C)	0.12	Ammonium nitrate	0.398
Copper	0.092	Potassium chlorate	0.197
Iodine	0.052	Sulfuric acid	0.340
Lead	0.306	Ethyl alcohol	0.581
Silver	0.056	Benzene	0.406

It should be pointed out that the specific heat of a substance is slightly different at different temperatures just as the density of solids and liquids is slightly different at different temperatures.

The specific heat is a complex property relating the number of calories required to change the temperature of 1 g. of an object and the temperature change brought about. As such it can be used as a conversion factor to relate heat energy per gram and temperature change of a given substance.

Example: 250 cal. of heat energy will raise the temperature of 50 g. of iron by 46.7°C. What is the specific heat of iron?

Answer: $$\text{Specific heat} = \frac{\text{Heat energy}}{\text{Mass} \times \text{Temperature change}}$$

$$= \frac{250 \text{ cal.}}{50 \text{ g.} \times 46.7°\text{C.}} = 0.107 \frac{\text{cal.}}{\text{g.-}°\text{C.}}$$

Example: The specific heat of water is 1.00 cal./g.-°C. How many calories are required to raise the temperature of 20 g. of water by 15°C.?

Answer: $$\text{Heat energy} = 15°\cancel{\text{C.}} \times \left(\frac{1 \text{ cal.}}{\text{g.-}°\cancel{\text{C.}}}\right) \times 20 \cancel{\text{g.}}$$

$$= 300 \text{ cal.}$$

Thus it will require 300 cal. to raise the temperature 15°C.

Do problems 6–9 at the end of the chapter.

5. Molar Heat Capacity—Law of Dulong and Petit

The molar heat capacity is defined as the energy required to raise the temperature of 1 mole of a substance by 1°C. It is frequently used since it employs the theoretical units of moles rather than the practical mass units of grams. We can relate it to the specific heat by using the conversion factor of molecular weight.

Definition:

Molar heat capacity = Specific heat × Molecular weight

Its units are calories per mole per degree Centigrade (cal./mole-°C.), and it is symbolized by C_p, the subscript p meaning that the property is measured at a fixed pressure, usually 1 atmosphere.

Example: The specific heat of water is 1.00 cal./g.-°C. What is its molar heat capacity?

Answer:

$$\text{Molar heat capacity of water} = \left(\frac{1.00 \text{ cal.}}{\cancel{\text{g.}}-°\text{C.}}\right) \times \left(\frac{18 \cancel{\text{g. } H_2O}}{1 \text{ mole } \cancel{H_2O}}\right)$$

$$= \frac{18 \text{ cal.}}{\text{mole-}°\text{C.}}$$

One of the interesting discoveries of the last century was that the molar heat capacity of metals is roughly 6 cal./mole-°C. This law, known as the Law of Dulong and Petit, is accurate to within about 10% for all metals. It was originally used to obtain crude atomic weights for the metallic elements.

Example: The specific heat of lead is 0.0306 cal./g-°C. What is its atomic weight?

Answer: From the Law of Dulong and Petit: molar heat capacity of lead = 6 cal./mole-°C. From the definition of molar heat capacity:

$$\text{Atomic weight of lead} = \frac{\text{Molar heat capacity}}{\text{Specific heat}}$$

$$= \frac{6.0 \ \text{cal./mole-}°C.}{0.0306 \ \text{cal./g.-}°C.} = 196 \ \text{g./mole}$$

which is pretty close to the true atomic weight of 207 g./mole.

Do problems 11–14 at the end of the chapter.

6. Heat Content

The fact that an object has the ability to change temperature indicates that it always possesses a certain amount of heat energy. At absolute zero, when it is incapable of being cooled further, it will have no more heat energy. However, substances contain another form of energy which is not capable of being removed simply by heating or cooling but can be released only when the atoms making up the substance are rearranged, as in a chemical reaction. This latent (hidden) chemical energy is also part of the total energy of the substance, just as the potential energy in a coiled spring must be counted as part of the energy of the spring. The sum of the heat energy and this internal chemical energy is called the *heat content* of a substance. There is no simple, direct way of measuring the total heat content of a substance, but we can measure the changes in heat content which occur when chemical changes take place, and these in turn can be related to the heat content. The heat content is also known as the *enthalpy*. Its symbol is H.

7. Heats of Formation

Since we can't have, or at least don't need, an absolute measure of the total heat energy or *heat content* of a substance, but merely a relative measure, we must establish a reference point or standard from which to measure this heat content for each substance. This can be done by measuring the heat content of a compound with respect to the elements which make it up. Thus we find that when carbon (graphite) reacts with oxygen (gas) at 20°C., carbon dioxide (gas) is produced and 94.0 kilocalories (kcal.) of heat are liberated for every mole of CO_2 (gas) produced. This thermal energy can be written in the balanced equation as one of the products:

$$C\ (s) + O_2\ (g) \rightarrow CO_2\ (g) + 94.0\ \text{kcal.}$$

The significance of this equation is that 1 mole of CO_2 (g) contains 94.0 kcal. less heat than the elements from which it is made. [Note that the parentheses refer to the physical state: (g), gas; (s), solid; (l), liquid.]

We will define the total heat content of a substance with reference to the elements from which it is made. To make an intensive property of it we will define it as the *heat content per mole* or molar heat content. By placing all free elements at zero on our reference scale, we see that the molar heat content of a substance is equal to minus the heat liberated or absorbed when 1 mole of the compound is formed from its elements. This quantity is also referred to as the heat of formation. Its usual symbol is H_f.

Definition:

Heat of formation = Molar heat content (Enthalpy)

(ΔH_f) = The heat absorbed [or minus (−) the heat liberated] when 1 mole of a compound is formed from the free elements (all substances being in their standard states)

Since the total heat energy depends slightly on temperature and pressure, we will take as our reference or standard set of conditions 25°C. and 1 atm. total pressure.*

From the example quoted, carbon dioxide at 25°C. and 1 atm. pressure has a heat of formation of −94.0 kcal./mole or, alternatively, a total heat content of −94.0 kcal./mole at 1 atm. pressure and 25°C.

Conversely this implies that if there were a way of breaking carbon dioxide back into its elements (the reverse reaction) it would require an amount of work equal to 94.0 kcal./mole.

In Table VIII are given heats of formation of a few compounds at standard conditions.

* In this chapter, all calculations of heats of chemical reaction should properly be called standard heats, since they refer to the heat change when the reaction occurs at 25°C. and 1 atm. pressure (standard conditions). It is possible to estimate from these standard heats what the heat changes would be at other temperatures if we know the molar heat capacities of each of the substances involved. From this latter information we can compute the heat contents of each of the substances at the different temperature and, from the difference in heat contents between reactants and products, the heat of reaction at that temperature.

TABLE VIII
Some Heats of Formation at Standard Conditions

Substance	Heat of Formation ΔH_f (kcal./mole)	Substance	Heat of Formation ΔH_f (kcal./mole)
CO (g)	-26.4	$NaCl$ (s)	-98.2
CO_2 (g)	-94.0	$KClO_3$ (s)	-93.5
H_2O (l)	-68.3	H_2SO_4 (l)	-193.9
SO_2 (g)	-70.9	$NaOH$ (s)	-102.0
HCl (g)	-22.0	CH_4 (g)	-17.9 (methane)
NO (g)	$+21.6$	C_2H_6 (g)	-20.2 (ethane)
NO_2 (g)	$+7.9$	C_2H_2 (g)	$+54.2$ (acetylene)
H_2S (g)	-4.8	C_6H_6 (l)	$+12.5$ (benzene)
NH_3 (g)	-11.0	H_2O (g)	-57.8
SO (g)	-94.6		

A negative heat of formation (or heat content) signifies that heat is given out when the compound is formed from its elements (the reaction is *exo*thermic). Alternatively, a positive heat of formation signifies that heat is absorbed when the compound is formed from its elements (i.e., the reaction is *endo*thermic).

8. Heats of Reaction

If we know the heat contents of all the substances participating in a reaction (i.e., their heats of formation), then we can calculate the heat of the reaction. This follows since, from the law of conservation of energy, the heat of the reaction must simply be the difference in heat contents of the products and the reactants.

Definition:

$$\text{Heat of reaction} = \left(\begin{array}{c}\text{Total heat content}\\ \text{of products}\end{array}\right) - \left(\begin{array}{c}\text{Total heat contents}\\ \text{of reactants}\end{array}\right)$$

This equation simply expresses an energy balance for the reaction. If the total heat contents of the products are less than the total heat contents of the reaction, then heat must have been liberated (exothermic reaction). If the total heat contents of the products are greater than the total heat contents of the reactants, then heat must have been absorbed during the reaction and the reaction is endothermic.

Example:

$$2CO\ (g) + O_2\ (g) \rightarrow 2CO_2\ (g)$$

What is the heat of this reaction, given the heats of formation? $\Delta H_f(CO) = -26.4$ kcal./mole; $\Delta H_f(CO_2) = -94.0$ kcal./mole. *Note:* $\Delta H_f(O_2) =$ zero, since it is a free element.

Answer: By definition:

$$\text{Heat of reaction} = \left(\begin{array}{c}\text{Total heat contents}\\ \text{of products}\end{array}\right) - \left(\begin{array}{c}\text{Total heat contents}\\ \text{of reactants}\end{array}\right)$$

$$\Delta H = [2\Delta H_f(CO_2)] - [2\Delta H_f(CO) + \Delta H_f(O_2)]$$

$$= (-2 \times 94.0) - [2(-26.4) + 0]$$

$$= (-188.0) - (-52.8)$$

$$= -135.2 \text{ kcal.}$$

This means that the reaction is exothermic, since the total heat contents of the products are *less* (observe the minus sign) than the total heat contents of the reactants. That is, heat is a product of the reaction.

In similar fashion, if we know the heat of a reaction and the heats of formation of all but one of the compounds present in the equation, we can calculate the heat of formation of this compound.

Example:
$$2C_2H_6 \ (g) + 7O_2 \ (g) \rightarrow 4CO_2 \ (g) + 6H_2O \ (l) + 746 \text{ kcal.}$$

If the heats of formation are: $\Delta H_f(CO_2) = -94.0$ kcal./mole and $\Delta H_f(H_2O) = -68.3$ kcal./mole, what is $\Delta H_f(C_2H_6)$?

Answer: From the definition:

$$\text{Heat of reaction} = \left(\begin{array}{c}\text{Total heat contents}\\ \text{of products}\end{array}\right) - \left(\begin{array}{c}\text{Total heat contents}\\ \text{of reactants}\end{array}\right)$$

But the heat of reaction $= -746$ kcal., since the reaction is exothermic (i.e., heat is given out). Thus

$$-746 \text{ kcal.} = [4\Delta H_f(CO_2) + 6\Delta H_f(H_2O)] - [2\Delta H_f(C_2H_6)]$$

Solving for $\Delta H_f(C_2H_6)$, we have

$$\Delta H_f(C_2H_6) = 2\Delta H_f(CO_2) + 3\Delta H_f(H_2O) + 373$$

and, on substitution,

$$\Delta H_f(C_2H_6) = 2(-94.0) + 3(-68.3) + 373$$

$$= -188.0 - 204.9 + 373$$

$$= -20 \text{ kcal.}$$

Example:
Solutions of HCl and NaOH will react to neutralize each other, as follows:
$$HCl(aq.) + NaOH(aq.) \rightarrow NaCl(aq.) + H_2O(l)$$
The symbol (aq.) refers to a water solution (aqueous) of the substance involved. If the standard heats of formation of the species are as follows, calculate the molar heat of neutralization.

Answer:

ΔH_f (HCl(aq.)) = -40.0 kcal./mole ΔH_f [NaCl (aq.)] = -97.3 kcal./mole
ΔH_f (NaOH(aq.)) = -112.2 kcal./mole ΔH_f [H$_2$O(l)]
        = -68.3 kcal./mole (Table VIII)

Heat of neutralization = Heat of reaction
= [ΔH_f(NaCl(aq.)) + ΔH_f(H$_2$O(l))] $-$ [ΔH_f(HCl(aq.)) + ΔH_f(NaOH(aq.))]
= [$-97.3 - 68.3$] $-$ [$-40.0 - 112.2$]
= $-165.6 - [-152.2]$ = -13.4 kcal./mole

Since the molar heat of neutralization is negative, the reaction is exothermic, i.e., heat is given out on neutralization.

Example:

When 1 mole of gas HCl at 25°C, is dissolved in 1 liter of water, 18.0 kcal. of heat are liberated. Using the data in Table VIII, calculate the molar heat of formation of HCl (aqueous).

Answer:

The reaction is:

$$\text{HCl}(g) \xrightarrow{\text{(water)}} \text{HCl (aq.)}$$

Heat of reaction = ΔH_f[HCl(aq.)] $-$ ΔH_f[HCl(g)]

or, on rearranging and solving for HCl(aq.):

$$\Delta H_f \text{[HCl(aq.)]} = \text{Heat of reaction} + \Delta H_f \text{[HCl}(g)\text{]}$$
$$= -18.0 - 22.0$$
$$= -40.0 \text{ kcal./mole}$$

Do problems 15 and 16 at the end of the chapter.

9. Heat Calculations

For calculating the amounts of heat liberated or absorbed when given amounts of a compound react in a chemical reaction, the balanced equation containing the heat of reaction can be used to obtain a heat-chemical conversion factor.

Example: The combustion of butane liberates energy according to the following:

$$2\text{C}_4\text{H}_{10} (g) + 13\text{O}_2 (g) \rightarrow 8\text{CO}_2 (g) + 10\text{H}_2\text{O} (g) + 1360 \text{ kcal.}$$

How much heat will be liberated by burning 10 g. of C$_4$H$_{10}$?

Answer:

$$10 \text{ g. C}_4\text{H}_{10} = 10 \text{ g. C}_4\text{H}_{10} \times \left(\frac{1 \text{ mole C}_4\text{H}_{10}}{58 \text{ g. C}_4\text{H}_{10}} \right) \times \left(\frac{1360 \text{ kcal.}}{2 \text{ moles C}_4\text{H}_{10}} \right)$$
$$= 117 \text{ kcal.}$$

Do problems 17 and 18 at the end of the chapter.

10. Additivity Laws for Molecular Properties; C_p and ΔH_f

Many properties of molecules can be looked on as arising from fixed contributions coming from each part of the molecule. Thus, the mass of a molecule is just the sum of the masses of the individual atoms present in it. This is an exact relation and a reflection of the law of Conservation of Mass. Scientists in studying the properties of molecules usually examine molecular properties to see if there are similar properties of atoms that are conserved in the molecules they form. It turns out that there are very few other properties such as mass, which are conserved. However, many molecular properties are approximately conserved.

Two such properties are the molar heat capacity, C_p, and the molar heat of formation, ΔH_f. If we write out the structure of a covalently bonded molecule, then we find that to each bond we can assign a value of C_p and ΔH_f, such that the value of C_p or ΔH_f is simply the sum of the contributions from each of these bonds.

Table IX lists values of C_p and ΔH_f for some bonds in gas-phase molecules, mostly organic. From these values, we can deduce the approximate values of ΔH_f and C_p for gas-phase molecules containing these bonds.

TABLE IX

SOME VALUES OF MOLAR HEAT CAPACITY, C_p, AND HEAT OF FORMATION, ΔH_f, FOR DIFFERENT BONDS IN COVALENT MOLECULES IN THE GAS-PHASE AT 25° C. AND 1 ATMOSPHERE

Bond	Molar Heat Capacity, C_p (cal./mole —°C)	Heat of Formation ΔH_f (kcal./mole)
C-H	1.74	−3.83
C-C	1.98	2.73
C-Cl	4.64	−7.4
C-Br	5.14	2.2
C-I	5.54	14.1
C-O	2.7	−12.0
O-H	2.7	−27.0
C-N	2.1	9.3
N-H	2.3	−2.6
S-H	3.2	−0.8
N-N	1.9	+32.4
C-S	3.4	6.7

Example: Calculate the molar heat capacity and heat of formation of the molecule C_2H_6 (ethane) at 25°C. and 1 atmosphere pressure.

Answer: To use this table, we must be able to write out the molecular structure in terms of bonds. For C_2H_6, it is:

$$\begin{array}{ccc} H & H \\ | & | \\ H-C-C-H \\ | & | \\ H & H \end{array}$$

The molecule has 6 C-H bonds and 1 C-C bond; hence from Table IX:

$$\begin{aligned} C_p &= 6 \times 1.74 + 1 \times 1.98 \\ &= 10.44 + 1.98 = 12.42 \rightarrow 12.4 \text{ cal./mole-°C.} \end{aligned}$$

$$\begin{aligned} \Delta H_f &= 6 \times (-3.83) + 1 \times 2.73 \\ &= -22.98 + 2.73 = -20.25 \rightarrow -20.3 \text{ kcal./mole} \end{aligned}$$

These can be compared with the measured values $C_p[C_2H_6(g)] = 12.7$ cal./mole-°C. and $\Delta H_f[C_2H_6(g)] = -20.2$ kcal./mole (see Table VIII).

Example: Calculate the molar heat capacity and heat of formation of ethyl alcohol (C_2H_5OH) gas.

Answer: The structure is:

$$\begin{array}{ccc} H & H \\ | & | \\ H-C-C-O-H \\ | & | \\ H & H \end{array}$$

which has 5 C-H bonds, 1 O-H bond, 1 C-C bond, and 1 C-O bond. Thus, from Table IX:

$$\begin{aligned} C_p &= 5 \times 1.74 + 1 \times 2.7 + 1 \times 1.98 + 1 \times 2.7 \\ &= 16.1 \text{ cal./mole-°C.} \end{aligned}$$

$$\begin{aligned} \Delta H_f &= 5 \times (-3.83) + 1 \times (-27.0) + 1 \times (2.73) + 1 \times (-12.0) \\ &= -55.4 \text{ kcal./mole} \end{aligned}$$

Measured values are: $C_p = 15.7$ cal./mole-°C.; $\Delta H_f = -56.2$ kcal./mole

---

Values of C_p computed from this Table IX are usually accurate to within $\pm$ 1.5 cal./mole-°C. and values of ΔH_f are usually accurate to within $\pm$ 2.5 kcal./mole.

11. Heat Changes Accompanying Physical Changes—Intermolecular Forces

Heat changes generally accompany physical changes as well as chemical changes. Table X lists some definitions of such heat changes. Remember, when heat change is negative, it means heat is liberated in the change.

TABLE X
DEFINITION OF HEATS OF PHYSICAL CHANGES

Term	Symbol	Definition
Heat of vaporization	ΔH_v	Heat *absorbed* when 1 mole of liquid or solid is changed to the gaseous state.
Heat of condensation	ΔH_c $(\Delta H_c = -\Delta H_v)$	Heat *liberated* when 1 mole of gas is condensed to a liquid or solid. Opposite of ΔH_v.
Heat of melting	ΔH_m	Heat absorbed when 1 mole of solid is melted.
Heat of solution	ΔH_s	Heat liberated *or* absorbed when 1 mole of substance (solute) is dissolved in a fixed quantity of solvent.

Before closing the chapter it is well to say something about the origin of these latent heat energies. Where do these energies of chemical change and physical change come from?

The answer to this question is to be found in the forces that act between molecules. The property that characterizes liquids and solids is their cohesiveness. It takes work to separate the molecules in a liquid or a solid. This work we call the energy of vaporization, and it acts to overcome the attractive forces between the molecules making up the liquid or solid. Conversely, where we permit a gas to condense to a liquid or solid, the attractive forces this time do work, and it is released as heat energy.

A chemical change involves a rearrangement of the atoms in one or more compounds or elements. If energy is required to overcome the attractive forces in order to make the rearrangement, then we say that heat is absorbed. If, on the other hand, the attractive forces are greater in the rearranged form (their products), then these intermolecular forces have performed work and heat energy is liberated.

When a solid is heated to a sufficiently high temperature, it will usually melt to a liquid. The heating is usually carried out at ambient pressures, i.e., 1 atmosphere. From some solids, however, vaporization will take place before melting can occur. Thus, solid CO_2 (dry ice) will not melt if heated but will vaporize directly to a gas. This vaporization process of transforming a solid directly to a gas is referred to as sublimation. The heat required for it is called the heat of sublimation, ΔH_{sub}.

A number of solids, when heated under 1 atmosphere external pressure, will sublime rather than melt. Acetylene and CO_2 are two examples. In the case of CO_2, if the external pressure is raised to 5.2 atmospheres or above, then melting will occur prior to vaporization. At 5.2 atmosphere pressure, CO_2 melts at $-56.6°C$. Acetylene (C_2H_2) melts at $-81.5°C$. if the pressure exceeds 1.5 atmospheres.

Any solid can be made to sublime at a sufficiently low pressure. Thus, ice will sublime directly to water vapor at $0°C$., if the pressure is below 4.5 torr (i.e., 0.0059 atm.). The heat of sublimation, ΔH_{sub}, is just equal to the heat of melting, ΔH_m, plus the heat of vaporizing the liquid:

$$\Delta H_{sub} = \Delta H_m + \Delta H_v$$

This is a consequence of the law of conservation of energy which states that energy can never be created or destroyed, but only transformed from one form to another:

A. Solid $\rightleftarrows$ Gas — ΔH_{sub}

B. $\begin{cases} \text{Solid} \rightleftarrows \text{Liquid} — \Delta H_m \\ \text{Liquid} \rightleftarrows \text{Gas} — \Delta H_v \end{cases}$

If we transform a solid into a liquid by adding energy (heating) in the amount ΔH_m kcal./mole, and then add an additional amount ΔH_v kcal./mole to change the liquid into a vapor, the final mole of vapor will have a heat content higher by the sum ($\Delta H_m + \Delta H_v$) kcal. than the original mole of solid. If we directly sublime (process A above) 1 mole of solid to 1 mole of vapor by adding ΔH_{sub} kcal., the final mole of vapor must have the same heat content as it did in process B above. Hence, $\Delta H_{sub} = \Delta H_m + \Delta H_v$.

Do problems 10, 19, and 20 at the end of the chapter.

12. Problems

1. If you push against a wall it does not move (usually), and thus, according

to our definition, no mechanical work is done. Nevertheless, you will become tired from this effort. Explain.

2. The force of gravity is 980 dynes/g. If it acts on a 10.0-g. mass during the time in which the body falls a distance of 1.00 m., having initially been at rest, calculate the increase in kinetic energy of the body. What is the final velocity of the body?

3. Make the following conversions:

(a) 500 cal. to ergs.   (c) 8 joules to calories.
(b) 6×10^8 ergs to kilocalories.   (d) 60 ergs to microjoules.

4. Make the following conversions in temperature scale:

(a) 150°C. to °K.   (g) −100°F. to °K.
(b) 0°K. to °F. (absolute zero).   (h) 50°F. to °R.
(c) 0°K. to °C.   (i) 10°R. to °F.
(d) 1000°K. to °C.   (j) 500°C. to °R.
(e) 500°K. to °F.   (k) 90°R. to °K.
(f) −40°C. to °F.

5. When the temperature of a body increases by 15°C., what does this increase correspond to in °K.? in °F.?

6. From Table VII, calculate the number of calories required to raise the temperature of 40 g. of copper from 100°C. to 150°C.

7. How many calories are required to raise the temperature of 1.0 lb. of water from 0°C. to the boiling point?

8. Compute the specific heat of water in units of ergs/lb.-°F.

9. Assuming that the specific heat of iron did not change, calculate how many calories of heat would have to be removed to cool 1.00 mole of Fe from 25°C. to 50°K?

10. Which would you think would require more energy to separate into its individual molecules, a substance at a high temperature or the same substance at a lower temperature? Explain your reasoning.

11. It takes 80 cal. to raise the temperature of 20 g. of wool by 16°C. What is the specific heat of wool?

12. An unknown metal has a specific heat of 0.150 cal./g. °C. What is the atomic weight of the metal, roughly? Could you guess from the Table of Atomic Weights which metal this might be?

13. From the knowledge of the atomic weight of nickel (58.7), what would you estimate its specific heat to be?

14. How many ergs are required to raise the temperature of 1 molecule of liquid water by 10°C.?

15. Calculate the heat of the following reactions (using the values for ΔH_f in Table VIII):

(a) $2NO + O_2 \rightarrow 2NO_2$
(b) $CH_4 + 2O_2 \rightarrow CO_2 + 2H_2O(l)$
(c) $C + H_2O(g) \rightarrow CO + H_2$
(d) $C_2H_2 + 2H_2 \rightarrow C_2H_6$
(e) $2Na + 2H_2O(l) \rightarrow 2NaOH(s) + H_2$
(f) $H_2S + 2O_2 \rightarrow H_2SO_4(l)$

16. For the following reactions, calculate the heat of formation of the single compound which in each case is not listed in Table VIII:

(a) $4C_3H_5(NO_3)_3 + 11O_2 \rightarrow 12CO_2 + 12NO_2 + 10H_2O + 1330$ kcal. (combustion of nitroglycerine).

(b) $6CO_2 + 6H_2O \rightarrow C_6H_{12}O_6$ (sugar) $+ 6O_2 - 676$ kcal. (photosynthesis of fructose).

(c) $3SO_2 + O_3 \rightarrow 3SO_3 + 105.6$ kcal.

(d) $2NaOH + H_2SO_4 \rightarrow Na_2SO_4 + 2H_2O + 30$ kcal.

17. For the reaction shown in problem 16(a), how many liters of oxygen at STP are required to produce 500 kcal. of heat?

18. How many kilocalories of heat energy are required by a plant to produce 50 g. of fructose [reaction 16(b) above]?

19. The heat of vaporization of water is 540 cal./g. How many calories are required to vaporize 200 g. of water? What is the molar heat of vaporization of water?

20. The heat of fusion of ice is 80 cal./g.

(a) What is the molar heat of fusion of ice?

(b) How many calories are required to melt 500 g. of ice?

(c) How many calories are required to transform 50 g. of water from ice at 0°C. to steam at 100°C. (See problem 19 for heat of vaporization.)

21. When H_2 (gas) and O_2 (gas) react to form H_2O (liquid), 68.3 kcal. of heat are liberated per mole of H_2O formed. Would this be more or less if we used liquid H_2 and liquid O_2 as our starting materials? Explain.

22. In an automobile engine, liquid hydrocarbons are burned in air to form CO_2 (gas) and H_2O (gas). Would more or less energy be produced in the combustion if the product H_2O were formed in the liquid state rather than as a gas? Explain.

Note: The heats of combustion of most hydrocarbons correspond to about 115 kcal. per mole of gas phase H_2O formed, whereas the heat of vaporization of water is 10 kcal./mole.

23. If a person goes on a diet to lose weight does he lose more weight by consuming very cold or very hot foods? Explain.

24. How much work is expended by a 150 lb. person who climbs a 5000 ft. mountain? (Consider only work against gravity.) Express your answer in ft-lb, in ergs, and in kcal.

25. The average velocity of an O_2 molecule at 25°C. is about 4.8×10^4 cm./sec. Calculate the total kinetic energy of a mole of O_2 gas at 25°C. in calories.

26. The heat of vaporization of benzene (C_6H_6) is 7.3 kcal./mole at 25°C. From the ΔH_f ($C_6H_6(l)$) given in Table VIII, calculate ΔH_f ($C_6H_6(g)$).

27. From the values in Table VIII, calculate the heat of the synthesis reaction:
$$2CO(g) + 5H_2(g) \rightarrow C_2H_6(g) + 2H_2O(l)$$

28. It is found that the heat of neutralization of a 0.1 molar solution of acetic acid with 0.1 molar NaOH is -11 kcal./mole. The heat of neutralization of the strong, completely ionized acid, 0.1 molar HCl is -13.4 kcal./mole. Can you offer a reason for the difference? Note that acetic acid is a weak acid.

29. From the data in Table IX, calculate C_p and ΔH_f for the following gas-phase molecules:

(a) $n - C_4H_{10}$ (structure is $CH_3 - CH_2 - CH_2 - CH_3$) butane

(b) $CH_3 - CH_2Cl$ (ethyl chloride)

(c) $CH_3 - NH_2$ (methylamine)

(d) $CHCl_3$ (chloroform)

30. Using the data in Tables VIII and IX only as needed, calculate the heat of the following reactions:

(a) $C_2H_6(g) + 7Cl_2(g) \rightarrow 2CCl_4(g) + 6HCl(g)$

(b) $CH_3SH(g) + 3O_2(g) \rightarrow CO_2(g) + SO_2(g) + 2H_2O(l)$

CHAPTER VI

The Properties of Gases

1. Measurement of Gases

As noted in Chapter II, the method employed for measuring a quantity of matter depends on the physical state of the matter. For solids, direct weighing is most convenient. For liquids, direct measurement of volume proves to be most convenient, but additional information is needed to convert from volume units to mass units, namely, the density of the liquid.

For gases, direct weighing is extremely difficult because gases are so many times lighter than the containers needed to hold them. For this reason, gases are measured out in terms of volume, and to convert to mass units additional information is needed giving the density of the gas.

The density of a substance is defined as the mass divided by the volume. The volume of a given quantity of liquid or solid depends very little on its temperature or pressure. On the other hand, the volume occupied by a given quantity of gas is much more sensitive to its temperature and pressure. Fortunately, there are laws which enable us to calculate, if we know the volume occupied by a given quantity of gas at one temperature and pressure, precisely what the volume will be at any other temperature and pressure. Thus it is possible from a knowledge of the density of a gas at any one pressure and temperature to compute the density of the same gas at any other temperature and pressure. This is a great convenience in measuring quantities of gas.

2. Pressure

The pressure exerted by a gas upon the walls of its container is the same at every part of the vessel. It may be measured by a pressure gauge or a mercury manometer. The latter is more frequently used in the laboratory, and the principle on which it is based is that the pressure exerted by a gas on a liquid in a U-tube can be counterbalanced by the weight of liquid in the other arm of the U-tube (Figure 1).

The unbalance in the arms of the manometer is caused by the pressure of the gas on one of these arms. We can measure the unbalance in units of length, and this provides a direct measure of the pressure exerted by the gas. Thus it is customary to speak of a pressure of 20 cm. Hg, meaning that the unbalance in the arms of the manometer (i.e., the difference between the two levels) is 20 cm., and the manometer fluid is mercury.

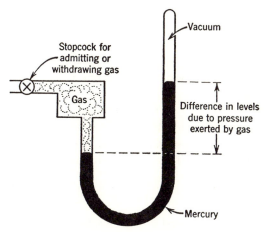

FIG. 1.   Mercury manometer.

These units can be related to the units of dynes by the properties of mercury (Hg). The pressure 20.0 cm. below the upper level of a tube full of mercury is simply the force per unit area due to the weight of the mercury above. The volume of a column of mercury 20.0 cm. high and having a cross-sectional area of 1 cm.^{2} is 20.0 cm.^{3}. Multiplying this by the density we obtain the mass:

$$20.0 \text{ cm.}^3 \text{ Hg} = 20.0 \text{ cm.}^3 \text{ Hg} \times \left(\frac{13.55 \text{ g. Hg}}{1 \text{ cm.}^3 \text{ Hg}} \right) = 271.0 \text{ g. Hg}$$

The force of gravity is 980 dynes/g., and so the total force of gravity acting on this column of mercury is

$$\text{Force} = 271.0 \text{ g.} \times \frac{980 \text{ dynes}}{\text{g.}}$$

$$= 265{,}900 \text{ dynes} = 2.66 \times 10^5 \text{ dynes}$$

We thus have the total pressure in terms of force per unit area.

$$20.0 \text{ cm. Hg pressure} = \left(\frac{20.0 \text{ cm.}^3 \text{ Hg}}{1 \text{ cm.}^2}\right) \times \left(\frac{13.55 \text{ g. Hg}}{1 \text{ cm.}^3 \text{ Hg}}\right) \times \left(\frac{980 \text{ dynes}}{1 \text{ g. Hg}}\right)$$

$$= 2.66 \times 10^5 \text{ dynes/cm.}^2$$

or $\qquad$ $1.00 \text{ cm. Hg} = 1.33 \times 10^4 \text{ dynes/cm.}^2$

Another unit for measurement of pressure is the atmosphere (atm.).

Definition: $\qquad$ 1 atm. = 76 cm. Hg (exact)

Then $\qquad$ $1 \text{ atm.} = 76 \text{ cm. Hg} \times \left(\frac{1.33 \times 10^4 \text{ dynes/cm.}^2}{1 \text{ cm. Hg}}\right)$

$$= 1.01 \times 10^6 \text{ dynes/cm.}^2$$

In honor of the Italian physicist Torricelli, who first designed the mercury barometer for measuring atmospheric pressure, the unit *torr* has been defined:

$$1 \text{ torr} = 1 \text{ mmHg}$$

Thus: $\qquad$ 1 atm. = 760 torr

The English system uses pounds force per square inch or simply pounds per square inch. The student should show that this can be related to atmospheres:

$$1 \text{ atm.} = 14.7 \text{ lb./sq. in.}$$

Do problems 1–6 at the end of the chapter.

3. Boyle's Law—Pressure and Volume of a Gas

If a given mass of gas (m) is kept in a vessel and its volume changed, then its pressure will change in such a manner that the product of the pressure (P) and volume (V) do not change. This is known as Boyle's law and is true only when the temperature (T) of the gas is kept fixed. Algebraically:

Boyle's Law: $\qquad$ $P \times V = \text{Constant}$ $\quad$ (if T and m are kept fixed)

The constant in this equation will, of course, be different for each temperature and for each different quantity of gas. It is, however, important to know that once these are fixed it is the same for each gas! That is, 10 cc. of hydrogen gas and 10 cc. of oxygen gas behave in the same way. They have the same constant.

This law permits us to calculate the volume (of a fixed mass of gas, kept at constant temperature) at any pressure if we know the volume of the gas at any one pressure.

Example: What volume will be occupied by 10 l. of a gas originally measured at a pressure of 15 lb./sq. in. if it is compressed to 90 lb./sq. in.?

Answer: Since, from Boyle's law, $P \times V$ after the compression must be the same as $P \times V$ before the compression:

$$P_{\text{final}} \times V_{\text{final}} = P_{\text{initial}} \times V_{\text{initial}}$$

and, solving for the final volume:

$$V_{\text{final}} = \frac{P_{\text{initial}}}{P_{\text{final}}} \times V_{\text{initial}}$$

and substituting:

$$V_{\text{final}} = \frac{15 \text{ lb./in.}^2}{90 \text{ lb./in.}^2} \times 10 \text{ l.}$$

$$= 1.67 \text{ l.} \rightarrow 1.7 \text{ l.}$$

Do problems 7 and 8 at the end of the chapter.

4. Charles' Law—Temperature and Volume

It is found by experiment that, if the temperature of a given quantity of gas (m) is changed, then its volume will change in a corresponding manner such that the ratio of the volume (V) to the *absolute temperature* (T) does not change. This is known as Charles' law and is true only if the pressure of the gas is kept fixed. Algebraically:

Charles' Law:

$$\frac{V}{T} = \text{Constant} \quad \text{(if } P \text{ and } m \text{ are kept fixed)}$$

Once again, the constant in this equation will be different for each different pressure and each different quantity of gas (m), but, once these two are fixed, the ratio does not change.

Example: A quantity of gas at 20°C. and 1 atm. pressure occupies a volume of 200 cc. What volume will it occupy at −40°C., the pressure being kept fixed?

Answer: From Charles' law, the ratio of the volume to *absolute* temperature must be the same before and after the experiment:

$$\frac{V_{\text{final}}}{T_{\text{final}}} = \frac{V_{\text{initial}}}{T_{\text{initial}}}$$

Solving for V_{final} (the final volume):

$$V_{\text{final}} = \left(\frac{T_{\text{final}}}{T_{\text{initial}}}\right) \times V_{\text{initial}}$$

and substituting:

$$V_{\text{final}} = \left(\frac{233°\text{K.}}{293°\text{K.}}\right) \times 200 \text{ cc.}$$

$$= 159 \text{ cc.}$$

Do problems 9 and 10 at the end of the chapter.

5. Combined Form of Gas Laws

In the usual experiment it is rare that the pressure and temperature are kept fixed, and it is important to have a law which tells us how the volume of a gas will change under these circumstances.

Charles' and Boyle's laws can be combined into a single law which states:

Combined Gas Law: The product of the pressure and volume of a given quantity of gas divided by the absolute temperature is a constant which does not change. Even though the individual properties may change in an experiment, this ratio does not.

Algebraically:

$$\frac{P \times V}{T} = \text{Constant} \qquad \text{(if } m \text{ is fixed)}$$

From this combined law we can calculate the way in which the volume or pressure or temperature changes if the initial conditions (P, V, T) are known and two of the final conditions are known (i.e., two of the three quantities, P, V, T).

Example: A quantity of gas has a volume of 180 cc. at 40°C. and a pressure of 80 cm. Hg. What will this volume be at standard conditions (0°C. and 76 cm. Hg)?

Answer: From the combined gas law:

$$\frac{P_{\text{final}} \times V_{\text{final}}}{T_{\text{final}}} = \frac{P_{\text{initial}} \times V_{\text{initial}}}{T_{\text{initial}}}$$

Solving for V_{final}:

$$V_{\text{final}} = V_{\text{initial}} \times \left(\frac{P_{\text{initial}}}{P_{\text{final}}}\right) \times \left(\frac{T_{\text{final}}}{T_{\text{initial}}}\right)$$

and, substituting, we have

$$V_{\text{final}} = 180 \text{ cc.} \times \left(\frac{80 \text{ cm. Hg}}{76 \text{ cm. Hg}}\right) \times \left(\frac{273°\text{K.}}{313°\text{K.}}\right)$$

$$= 165 \text{ cc. STP}$$

Note: In the preceding examples, it should be noted that cooling a gas causes it to contract and heating causes it to expand. Similarly, compressing a gas reduces its volume and decreasing its pressure increases the volume. These rules provide a quick check on possible errors.

Thus, in the above example we must cool the gas from 40°C. (313°K.) to 0°C. (273°K.). This should reduce its volume, and accordingly we observe that the initial volume of 180 cc. is multiplied by a fraction ($273/313$) which is less than 1.

Also we must expand the gas to change its pressure from 80 cm. Hg to 76 cm. Hg. Again observe that the initial volume of 180 cc. is multiplied by the fraction ($80/76$) which is greater than 1.

Do problems 11, 12, and 13 at the end of the chapter.

6. Ideal Gas Law

We may now ask, how do the variables P, V, and T behave when the mass of gas is not held constant? The answer to this is found experimentally, and we find that the ratio PV/T is proportional to the mass (m) of the gas taken. Algebraically:

Gas Law:

$$\frac{P \times V}{T \times m} = \text{Constant}$$

This constant is now a fixed number for each gas but different for different gases. It is found, however, that it is inversely proportional to the molecular weight of the gas. Thus the constant for oxygen gas ($O_2 = 32$) is $\frac{1}{16}$ as large as the constant for hydrogen ($H_2 = 2$). This permits us to write an equation which includes the molecular weight (M) of the gas.

Ideal Gas Law:

$$\frac{P \times V \times M}{T \times m} = \text{Constant} = R$$

In this last equation the constant is now the *same* for every gas. It is known as the *universal gas constant* and given the symbol R. The ideal gas law, as it is called, now tells us that the product of the pressure, volume, and molecular weight of a gas, divided by the absolute temperature and mass of the gas, is a number R, which is the same for all gases. R can be calculated experimentally from the measurement of the properties of any gas.

Example: 32 g. of oxygen gas (O_2) at 0°C. and 1 atm. pressure is found to occupy a volume of 22.4 l. Calculate the value of R.

Answer:

$$R = \frac{P \times V \times M}{T \times m}$$

and on substitution:

$$R = \frac{1 \text{ atm.} \times 22.4 \text{ l.} \times 32 \text{ g.}/\text{mole}}{273°\text{K.} \times 32 \text{ g.}}$$

$$= 0.0821 \frac{\text{l.-atm.}}{\text{mole-}°\text{K.}}$$

or, converting to different volume units:

$$R = 0.0821 \frac{\text{l.-atm.}}{\text{mole-}°\text{K.}} \times \left(\frac{1000 \text{ cc.}}{1 \text{ l.}} \right) = 82.1 \frac{\text{cc.-atm.}}{\text{mole-}°\text{K.}}$$

R can be expressed in many other units as well. In addition to the ones shown in the example, which are most frequently used, we have $R = 1.99$ cal./mole-°K. and $R = 8.31 \times 10^7$ ergs/mole-°K.

The ideal gas law equation can be rearranged and written as follows:

$$PV = \frac{m}{M} RT$$

but, since the mass (m) of a gas divided by its molecular weight (M) is simply the number of moles of the gas, $n = m/M$. This equation can be written in its more usual form:

$$PV = nRT$$

Gases whose P, V, T, M behavior follows these last two equations are called "Ideal Gases." Deviations from the ratio $PV/T = R$ for a mole of ideal gas is referred to as a deviation from ideality.

When gases behave non-ideally, their deviations from the ideal gas law are divided into two classes. Gases, whose pressure (at a fixed T, V, and n) is less than that calculated for the ideal gas law, are said to show negative deviations from ideal behavior. Gases, whose pressures are higher than ideal, are said to show positive deviations.

In practice, it is found that any gas will show negative deviations from the ideal gas law at temperatures below 1.5 times the normal boiling temperature of its liquid. At higher temperatures and high pressures (above 20 atmospheres), gases tend to show positive deviations. All temperatures are in °K.

Do problem 14 at the end of the chapter.

7. Molecular Weight of Gases

The molecular weight of a gas may be calculated if we know the volume occupied by a given mass of the gas (i.e., the density of the gas).

This calculation is possible since we know that 1 mole of a gas will occupy a volume of 22.4 l. at STP. If we know the volume occupied at any other temperature and pressure, the combined form of the gas laws permits us to calculate the volume at STP, and using the molar density as a conversion factor we can calculate the molecular weight.

Example: 5.75 g. of a gas occupy a volume of 3.40 l. at a temperature of 50°C. and a pressure of 0.94 atm. What is its molecular weight?

Answer: Using the combined form of the gas laws:

$$V_{\text{STP}} = 3.40 \text{ l.} \times \left(\frac{273°\text{K.}}{323°\text{K.}}\right) \times \left(\frac{0.94 \text{ atm.}}{1.00 \text{ atm.}}\right)$$

$$= 2.70 \text{ l. STP}$$

From the given data: 5.75 g. = 2.70 l. STP. This provides a conversion factor:

$$1 \text{ mole} = 22.4 \text{ l. STP} \times \left(\frac{5.75 \text{ g.}}{2.70 \text{ l. STP}} \right)$$

$$= 47.6 \text{ g. (molecular weight)}$$

The ideal gas law, however, allows us to perform the same calculations in one step. The equation representing the ideal gas law has in it six quantities: M, m, R, T, P, V. If we know five of these we can always solve for the sixth. Since R is a known constant, this leaves only four experimental properties to be measured in order to compute the unknown. The following example illustrates the application of this equation to the problem already solved.

Example: Use the data from the previous example to calculate the molecular weight of the gas, making use of the ideal gas law.

Answer: Solving the ideal gas law equation for the molecular weight:

$$M = \frac{m \times R \times T}{P \times V}$$

Substituting the data:

$$M = \frac{5.75 \text{ g.}}{0.94 \text{ atm.}} \times 0.0821 \frac{\text{l. atm.}}{\text{mole-}°\text{K.}} \times \frac{323°\text{K.}}{3.40 \text{ l.}}$$

$$= \frac{5.75 \times 0.0821 \times 323}{0.94 \times 3.40} \frac{\text{g.}}{\text{mole}}$$

$$= 47.6 \text{ g./mole}$$

(which checks with the previous calculations)

Do problems 15 and 16 at the end of the chapter.

8. Dalton's Law—Mixtures of Gases

Quite frequently it is convenient to collect gases in vessels above the surfaces of liquids such as water. The gases that we collect will not be pure but will consist of a mixture of the original gas plus the vapor of the liquid used to confine it. For mercury the amount of mercury vapor included is so small as to be negligible. However, for water, the amount of water vapor is quite significant. How can we estimate the composition of a gas collected above water?

The answer to this question is provided by Dalton's law.

Dalton's Law: The total pressure exerted by a mixture of gases is simply the sum of the partial pressures which each gas would exert if the others were not present.

Now if a gas is collected above water and the result is a mixture of water vapor with the original gas, the total pressure of the mixture is equal to the sum of the pressures exerted by the original gas and by the water vapor, each acting as if the others were not there.   If we could know what pressure the water vapor exerted, we could subtract this from the total, observed pressure, and the difference would then be the true pressure exerted by the original gas in the vessel.

This information about the pressure of water vapor can be obtained from a table of vapor pressures for water.

Example: A sample of oxygen is collected by displacing water from an inverted tube. The temperature is 25°C., the pressure is 750 torr (barometric pressure), and the volume occupied is 280 cc.   What is the true volume of oxygen at STP?

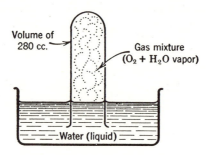

Volume of 280 cc.

Gas mixture $(O_2 + H_2O$ vapor)

Water (liquid)

Answer: From the table of vapor pressures we find the vapor pressure of water to be 23.5 torr at 25°C.   The partial pressure of oxygen in the tube is thus equal to $750 - 23.5 = 726.5$ torr.   Now, using the gas laws:

$$V_{STP} = V_{initial} \times \left(\frac{P_{initial}}{P_{final}}\right) \times \left(\frac{T_{final}}{T_{initial}}\right)$$

$$= 280 \text{ cc.} \times \left(\frac{726.5 \text{ torr}}{760 \text{ torr}}\right) \times \left(\frac{273°\text{K.}}{298°\text{K.}}\right)$$

$$= 245 \text{ cc. STP}$$

Do problems 17 and 18 at the end of the chapter.

9. Gas Densities—Molecular Weights

The density of a gas is given by the ratio of its mass to its volume:

$$d = \frac{m}{V}$$

But the volume of a gas depends on its temperature and pressure. Using the ideal gas law:

$$PV = \frac{m}{M} RT$$

we can rearrange the equation to obtain the ratio m/V:

$$\boxed{\frac{m}{V} = \frac{MP}{RT} = d \quad \text{(density)}}$$

Thus the ideal gas law gives us a direct relation between the density of a gas and its other properties.

In particular, if we know the density of a gas at a given pressure and temperature we can calculate its molecular weight.

Example: The density of a certain gas at 30°C. and 1.3 atm. pressure is 0.027 g./cc. What is its molecular weight?

Answer: Rearranging the above equation to solve for M:

$$M = \frac{dRT}{P}$$

Substituting:

$$M = \frac{\left(\dfrac{0.027 \text{ g.}}{\text{cc.}}\right) \times \left(\dfrac{82.1 \text{ cc.-atm.}}{\text{mole-°K.}}\right) \times 303°\text{K.}}{1.3 \text{ atm.}}$$

$$= \frac{0.027 \times 82.1 \times 303}{1.3} \quad \frac{\text{g.}}{\text{mole}}$$

$$= 516 \text{ g./mole} \ \rightarrow\ 520 \text{ g./mole}$$

This same formula gives us a method for obtaining molecular weights by a comparison of the densities of two gases, measured at the same temperature and pressure. If d_1 refers to gas 1 and d_2 refers to gas 2 we have:

$$d_1 = \frac{M_1 P}{RT}; \quad d_2 = \frac{M_2 P}{RT} \qquad \text{(same } P \text{ and } T\text{)}$$

Dividing these two equations, we have (P, T, and R cancel):

$$\frac{d_1}{d_2} = \frac{M_1}{M_2}$$

This tells us that the densities of two different gases (measured at the same T and P) will be in the ratio of their molecular weights. If we know the molecular weight of one of the gases, this equation then permits us to calculate the molecular weight of the other.

Example: The density of a certain gas is 1.64 g./l. At the same temperature and pressure, oxygen (O_2) has the density of 1.45 g./l. What is the molecular weight of the gas?

Answer: From our equation, solving for M_1:

$$M_1 = M_2\left(\frac{d_1}{d_2}\right) \quad \text{(let subscript 2 represent oxygen)}$$

$$= \left(\frac{32 \text{ g. } O_2}{\text{mole}}\right) \times \left(\frac{1.64 \text{ g./l.}}{1.45 \text{ g. } O_2/\text{l.}}\right)$$

$$= 36.2 \text{ g./mole}$$

Finally, we can use our original equation to obtain the density of a gas at any temperature and pressure, if we know its density at a given temperature and pressure.

Let d_1 be the measured density of a gas at T_1 and P_1 and d_2 the desired density at different conditions T_2 and P_2. We can write two equations:

$$d_1 = \frac{MP_1}{RT_1}; \quad d_2 = \frac{MP_2}{RT_2} \quad \text{(M is the same, since the gas is the same)}$$

Dividing these two equations we have:

$$\frac{d_2}{d_1} = \left(\frac{P_2}{P_1}\right) \times \left(\frac{T_1}{T_2}\right) \quad \text{(M and R cancel)}$$

Thus we have derived an equation giving the change in density of a gas with temperature and pressure.

Example: The density of a certain gas is 1.85 g./l. at 80 cm. Hg and 50°C. What is it at STP?

Answer: Solving the above equation for d_2:

$$d_2 = d_1 \times \left(\frac{P_2}{P_1}\right) \times \left(\frac{T_1}{T_2}\right)$$

Letting subscript 1 refer to given conditions and 2 to STP:

$$d_2 = \left(\frac{1.85 \text{ g.}}{\text{l.}}\right) \times \left(\frac{76 \text{ cm. Hg}}{80 \text{ cm. Hg}}\right) \times \left(\frac{323°\text{K.}}{273°\text{K.}}\right) = 2.08 \frac{\text{g.}}{\text{l. STP}}$$

Note: We can check the calculation by observing that going from 80 cm. Hg to 76 cm. Hg (standard pressure) represents an expansion in volume and thus a decrease in density (ratio 76/80 is less than 1). Similarly going from 50°C. to 0°C. represents a contraction in volume on cooling and thus an increase in density (ratio 323/273 is greater than 1).

10. Molar Heat Capacity of Gases—Molecular Structure

When 3 cal. of heat energy are added to 1 mole of a monatomic gas such as helium or neon, its temperature rises by precisely 1°C. This is found to be true independent of the initial temperature or pressure within very broad limits. The only restriction is that the heating takes place in a rigid vessel so that the volume of the gas is kept constant.

This observation is represented in the statement that C_v, the molar heat capacity (at constant volume) equals 3 cal./mole-°C. for monatomic gases.

For diatomic (or linear) gases such as N_2, O_2, CO_2, etc., we find experimentally that $C_v = 5$ cal./mole-°C., whereas for polyatomic, non-linear gases such as CH_4, NH_3, NO_2, etc., C_v is usually in excess of 6 cal./mole-°C.

These very interesting facts give us a quite unexpected and extraordinary method of inferring information about the structure of gas molecules just from the very simple observation of the quantity of heat which is required to raise their temperatures by 1°C.

Example: It is found that it takes 3.9 cal. of heat energy to raise the temperature of 5.6 l. of an unknown gas from 119°C. to 126°C. The pressure of the gas contained in a glass flask was initially 600 torr.

Calculate C_v for the gas and state what information this value gives about its structure.

Answer: C_v has the units of cal./mole-°C. This problem belongs to Case B (Chapter I, section 13), namely that of determining C_v, a complex property. We are given the number of calories and the change in temperature so that we need to compute number of moles.   Converting 5.6 l. gas to STP:

$$5.6 \text{ l. gas} = 5.6 \text{ l. gas} \times \left(\frac{273°\text{K.}}{392°\text{K.}}\right) \times \left(\frac{600 \text{ torr}}{760 \text{ torr}}\right)$$

$$= 3.1 \text{ l. STP}$$

$$= 3.1 \text{ l. STP} \times \left(\frac{1 \text{ mole}}{22.4 \text{ l. STP}}\right) = 0.138 \text{ moles}$$

$$C_v = \frac{\text{Calories}}{\text{Moles} \times \text{change in °C.}} = \frac{3.9 \text{ cal.}}{0.138 \text{ mole} \times 7°\text{C}}$$

$$= 4.1 \text{ cal./mole-°C.}$$

Since this lies between the low values of 3.0 for monatomic gases and 5.0 for diatomic (or linear) molecules, we infer that the gas is a mixture containing at least $(4.1 - 3.0)/(5.0 - 3.0) = 0.55$ mole fraction of a monatomic gas.

For NO_2, C_v at room temperature is about 6.1 cal./mole-°C. whereas for CO_2, N_2O, and C_2H_2 it is 5.0 cal./mole-°C.   This is sufficient information to conclude that NO_2 is triangular in shape, and CO_2, N_2O and C_2H_2 (acetylene) must be linear!

These observations may be deduced from the law of equipartition of energy which states that every independent mode of motion of a molecule at thermal equilibrium tends to have a kinetic energy associated with it of $\frac{1}{2} RT$ (R is the universal gas constant $= 1.99$ cal./mole-°C.).   A monatomic gas such as He or Ne has only translational motion in three dimensions in a flask and thus $3 \times \frac{1}{2}RT = \frac{3}{2}RT$, translational energy.   We see that if its temperature increases by 1°C. its kinetic (translational) energy increases by $\frac{3}{2}R = 3$ cal./mole. For linear molecules such as CO_2 there are two independent axes of rotation and hence $2 \times \frac{1}{2}RT$ rotational energy.   A 1°C. increase in temperature will thus increase this by $2 \times \frac{1}{2}R = 2$ cal./mole.   Thus a total of 5 cal. is needed to heat 1 mole of CO_2 by 1°C.   For non-linear molecules we have three rotational axes and hence a total heat capacity of 6 cal./mole-°C.

Note: At low temperatures, the internal vibrations of molecules as well as their electronic motions are frozen in and do not share in the equipartition.   At sufficiently high temperatures ($\sim$1000°K for vibrations, $\sim$5000°K for electronic motion), this is no longer true.

11. Molar Heat Capacity of Gases—Constant Volume and Constant Pressure

For an ideal gas, the molar heat capacity at constant volume C_v is related to the molar heat capacity at constant pressure, C_p (see Table IX, pg. 00), by the relation:

$$C_p = C_v + R = C_v + 2.0 \text{ cal./mole-}°\text{C.}$$

The reason that C_p is greater than C_v is that in heating a gas at constant volume, its pressure increases. If we do the experiment with the gas in a cylinder with a movable piston, we can allow the gas to constantly expand against the force of the external atmosphere so that the pressure remains fixed. In this case, the gas pressure remains constant but the gas does work against the pressure of the atmosphere and it will become colder as a result if no heat energy is provided to compensate this extra work done by the expanding gas.

This extra energy can be computed from the ideal gas laws. If a gas expands against a constant external pressure, P, then it will do work (lose energy) exactly equal to P times the change in volume:

$$\text{Work Done} = P \left(V_{\text{final}} - V_{\text{initial}} \right)$$

But from the ideal gas law, for 1 mole of a gas:

$$V_{\text{final}} = \frac{RT_{\text{final}}}{P}; \; V_{\text{initial}} = \frac{RT_{\text{initial}}}{P}$$

$$\therefore \text{Work Done} = P \left(\frac{RT_{\text{final}}}{P} - \frac{RT_{\text{initial}}}{P} \right) = R \left(T_{\text{final}} - T_{\text{initial}} \right)$$

But in determining C_p or C_v, we add the amount of heat to 1 mole, just necessary to raise the temperature by 1°C., hence, $T_{\text{final}} - T_{\text{initial}} = 1°\text{C.}$ and;

$$\text{Work Done} = R(1°\text{C.}) = 2 \text{ cal./mole}$$

12. Problems

1. The difference between two arms of a mercury manometer is 40 cm. Hg. If the lower arm is exposed to a gas while the upper arm is evacuated, what is the pressure of the gas in atmospheres?

2. If the difference between the two arms in the preceding problem were 10 in., calculate the pressure in centimeters Hg; in atmospheres.

3. If a gas is confined by a water instead of a mercury manometer and the water level exposed to the gas is 90 cm. lower than the other arm which is exposed to the air (whose pressure is 75.0 cm. Hg), calculate the gas pressure in centimeters Hg; in atmospheres. (Density Hg = 13.55 g./cc.)

4. Make the following conversions:

  (a) 250 mm. Hg to dynes per square centimeters.
  (b) 35 cm. Hg to atmospheres.
  (c) 3.0 lb./sq. in. to centimeters Hg.
  (d) 60 dynes/cm.^{2} to atmospheres.

5. A block of iron rests on a base whose area is 3 cm.^{2}. If the volume of the block is 25 cm.^{3} and the density of iron is 7.8 g./cm.^{3}, calculate the pressure on the base in dynes per square centimeter.

6. A phonograph pickup weighs 2 oz. It rests on a needle whose tip has a contact area of 0.045 sq. mm. Calculate the pressure exerted by the pickup on a record in pounds per square inch.

7. What is the volume occupied by 10 l. of gas at 76 cm. Hg after it has been compressed at constant temperature to 5 atm.?

8. What pressure is required to compress 180 cc. of gas at constant temperature to a volume of 24 cc. if the initial pressure is 30 cm. Hg?

9. 23 l. of gas is heated from 30°C. to 150°C. at constant pressure. What is its final volume?

10. To what temperature must 280 cc. of CO_2 be cooled for its volume to decrease to 130 cc.? The initial temperature is −20°C., and the gas is kept at constant pressure.

11. What is the volume of 600 cc. STP of N_2 at 45°C. and 2.0 atm. pressure?

12. 720 cc. of nitrogen gas is collected over mercury at −10°C. and 25.0 cm. Hg pressure. What is this volume at STP?

13. In each of the following problems, what information is lacking to calculate the answer?

  (a) A volume of gas is compressed from 1.0 atm. to 10.0 atm. at constant temperature. What is its final volume?
  (b) What volume will 15 cc. of gas occupy when cooled to −30°C. at constant pressure?
  (c) 80 cc. of gas at −38°C. is heated to 40°C. and 80 mm. Hg pressure. What was the initial pressure?

14. Using conversion factors, convert the value of R from 82.1 cc.-atm./mole-°K. to units of:

  (a) cc.-mm. Hg/mole-°K.
  (b) l.-lb./sq. in./mole-°K.
  (c) ergs/mole-°K.

15. 820 cc. of an unknown gas at 35°C. and 80 cm. Hg is found to weigh 2.46 g. What is its molecular weight?

16. What volume will be occupied by 1.5 g. of NO gas at 75°C. and 300 mm. Hg pressure?

17. 1.47 l. of a gas is collected over water at 30°C. and barometric pressure of 744 torr. If the gas has a weight of 1.98 g. and the vapor pressure of water at 30°C. is 32 mm. Hg, what is the molecular weight of the gas?

18. 250 cc. of a gas is collected over acetone at -10°C. and 85 cm. Hg pressure. If the gas weighs 1.34 g. and the vapor pressure of acetone at -10°C. is 39 mm. Hg, what is the molecular weight of the gas?

19. The density of a gas is found to be 2.07 g./l. at 30°C. and 2.00 atm. pressure. What is its density at STP?

20. The density of Cl_2 gas is 3.17 mg./cc. at STP. What is its density at 100°C. and 70.0 cm. Hg pressure?

21. The pressure of the atmosphere 100 miles above the earth is about 2×10^{-6} torr, and the temperature is about -180°C. How many molecules are there in 1 cc. of gas at this altitude?

22. What is the density of SO_2 gas at STP? at 40°C. and 80.0 cm. Hg pressure?

23. It is found that an unknown gas has a density 2.5 times as great as that of oxygen (O_2) at the same T and P. What is its molecular weight?

24. At a given T and P, CO_2 has a density which is 0.758 that of a gas X. What is the molecular weight of X? What is the density of X at STP?

25. The composition of dry air is about 78 mole per cent N_2, 21 mole per cent O_2 and 1 mole per cent Ar. When the barometric pressure is 750 torr, calculate the partial pressures exerted by these three components in dry air.

26. A 1500 cc. flask contains 400 mg. O_2 and 60 mg. H_2 at 100°C.

    (a) What is the total pressure in the flask?

    (b) If the mixture is permitted to react to form H_2O (gas) at the same temperature, what materials will be left and what will be their partial pressures?

27. Which has the larger value of C_v, He or HCl? Explain.

28. How many calories of heat are required to raise the temperature of 5 lb. of O_2 gas from 60°C. to 120°C. at constant volume?

29. 11.2 l. STP of an unknown gas require 25 cal. to raise its temperature by 10°C. Calculate C_v for the gas. What can you conclude about its structure?

30. A mixture of gases consists of 50 mole per cent He and 50 mole per cent H_2. Calculate C_v for this mixture. What is the average molecular weight of the mixture?

31. At room temperature CCl_4 gas has a C_v much in excess of 6 cal./mole-°C. How can one account for such behavior?

32. 100 cal. of heat energy are added to 4.0 l. STP of CO_2 gas held in a rigid flask. Calculate the final temperature and pressure of the gas.

33. How much heat must be extracted from 10 g. of Ne gas held in a glass flask (initially at 40°C.) to reduce its temperature to 80°K?

34. When a gas is permitted to expand by allowing it to slowly move a piston which is confining it, the gas is found to be colder. (a) Explain why this is to be expected. (b) On the basis of this observation explain whether or not it will take more or less heat to raise the temperature of 1 mole of He gas by 1°C. if the process is conducted at constant volume (C_v) or constant pressure (C_p).

Note: The quantity of heat required to change the temperature of 1 mole of a gas by 1°C. under such conditions that the pressure is kept constant is known as the molar heat capacity at constant pressure C_p.

35. At 65°C. pure PCl_5 gas is present in a flask at a pressure of 200 torr. At 200°C. this is completely dissociated into $PCl_3 + Cl_2$. Calculate the pressure in the flask at 200°C.

36. At 400°C. a flask contains 26 torr pressure of NO and 60 torr pressure of Chlorine. At temperatures below 200°C. these two react completely with each other to form NOCl. Calculate the final pressure in this flask at 100°C. What species are present?

37. A 1.00-l. flask contains at 30°C. equal molar amounts of NO_2 gas and N_2O_4 gas. If the total mass of gas in the flask is 1.60 g. calculate the total pressure.

38. If the heating of the gas in problem 32 is carried out in a cylinder with movable piston so that the gas is allowed to expand at a constant external pressure of 1 atmosphere, calculate the final temperature and volume of the gas.

39. How many calories of heat are required to raise the temperature of 30 l. of N_2 from 20°C. to 120°C. at a constant pressure of ½ atmosphere? What is the final volume of the gas?

CHAPTER VII

The Concept of Combining Power—Valence

1. The Equivalent Weights of Elements

In the present chapter we shall investigate the concept of "valence" which is so frequently used in chemistry.

It was early discovered that elements seem to combine with each other in weight ratios which, in many cases, were simply related to their atomic weight. Thus it was found that 8 g. of oxygen (O) will react with precisely 20 g. of calcium (Ca) to form calcium oxide. It was further found that 8 g. of oxygen (O) would react with precisely 1.0 g. of hydrogen to form water. Then when the reaction of calcium with hydrogen was investigated it was discovered that 20 g. of calcium would react with just 1.0 g. of hydrogen to form calcium hydride. Many more illustrations of this type of relation could be given for other elements.

We can compile from such evidence an experimental table of weights for each element which gives that weight of the element which will combine with 8.00 g. of oxygen. These weights can be measured very precisely. The table is known as the Table of Combining Weights, or, according to current practice, the Table of Equivalent Weights. Table XI contains the equivalent weights of a few common elements.

TABLE XI
EQUIVALENT WEIGHTS OF A FEW COMMON ELEMENTS

Element	Equivalent Weight (grams)	Element	Equivalent Weight (grams)
Oxygen	8.000*	Aluminum	9.00
Hydrogen	1.008	Calcium	20.04
Sulfur	16.000	Zinc	32.7
Carbon	3.000	Sodium	23.00
Chlorine	35.46	Potassium	39.10
Bromine	79.92	Lead	103.61

* 8.000 g. of oxygen is the standard unit for the table.

The usefulness of such a table is that in most cases, if we want to know what weight of one element will react with a given weight of another element, these weights will be in the ratio of the equivalent weights.

Thus from the table we can predict that 1.008 g. of hydrogen will react with 16.000 g. of sulfur; 32.7 g. of zinc will react with 35.46 g. of chlorine, etc.

From these equivalent weights chemical weight conversion factors can be obtained which can be used to determine the weight of one element that will react with another.

Example: What weight of aluminum (Al) will combine with 5.00 g. of chlorine (Cl)?

Answer: From the table:

$$9.00 \text{ g. Al} = 35.46 \text{ g. Cl}$$

Then:

$$5.00 \text{ g. Cl} = 5.00 \text{ g. Cl} \times \left(\frac{9.00 \text{ g. Al}}{35.46 \text{ g. Cl}} \right)$$

$$= 1.27 \text{ g. Al}$$

Example: 7.95 g. of an element X will combine with 3.47 g. of oxygen. What is the equivalent weight of the element?

Answer: By equivalent weight, we mean that weight which will combine with 8.000 g. O.

$$8.000 \text{ g. O} = 8.000 \text{ g. O} \times \left(\frac{7.95 \text{ g. X}}{3.47 \text{ g. O}} \right)$$

$$= 18.32 \text{ g. X}$$

The significance of these results is that 1 equivalent weight of an element has precisely the same *capacity* for chemical combination as 1 equivalent weight of any other element, and all of these have the same chemical combining power as 8.000 g. of oxygen (standard).

Do problems 1 and 2 at the end of the chapter.

2. The Equivalent—Units

We can define the following useful unit:

Definition:

1 equivalent of an element = The equivalent weight of the
element in grams

From the preceding discussion we see that the unit "equivalent" is well chosen since 1 equivalent of one element is indeed equal (or equivalent) in *chemical combining power* to 1 equivalent of any other element.

Conversely, we may conclude, when two elements react to form a compound, equal numbers of equivalents of both must react.

Example: How many equivalents of zinc (Zn) will react with 20 g. of bromine (Br)?

Answer: From Table XI:

$$20 \text{ g. Br} = 20 \text{ g. Br} \times \left(\frac{1 \text{ eq. Br}}{79.92 \text{ g. Br}} \right)$$

$$= 0.25 \text{ eq. Br}$$

By the principle of equivalence, this will require precisely *0.250 eq. Zn,* or, in grams:

$$0.25 \text{ eq. Zn} = 0.25 \text{ eq. Zn} \times \left(\frac{32.7 \text{ g. Zn}}{1 \text{ eq. Zn}} \right) = 8.2 \text{ g. Zn}$$

It would thus appear that the Table of Equivalent Weights gives all the information needed to perform calculations for chemical reactions. This statement is subject to some restriction as we shall see later.

When two different elements react to form a compound, the latter will contain two elements and so is called a binary compound (e.g., $NaCl$, HCl, $ZnCl_2$, etc.). But, by the principle of equivalence, a binary compound will contain equal numbers of equivalents of each element present in it. Thus we can define the equivalent weight of a binary compound:

Definition:

The equivalent weight of   = That weight of the compound which
     a binary compound   contains 1 equivalent of each element

Similarly we can definite a unit:

Definition:

1 equivalent of a compound   = The equivalent weight in grams

---

Example: What is the equivalent weight of sodium chloride?

Answer: We don't need to know the formula! From the table and the definition:

Eq. wt. of sodium chloride = Eq. wt. of Na + Eq. wt. of Cl
                     = 23.00 g. + 35.46 g.
                     = 58.46 g.

In a similar fashion we can obtain the equivalent weight of any binary compound by adding the equivalent weights of the elements present in it. For this we do not need to know the formula! Thus the equivalent weight of calcium oxide is 28.04 g.; of aluminum bromide, 88.92 g.; of potassium sulfide, 55.10 g.

3. Radicals

The chemical behavior of most inorganic compounds containing more than two elements is such that they may be considered as being composed of only two groups, one of which may be an element and the other a complex group containing more than one element. Thus calcium sulfate ($CaSO_4$) behaves chemically as though it were made of the two groups, calcium (Ca) and sulfate (SO_4). The sulfate (SO_4) preserves its identity in many chemical reactions of the compound $CaSO_4$. Such groups are known as radicals, and their equivalent weights may be obtained experimentally by determining the weight of the group that will combine with 1 equivalent of an element whose equivalent weight is known.

Thus it can be shown by direct analysis that 20 g. of calcium (1 equivalent) will combine with 48 g. of sulfate (SO_4), and so the equivalent weight of the sulfate radical (SO_4) is 48 g. Table XII lists some of the more common radicals and their equivalent weights.

TABLE XII
EQUIVALENT WEIGHTS OF SOME RADICALS

Radical	Formula	Equivalent Weight (grams)
Sulfate	SO_4	48.0
Nitrate	NO_3	62.0
Ammonium	NH_4	18.0
Carbonate	CO_3	30.0
Chromate	CrO_4	58.0
Chlorate	ClO_3	83.5
Hydroxide	OH	17.0

For compounds containing such radicals, 1 equivalent of the compound will contain 1 equivalent of the element and 1 equivalent of the radical. Thus the equivalent weight of sodium sulfate is 71.0 g.; of ammonium chloride, 53.5 g.; of aluminum chromate, 67.0 g.; etc.

Do problems 3 and 4 at the end of the chapter.

4. The Principle of Equivalence—Chemical Reactions

When reactions occur between elements or compounds, the principle of equivalence predicts that 1 equivalent of each substance will react and 1 equivalent of each product will be found.

Thus when zinc reacts with sulfuric acid (hydrogen sulfate) we can represent the reaction as:

1 eq. zinc + 1 eq. sulfuric acid → 1 eq. zinc sulfate + 1 eq. hydrogen

Similar equations can be written for all such reactions. Only for those reactions in which radicals are destroyed or lose their identity does this principle fail or, rather, need modification.

It appears, then, that if we are provided with a Table of Equivalent Weights we can dispense with the writing and balancing of chemical equations since the principle of equivalence tells us precisely the weight relationships in all chemical reactions, namely, 1 equivalent per 1 equivalent for all substances in the reaction.

Example: Ammonium sulfate reacts with calcium chloride to produce ammonium chloride and a precipitate of insoluble calcium sulfate. How much calcium sulfate will be produced from the reaction of 14.0 g. of ammonium sulfate?

Answer: The student should check the origin of all the conversion factors:

$$14.0 \text{ g. ammonium sulfate} = 14.0 \text{ g. amm. sulf.} \times \left(\frac{1 \text{ eq. amm. sulf.}}{66.0 \text{ g. amm. sulf.}} \right)$$

$$\times \left(\frac{1 \text{ eq. cal. sulf.}}{1 \text{ eq. amm. sulf.}} \right) \times \left(\frac{68 \text{ g. cal. sulf.}}{1 \text{ eq. cal. sulf.}} \right)$$

$$= 14.4 \text{ g. calcium sulfate}$$

Do problem 5 at the end of the chapter.

5. Combining Power—The Unit of Valence

We can now ask, how is this new system of units, *equivalents,* which are so useful in reactions, related to our theoretical units of moles?

If we compare the equivalent weights listed in Table XI and Table XII with the molecular weights of the substances, we will find that the molecular weights are always some simple multiple of the equivalent weights. This is shown in Table XIII.

TABLE XIII
SOME EQUIVALENT AND MOLECULAR WEIGHTS

Substance	Equivalent Weight (g./eq.)	Molecular Weight (g./mole)	Ratio of Mol. Wt./Eq. Wt.
Oxygen (O)	8.000	16.00	2
Carbon (C)	3.00	12.00	4
Calcium (Ca)	20.00	40.00	2
Nitrate (NO_3)	62.00	62.00	1
Sulfate (SO_4)	48.00	96.0	2
Calcium sulfate ($CaSO_4$)	68.0	136.0	2

This ratio of the molecular weight of an element or group to the equivalent weight of the same element or group has the units of equivalents per mole and tells us how many equivalent weights or combining powers are present in one mole of the element or group.

$$\text{Ratio of molecular weight divided by equivalent weight} = \frac{\text{Grams/Mole}}{\text{Grams/Equivalent}} = \frac{\text{Equivalents}}{\text{Mole}}$$

Because of the standards used for oxygen (1 eq. O = 8 g. O; 1 mole O = 16 g. O), it turns out that this ratio is never less than 1 and always turns out to be a whole number. This number is called the *valence* of the element or group, and we see now the precise meaning of the term:

Definition: Valence is the number of equivalents of an element or compound present in 1 mole of the compound.

or

$$\text{Valence} = \frac{\text{Equivalents}}{\text{Mole}}$$

The valence of an element or group is thus a quantity which measures the capacity of the element or group for chemical combination.

When we say that oxygen (O) has a valence of 2 and hydrogen (H) has a valence of 1, we mean that the combining power of 1 mole of O is twice as great as the combining power of 1 mole of H, and, in particular, 1 mole of O can combine with 2 moles of H.

6. Valence and Chemical Formulae

If we know the valences of elements or radicals, we can write chemical formulae for the compounds they form. Thus the valence of aluminum (Al) is 3, that of chlorine (Cl) is 1. Then, 1 mole of aluminum can combine with 3 moles of chlorine and the compound aluminum chloride will have the formula, $AlCl_3$. Similarly the valence of sulfate (SO_4) is 2. Then 2 moles of Al will combine with 3 moles of SO_4, since 2 moles of Al have a combining power of 6, and 3 moles of SO_4 have a combining power of 6. The formula is then $Al_2(SO_4)_3$. From a table of valences the student can write the formula for any combination of elements and radicals shown. (*Note:* This is, however, no guarantee that the compound written exists. If it does, however, the formula will be correct.)

Similarly, if we know the formula (from analysis) of a compound and the valence of one of the elements, we can always deduce the valence of the other.

Example: The formula of chromium (III) sulfate is $Cr_2(SO_4)_3$. What is the valence of chromium? The valence of SO_4 is 2.

Answer: The SO_4 group has a valence of 2. Three SO_4 groups represent 6 equivalents. There must be 6 equivalents of Cr combined with this. Since there are 2 moles of Cr, each mole of Cr must contain 3 equivalents. Thus the valence of Cr is 3.

Do problem 10 at the end of the chapter.

7. Calculation of Equivalent Weights

Because it is much simpler to remember the valences of elements and radicals rather than their combining weights, not much use is made of the Table of Equivalent Weights. Instead the equivalent weights of elements and radicals are calculated from the molecular weights and the known valences. For this purpose the definition of valence is used:

Definition:

$$\text{Equivalent weight} = \frac{\text{Molecular weight}}{\text{Valence}}$$

Example: The molecular weight of tin (Sn) is 118.7 g./mole. Its valence is 4. What is its equivalent weight?

Answer:

$$\text{Eq. wt. Sn} = \frac{118.7 \text{ g./mole}}{4 \text{ eq./mole}} = 29.68 \text{ g./eq.}$$

Similar calculations can be made for radicals.

Example: The molecular weight of orthosilicate (SiO_4) is 92 g./mole. Its valence is 4 eq./mole. What is its equivalent weight?

Answer:

$$\text{Eq. wt. SiO}_4 = \left(\frac{92 \text{ g.}}{\text{mole}} \right) \times \left(\frac{1 \text{ mole}}{4 \text{ eq.}} \right)$$
$$= 23 \text{ g./eq.}$$

Observe that the property of valence provides a conversion factor to convert from equivalents to moles. With these rules it is also possible to compute the equivalent weight of a compound from its molecular weight. Thus 1 mole of calcium chloride ($CaCl_2$) will weigh

111.0 g. and contain 2 equivalents of Ca and 2 equivalents of Cl. One-half of a mole will contain 1 equivalent of each group, and so the equivalent weight of $CaCl_2$ is 55.5 g./eq.

Table XIV shows some compounds and their equivalent weight:

TABLE XIV
EQUIVALENT WEIGHTS OF SOME COMPOUNDS

Compound	Molecular Weight (grams/mole)	Number of Equivalents in 1 mole	Equivalent Weight (grams/equiv.)
NaCl	58.5	1	58.5
Na_2SO_4	142	2	71
$ZnCl_2$	136.4	2	68.2
$ZnSO_4$	161.4	2	80.7
$AlCl_3$	133.5	3	44.5
$AlPO_4$	122	3	40.7
$Al_2(SO_4)_3$	342	6	57

8. Some Complications—Multivalence

Having thus developed what seems to be a super-elegant way of doing calculations of chemical reactions, a way that avoids the need for balanced equations, we must now raise the disappointing specter of complication. It is unfortunate, but true, that many, actually a majority, of the elements display the property of multivalence. That is, they may have one valence in a given set of reactions and another valence in a different set of reactions.

Even our standard-bearer, oxygen, has the valence of 2 in water (H_2O) but can form another compound with hydrogen in which it has the valence of 1, namely hydrogen peroxide (H_2O_2). Nitrogen can form five different oxides, each showing a different valence.

This means that we shall have to exercise due care in discussing equivalents and equivalent weights, since we must be sure of the particular valence state that is involved. This complication is almost great enough to cause us to discard the concept of equivalence entirely. However, its usefulness in titrations and ionic reactions is so great that it has persisted, and so we shall continue to apply it but be warned of the confusion and ambiguity that may arise.

9. Summary

1. *The equivalent weight of an element or radical is that weight which combines with 8.000 g. of oxygen.*

2. *The unit, equivalent, is defined as the equivalent weight in grams.*

3. *The law of combining weights states that 1 equivalent weight of any element will combine with precisely 1 equivalent weight of another element or compound.*

4. *By the equivalent weight of a compound we shall mean that weight which contains 1 equivalent of each element or radical.*

5. *The valence of a radical or element is defined as the number of equivalent weights present in 1 mole of the element or radical. It has the units of equivalents per mole and may be used as a property conversion factor to relate equivalents and moles.*

6. *The equivalent weight of an element or radical may be calculated from its molecular weight and its valence.*

$$Equivalent\ weight = Molecular\ weight/Valence$$

7. *From the principle of reacting equivalence contained in the law of combining weights we can say that, when elements or compounds react, equal numbers of equivalents will always react with each other to produce equal numbers of equivalents of each product. This allows us to perform calculations without the use of balanced equations.*

8. *The property of multivalence exhibited by most elements indicates that caution be used in dealing with the above properties.*

10. Problems

1. 4.25 g. of an element X combines with oxygen to form 5.40 g. of an oxide. What is the equivalent weight of X?

2. 1.08 g. of a metal oxide, on heating, decomposes to give the pure metal and 56.0 cc. STP of O_2 gas. What is the equivalent weight of the metal?

3. 2.94 g. of nickel (valence = 2) combines with an element X to form 4.49 g. of a compound. What is the equivalent weight of X?

4. Using the Table of Atomic Weights and known valences, make the following conversions:

(a) 2 moles of $FeCl_3$ to equivalents.

(b) 25.6 g. of zinc to equivalents.

(c) 13 mmoles of Pb to milliequivalents.

(d) 6.5 g. of $CaSO_4$ to equivalents.

(e) 10 l. STP of gaseous CCl_4 to equivalents.

(f) 0.45 eq. of H_2S to moles.

(g) 0.64 eq. of $AlBr_3$ to grams.

(h) 0.20 eq. of $CuSO_4$ to grams.

(i) 340 cc. STP of oxygen gas (O_2) to equivalents.

(j) 1.40 meq. of Cl_2 gas to cubic centimeters STP.

5. Using the principle of equivalence, calculate (without chemical equations):

(a) How many grams of $CuSO_4$ will react with 2.4 g. of Na_2S to produce CuS and Na_2SO_4?

(b) How many grams of NaOH will react with 30 g. of $Al_2(SO_4)_3$?

(c) How many grams of $Ca_3(PO_4)_2$ can be made from 20.0 g. of $Ca(NO_3)_2$ by the reaction with H_3PO_4?

(d) How many grams of sulfuric acid (H_2SO_4) are needed to react with 20 g. of NaOH?

(e) How many grams of Zn will react with 11 g. of NaOH to produce hydrogen and sodium zincate?

6. 14.2 g. of an unknown acid X are neutralized by precisely 12.0 g. of NaOH. What is the equivalent weight of the acid?

7. 24.5 mmoles of $Ca(OH)_2$ will exactly neutralize 1.37 g. of an acid. What is the equivalent weight of the acid?

8. When 18.7 g. of an unknown acid react with zinc metal, exactly 800 cc. STP of H_2 gas are evolved. What is the equivalent weight of the acid?

9. When 3.5 g. of an unknown metal react with an acid, exactly 250 cc. STP of H_2 gas are produced. What is the equivalent weight of the metal?

10. Given the valence of oxygen as 2, calculate the valence of the other element in each of the following compounds:

(a) N_2O.	(e) SO_3.	(i) Tl_2O_3.
(b) NO_2.	(f) CO.	(j) PtO_2.
(c) N_2O_3.	(g) Mn_2O_7.	(k) Fe_3O_4.
(d) P_2O_5.	(h) ClO_2.	

CHAPTER VIII

Measurement of Solutions

1. Solutions—Concentration Units

It is generally clumsy to carry out reactions between solids or gases. First, it is difficult to mix the solids adequately; and second, it is difficult to handle gases in the laboratory without a great deal of equipment. On the other hand, liquids mix quite readily, are easily handled, and can be measured out accurately and quickly by volumetric equipment.

For these reasons it is always preferable to carry out chemical reactions in the liquid state. But many substances are solids or gases! This difficulty is circumvented by finding a liquid capable of dissolving them. Thus sodium chloride is most frequently used in chemical reactions in the form of a solution of sodium chloride in water. The liquid is called the *solvent*. The substance dissolved in the liquid is called the *solute*. (This distinction sometimes appears quite arbitrary as when, for example, alcohol, itself a liquid, is dissolved in an equal volume of water. Either substance could then be called the solute.)

In order to do quantitative work with solutions, if we are going to measure the solution in terms of volume, we must know the amount of solute contained per unit of volume of solution. This property is known as the concentration of the solution.

There are a number of different units employed in specifying the concentration of solute present in a given quantity of solution. These are listed in Table XV.

Do problems 1 and 2 at the end of the chapter.

2. Conversion of Units

It can be seen that there are two types of units for expressing the concentration of a solution. One is in units of the weight of the solution and requires that the solution be weighed when dispensing (e.g., molality, weight per cent). This type of unit is used only in

TABLE XV

UNITS USED IN EXPRESSING CONCENTRATION OF A SOLUTION

Name	Definition	Most Frequently Used Units	Property of Solution Measured When Dispensing
Weight percent	Weight units of solute contained in 100 weight units of solution	$\dfrac{\text{Grams of solute}}{\text{100 grams of solution}}$	Weight of solution
Weight concentration	Weight of solute contained in a unit volume of solution	$\dfrac{\text{Grams of solute}}{\text{Liters of solution}}$	Volume of solution
Molarity (M)	Number of moles of solute contained in 1 l. of solution	$\dfrac{\text{Moles of solute}}{\text{Liters of solution}}$	Volume of solution
Normality (N)	Number of equivalents of solute contained in 1 l. of solution	$\dfrac{\text{Equivalent of solute}}{\text{Liters of solution}}$	Volume of solution
Molality	Number of moles of solute per kilogram of solvent	$\dfrac{\text{Moles of solute}}{\text{Kilograms of solvent}}$	Weight of solution

experiments requiring great accuracy. The second type of unit is in terms of the volume of the solution and is in much more common usage.

Before proceeding to the analysis of problems, let us first say a word about the relations between the three concentration units which refer to volume of solutions and are most popular.

The three units—for weight concentration, for molarity, and for normality—all have reference to the quantity of solute per liter of solution. We can use the appropriate conversion factors to convert from any one of these to the other.

Example: The concentration of a solution is given as 40.0 g. of NaCl per liter of solution (abbreviation: 40.0 g. NaCl/l.). Convert this to units of molarity and normality.

Answer: To convert to units of molarity, we use the relation:

$$1 \text{ mole NaCl} = 58.5 \text{ g. NaCl}$$

Then:

$$\frac{40.0 \text{ g. NaCl}}{1.} = \left(\frac{40.0 \text{ g. NaCl}}{1.}\right) \times \left(\frac{1 \text{ mole NaCl}}{58.5 \text{ g. NaCl}}\right)$$

$$= \frac{0.684 \text{ mole NaCl}}{1.} \text{ or } 0.684 \text{ } M \text{ NaCl}$$

For normality we need the relation:

$$58.5 \text{ g. NaCl} = 1 \text{ mole NaCl} = 1 \text{ eq. NaCl}$$

Then:

$$\frac{40.0 \text{ g. NaCl}}{1.} = \left(\frac{40.0 \text{ g. NaCl}}{1.}\right) \times \left(\frac{1 \text{ eq. NaCl}}{58.5 \text{ g. NaCl}}\right)$$

$$= \frac{0.684 \text{ eq. NaCl}}{1.} \text{ or } 0.684 \text{ } N \text{ NaCl}$$

Example: A solution of $ZnSO_4$ has a concentration of 0.70 mole $ZnSO_4$ per liter of solution (abbreviation: 0.70 M $ZnSO_4$). Express this in units of weight concentration and normality.

Answer:

(A) $$\frac{0.70 \text{ mole ZnSO}_4}{1.} = \left(\frac{0.70 \text{ mole ZnSO}_4}{1.}\right) \times \left(\frac{161.4 \text{ g. ZnSO}_4}{1 \text{ mole ZnSO}_4}\right)$$

$$= \frac{113 \text{ g. ZnSO}_4}{1.}$$

(B) $$\frac{0.70 \text{ mole ZnSO}_4}{1.} = \left(\frac{0.70 \text{ mole ZnSO}_4}{1.}\right) \times \left(\frac{2 \text{ eq. ZnSO}_4}{1 \text{ mole ZnSO}_4}\right)$$

$$= \frac{1.40 \text{ eq. ZnSO}_4}{1.} \text{ or } 1.40 \text{ } N \text{ ZnSO}_4$$

Do problem 3 at the end of the chapter.

3. Interpretation of Concentration Units

Concentration is an intensive property of a solution and in many respects is similar to density. Just as the density of a substance expresses a relation between the property of mass and the property of volume for the particular substance, so concentration expresses a relation between the property, the mass of the solute (present in a given quantity of the solution), and the property of the volume of the solution.

We can use concentration units to give conversion factors between these two properties. This is perhaps most easily seen from the definition in algebraic form:

Definition: Molarity $(M) = \dfrac{\text{Moles of solute}}{\text{Liters of solution}} = \dfrac{\text{Millimoles solute}}{\text{Milliliters solution}}$

Definition: Normality $(N) = \dfrac{\text{Equivalents of solute}}{\text{Liters of solution}}$

$= \dfrac{\text{Milliequivalents of solute}}{\text{Milliliters of solution}}$

Definition: Weight concentration $(C) = \dfrac{\text{Grams of solute}}{\text{Volume of solution}}$

Each equation relates three properties. If we know two of these we can always solve for the third. However, the method of conversion factors is so much simpler that we shall use it to illustrate problems:

Example: How many moles of HCl are there in 1.5 l. of a 2.0 M solution? How many grams of HCl?

Answer:

(A)　　　$1.5 \text{ l. soln.} = 1.5 \text{ l. soln.} \times \left(\dfrac{2.0 \text{ moles HCl}}{1 \text{ l. soln.}}\right)$

　　　　　$= 3.0 \text{ moles HCl}$

(B)　　　$1.5 \text{ l. soln.} = 1.5 \text{ l. soln.} \times \left(\dfrac{2.0 \text{ moles HCl}}{1 \text{ l. soln.}}\right) \times \left(\dfrac{36.5 \text{ g. HCl}}{1 \text{ mole HCl}}\right)$

　　　　　$= 110 \text{ g. HCl}$

Example: What volume of a 0.64 N solution of H_2SO_4 will contain 13.0 g. of H_2SO_4? What volume will contain 0.25 mole of H_2SO_4?

Answer:

(A)　$13.0 \text{ g. } H_2SO_4 = 13.0 \text{ g. } H_2SO_4 \times \left(\dfrac{1 \text{ mole } H_2SO_4}{98 \text{ g. } H_2SO_4}\right) \times \left(\dfrac{2 \text{ eq. } H_2SO_4}{1 \text{ mole } H_2SO_4}\right)$

　　　　　　　　　　　　　　　　　　　　　　　　$\times \left(\dfrac{1 \text{ l. soln.}}{0.64 \text{ eq. } H_2SO_4}\right)$

　　　$= \dfrac{13.0 \times 2}{98 \times 0.64} \text{ l. soln.}$

　　　$= 0.414 \text{ l. soln.}$

Note the logic of our procedure. We were given the property of mass in grams and asked to find the property of volume of solution, being also given the concentration in equivalents per liter. As indicated by the successive conversions, we converted mass to moles, moles to equivalents, and, finally, equivalents to liters of solution.

(B)　$0.25 \text{ mole } H_2SO_4 = 0.25 \text{ mole } H_2SO_4 \times \left(\dfrac{2 \text{ eq. } H_2SO_4}{1 \text{ mole } H_2SO_4}\right)$

　　　　　　　　　　　　　　　　　　　　$\times \left(\dfrac{1 \text{ l. soln.}}{0.64 \text{ eq. } H_2SO_4}\right)$

　　　$= 0.78 \text{ l. soln.}$

Example: How would you make 300 cc. of a 2.2 M solution of $AlCl_3$?
Answer:

$$300 \text{ cc. soln.} = 300 \text{ cc. soln.} \times \left(\frac{2.2 \text{ moles } AlCl_3}{1 \text{ l. soln.}}\right) \times \left(\frac{1 \text{ l.}}{1000 \text{ cc.}}\right)$$

$$\times \left(\frac{133.5 \text{ g. } AlCl_3}{1 \text{ mole } AlCl_3}\right)$$

$$= 88 \text{ g. } AlCl_3$$

Thus we would take 88 g. of $AlCl_3$ and add water until the total volume of the solution was 300 cc.

Do problems 4–10 at the end of the chapter.

4. Dilution

In laboratory practice we quite frequently are given stock solutions of certain concentrations and are required to make dilute solutions from them. In addition, in the course of reactions we will mix two solutions. This will result in diluting the concentration of each solution. It is, therefore, of interest to see how these changes, which are changes in volume, affect the concentration.

Example: A laboratory bottle is labeled 12.0 M HCl. How would you make from this 20 cc. of a 3.0 M HCl solution?

Answer: It is well to visualize first the procedure we must follow. We are going to take a certain volume of the concentrated 12.0 M HCl solution and add to it enough water to make 20 cc. of the 3.0 M solution. Thus we want to know how many cubic centimeters of the 12.0 M solution to start with.

In 20 cc. of 3.0 M HCl there are

$$20 \text{ cc. soln.} = 20 \text{ cc. soln.} \times \left(\frac{3.0 \text{ moles HCl}}{1 \text{ l. soln.}}\right) \times \left(\frac{1 \text{ l.}}{1000 \text{ cc.}}\right)$$

$$= 0.060 \text{ mole HCl}$$

In order to get 0.060 mole HCl we need to take

$$0.060 \text{ mole HCl} = 0.060 \text{ mole HCl} \times \left(\frac{1 \text{ l. soln.}}{12.0 \text{ moles HCl}}\right) \times \left(\frac{1000 \text{ cc.}}{1 \text{ l.}}\right)$$

$$= 5.0 \text{ cc. solution.}$$

Thus if we take 5.0 cc. of the stock solution and add enough water (about 1𝟧 cc.) to make 20 cc. of solution we will have a concentration of 3.0 M HCl.

Note that both parts of this problem could have been combined as a single series of conversions.

Example: How would you make 24 cc. of a 0.25 M solution of H_2SO_4, starting with 6.0 M H_2SO_4?

Answer: Let us label these solutions A and B to avoid confusion.

$$24 \text{ cc. soln. A} = 24 \text{ cc. soln. A} \times \left(\frac{0.25 \text{ mole } H_2SO_4}{1000 \text{ cc. soln. A}}\right) \times \left(\frac{1000 \text{ cc. soln. B}}{6.0 \text{ moles } H_2SO_4}\right)$$

$$= 1.00 \text{ cc. solution B}$$

Thus we take 1.00 cc. of solution B (6.0 M H_2SO_4) and dilute to 24 cc. with water.

Example: 50 cc. of a 3.0 M solution of HCl is mixed with 70 cc. of a 4.0 M solution of KNO_3. What is the final concentration of HCl and KNO_3 in the mixture? (Assume no contraction of volume on mixing.*)

Answer: Let us label the mixture, solution C. Its volume is $50 + 70 = 120$ cc. By definition:

$$\frac{\text{Molarity of HCl}}{\text{in mixture}} = \frac{\text{moles HCl}}{\text{l. soln. C}} = \frac{50 \text{ cc. soln. A} \times \left(\frac{3.0 \text{ moles HCl}}{1 \text{ l. soln. A}}\right) \times \left(\frac{1 \text{ l.}}{1000 \text{ cc.}}\right)}{120 \text{ cc. soln. C} \times \left(\frac{1 \text{ l.}}{1000 \text{ cc.}}\right)}$$

$$= \frac{1.25 \text{ moles HCl}}{\text{l. soln. C}} = 1.25 \; M \text{ HCl}$$

$$\frac{\text{Molarity of } KNO_3}{\text{in mixture}} = \frac{\text{moles } KNO_3}{\text{l. soln. C}}$$

$$= \frac{70 \text{ cc. soln. B} \times \left(\frac{4.0 \text{ moles } KNO_3}{1 \text{ l. soln. B}}\right) \times \left(\frac{1 \text{ l.}}{1000 \text{ cc.}}\right)}{120 \text{ cc. soln. C} \left(\frac{1 \text{ l.}}{1000 \text{ cc.}}\right)}$$

$$= 2.33 \frac{\text{moles } KNO_3}{\text{l. soln. C}} \rightarrow 2.3 \; M \; KNO_3$$

* Note that on dilution of a solution the original concentration is diminished in the ratio of the initial to the final volume. In the above example the final M HCl $= {}^{50}\!/_{120} \times$ (original M HCl).

Do problems 11–17 at the end of the chapter.

5. Chemical Reactions Involving Solutions

In Chapter IV we discussed the method of solving problems involving chemical reactions. The problems discussed, however, involved only the reaction of pure substances. How may we extend the methods to include solutions? The answer is that we will follow the same procedure but now add to it the methods just discussed for converting from volume of a solution to quantity of solute present.

Example:

$$3Cu + 8HNO_3 \rightarrow 3Cu(NO_3)_2 + 2NO\uparrow + 4H_2O$$

How many grams of copper may be dissolved in 150 cc. of 4 M HNO_3?

Answer:

$$150 \text{ cc. soln.} = 150 \text{ cc. soln.} \times \left(\frac{1 \text{ l.}}{1000 \text{ cc.}}\right) \times \left(\frac{4 \text{ moles } HNO_3}{1 \text{ l. soln.}}\right)$$

$$\times \left(\frac{3 \text{ moles } Cu}{8 \text{ moles } HNO_3}\right) \times \left(\frac{63.6 \text{ g. Cu}}{1 \text{ mole Cu}}\right)$$

$$= \frac{150 \times 4 \times 63.6 \times 3}{1000 \times 8} = 14.3 \text{ g. Cu}$$

Example:

$$3Cl_2 + 6NaOH \rightarrow 5NaCl + NaClO_3 + 3H_2O$$

How many liters STP of Cl_2 gas will react with 75 cc. of 1.6 M NaOH?

Answer:

$$75 \text{ cc. soln.} = 75 \text{ cc. soln.} \times \left(\frac{1 \text{ l.}}{1000 \text{ cc.}}\right) \times \left(\frac{1.6 \text{ moles NaOH}}{1 \text{ l. soln.}}\right)$$

$$\times \left(\frac{3 \text{ moles } Cl_2}{6 \text{ moles NaOH}}\right) \times \left(\frac{22.4 \text{ l. STP } Cl_2}{1 \text{ mole } Cl_2}\right)$$

$$= \frac{75 \times 1.6 \times 3 \times 22.4}{6 \times 1000} \text{ l. STP } Cl_2$$

$$= 1.34 \text{ l. STP } Cl_2$$

6. Reactions between Solutions—Principle of Equivalence

If we are dealing with the reaction between two solutions, we may use the procedure outlined in section 5.

Example:

$$2FeCl_3 + 3Ag_2SO_4 \rightarrow 6AgCl\downarrow + Fe_2(SO_4)_3$$

How many cubic centimeters of 0.20 M Ag_2SO_4 will react with 68 cc. of 0.65 M $FeCl_3$?

Answer: Label the $FeCl_3$ solution A and the Ag_2SO_4 solution B.

$$68 \text{ cc. soln. A} = 68 \text{ cc. soln. A} \times \left(\frac{1 \text{ l.}}{1000 \text{ cc.}}\right) \times \left(\frac{0.65 \text{ mole } FeCl_3}{1 \text{ l. soln. A}}\right)$$

$$\times \left(\frac{3 \text{ moles } Ag_2SO_4}{2 \text{ moles } FeCl_3}\right) \times \left(\frac{1 \text{ l. soln. B}}{0.20 \text{ mole } Ag_2SO_4}\right) \times \left(\frac{1000 \text{ cc.}}{1 \text{ l.}}\right)$$

$$= \frac{68 \times 0.65 \times 3 \times 1000}{1000 \times 2 \times 0.20} \text{ cc. soln. B} = 331 \text{ cc. soln. B}$$

$$\rightarrow 330 \text{ cc. soln. B}$$

However, when dealing with solutions we can also use the much more powerful methods outlined in the section dealing with equivalents (Chapter VII). The principle of equivalence tells us that equal volumes of solutions having the same normality will have the same capacity for chemical reaction since they will have equal numbers of equivalents in equal volumes.

Thus, in dealing with reactions between two solutions, if we convert their concentrations to normalities, we can dispense with a balanced chemical equation.

To illustrate this let us consider the previous example:

Example: How many cubic centimeters of 0.20 M Ag_2SO_4 will react with 68 cc. of 0.65 M $FeCl_3$?

Answer:

$$0.65 \ M \ FeCl_3 = \left(\frac{0.65 \ \text{mole } FeCl_3}{1. \ \text{soln.}}\right) \times \left(\frac{3 \ \text{eq. } FeCl_3}{1 \ \text{mole } FeCl_3}\right)$$

$$= \frac{1.95 \ \text{eq. } FeCl_3}{1. \ \text{soln.}} = 1.95 \ N \ FeCl_3$$

Similarly:

$$0.20 \ M \ Ag_2SO_4 = 0.40 \ N \ Ag_2SO_4$$

Then:

$$68 \text{ cc. soln. A} = 68 \ \text{cc. soln. A} \times \left(\frac{1 \ \text{l.}}{1000 \ \text{cc.}}\right) \times \left(\frac{1.95 \ \text{eq. } FeCl_3}{1 \ \text{l. soln. A}}\right)$$

$$\times \left(\frac{1 \ \text{eq. } Ag_2SO_4}{1 \ \text{eq. } FeCl_3}\right) \times \left(\frac{1 \ \text{l. soln. B}}{0.40 \ \text{eq. } Ag_2SO_4}\right) \times \left(\frac{1000 \ \text{cc.}}{1 \ \text{l.}}\right)$$

$$= \frac{68 \times 1.95 \times 1000}{1000 \times 0.40} \ \text{cc. soln. B}$$

$$= 331 \text{ cc. soln. B} \rightarrow 330 \text{ cc. soln. B}$$

Thus by means of the principle of equivalence we have achieved the same result without a balanced equation. Note that in the above example we could have combined the conversion of concentrations units with the other steps.

The student will probably by now have noticed that we generally deal in the laboratory with cubic centimeters (or milliliters) of solutions rather than with such large quantities as liters. Also, we are much more apt to use millimoles or milliequivalents of material rather than such large amounts as moles. Is it not possible to use these smaller units for calculations? The answer is, of course, yes, and it can easily be seen that a concentration of 1 mole/l. is the same as

1 millimole/milliliter (1 mmole/ml.). Similarly, a 2 N solution may be taken to mean 2 equivalents/liter (2 eq./l.) or 2 milliequivalents/milliliter (2 meq./ml.).

These latter units are more convenient than the other units and consequently widely used for most laboratory work. (*Note:* 1 ml. may be taken as equal to 1 cc.)

Example: How many milliliters of a 3.4 M solution of $Ba(NO_3)_2$ will react with 60 ml. of a 2.4 M solution of Na_3PO_4?

Answer: We will use the principle of equivalence and combine all steps: Call the $Ba(NO_3)_2$ solution A, and the Na_3PO_4 solution B.

$$60 \text{ ml. soln. B} = 60 \text{ ml. soln. B} \times \left(\frac{2.4 \text{ mmoles } Na_3PO_4}{1 \text{ ml. soln. B}} \right) \times \left(\frac{3 \text{ meq. } Na_3PO_4}{1 \text{ mmole } Na_3PO_4} \right)$$

$$\times \left(\frac{1 \text{ meq. } Ba(NO_3)_2}{1 \text{ meq. } Na_3PO_4} \right) \times \left(\frac{1 \text{ mmole } Ba(NO_3)_2}{2 \text{ meq. } Ba(NO_3)_2} \right) \times \left(\frac{1 \text{ ml. soln. A}}{3.4 \text{ mmole } Ba(NO_3)_2} \right)$$

$$= \frac{60 \times 2.4 \times 3}{2 \times 3.4} \text{ ml. soln. A}$$

$$= 64 \text{ ml. soln. A}$$

Do problems 18, 19, and 20 at the end of the chapter.

7. Titration of Acids and Bases

One of the most important of the general reactions in chemistry is the reaction of an acid and base to produce a salt and water. These reactions may be performed in solution, using a colored dye to indicate when neutralization has been achieved. When such neutralizations are performed extremely accurately, with burets to measure volumes they are referred to as titrations. We may apply the methods of the preceding chapter to such calculations.

Example: How many milliliters of 0.3 N H_2SO_4 are required to neutralize 40 ml. of 0.6 N NaOH?

Answer: Call the acid A, and the base solution B.

$$40 \text{ ml. soln. B} = 40 \text{ ml. soln. B} \times \left(\frac{0.6 \text{ meq. NaOH}}{1 \text{ ml. soln. B}} \right) \times \left(\frac{1 \text{ meq. } H_2SO_4}{1 \text{ meq. NaOH}} \right)$$

$$\times \left(\frac{1 \text{ ml. soln. A}}{0.3 \text{ meq. } H_2SO_4} \right)$$

$$= \frac{40 \times 0.6}{0.3} \text{ ml. soln. A}$$

$$= 80 \text{ ml. acid}$$

If the concentrations are given in terms of molarities, we may first convert to normalities or combine the conversions with the problem.

Example: How many milliliters of 0.46 M H_3PO_4 are needed to neutralize 60 ml. of 0.62 N NaOH?

Answer:

$$60 \text{ ml. soln. B} = 60 \text{ ml. soln. B} \times \left(\frac{0.62 \text{ meq. NaOH}}{1 \text{ ml. soln. B}} \right) \times \left(\frac{1 \text{ meq. } H_3PO_4}{1 \text{ meq. NaOH}} \right)$$

$$\times \left(\frac{1 \text{ mmole } H_3PO_4}{3 \text{ meq. } H_3PO_4} \right) \times \left(\frac{1 \text{ ml. soln. A}}{0.46 \text{ mmole } H_3PO_4} \right)$$

$$= \frac{60 \times 0.62}{3 \times 0.46} \text{ ml. soln. A}$$

$$= 26.9 \text{ ml. acid} \rightarrow 27 \text{ ml. acid}$$

It is interesting to observe that in all such problems we can from the principle of equivalence compute the quantity of each substance produced in the reaction. Thus, if we titrate an acid with a base we will produce a salt and water. For each milliequivalent of acid we will require 1 meq. equivalent of base and we will produce 1 meq. of salt and 1 meq. of water (considered as H—OH; 18 g./eq.).

Example: In the preceding example, the neutralization of 60 ml. of 0.62 N NaOH by 0.46 M H_3PO_4, how many equivalents of salt (Na_3PO_4) are produced? What is its weight?

Answer: Using the principle of equivalence to convert NaOH to Na_3PO_4:

$$60 \text{ ml. soln. B} = 60 \text{ ml. soln. B} \times \left(\frac{0.62 \text{ meq. NaOH}}{1 \text{ ml. soln. B}} \right) \times \left(\frac{1 \text{ meq. } Na_3PO_4}{1 \text{ meq. NaOH}} \right)$$

$$= 37.2 \text{ meq. } Na_3PO_4$$

To convert to weight we first convert to millimoles and then to weight units:

$$= 37.2 \text{ meq. } Na_3PO_4 \times \left(\frac{1 \text{ mmole } Na_3PO_4}{3 \text{ meq. } Na_3PO_4} \right)$$

$$\times \left(\frac{164 \text{ mg. } Na_3PO_4}{1 \text{ mmole } Na_3PO_4} \right) \times \left(\frac{1 \text{ g.}}{1000 \text{ mg.}} \right)$$

$$= 2.03 \text{ g. } Na_3PO_4$$

Do problems 21–24 at the end of the chapter.

8. Ionic Solutions

Compounds, such as ethyl alcohol (C_2H_5OH), acetone (CH_3COCH_3), and ammonia (NH_3) are covalently bound and hence are made up of molecules. When we dissolve them in water, or other solvents, the solutions can be considered as essentially binary mixtures. However, salts, such as $NaCl$, K_2SO_4, $CaCl_2$, etc., are not made up of molecules but are instead mixtures of positive and negative ions in fixed and definite proportions. When we dissolve these salts in ionizing solvents such as water, they become completely dissociated into their individual ions.

Thus, a 1-molar solution of $NaCl$ contains 1 mole Na^+ ions/liter soln. and 1 mole Cl^- ions/liter soln. A 0.15 molar $CaCl_2$ contains 0.15 mole Ca^{2+} ions/liter soln. and 0.30 mole Cl^- ions/liter soln. A 0.25 M $Al_2(SO_4)_3$ solution contains 0.50 mole Al^{3+} ions/liter soln. and 0.75 mole $(SO_4)^=$/liter soln. We must remember when dealing with salt solutions to always consider the positive and negative ions separately.

Example: 20 ml. of 0.60 M NaCl is mixed with 40 ml. of 0.80 M KCl. What ions are present in the final solution and what are their concentrations?

Answer: The 0.60 M NaCl soln. (A) contains 0.60 M Na^+ and 0.60 M Cl^-. The 0.80 M KCl soln. (B) contains 0.80 M K^+ and 0.80 M Cl^-. The concentrations of each ion in the final solution (C), whose volume will be 60 ml., is obtained by taking the total mmoles of each species present and dividing by the total ml. of the soln.

$$\text{Conc. } (Na^+) = \frac{\text{mmole } Na^+}{\text{ml. soln. C}}$$

$$= \frac{20 \text{ ml. soln. A} \times \left(\dfrac{0.60 \text{ mmole } Na^+}{1 \text{ ml. soln. A}}\right)}{60 \text{ ml. soln. C}} = 0.20 \ M \ Na^+$$

$$\text{Conc. } (K^+) = \frac{\text{mmoles } K^+}{\text{ml. soln. C}}$$

$$= \frac{40 \text{ ml. soln. B} \times \left(\dfrac{0.80 \text{ mmole } K^+}{1 \text{ ml. soln. B}}\right)}{60 \text{ ml. soln. C}} = 0.53 \ M \ K^+$$

$$\text{Conc. } (Cl^-) = \frac{\text{mmoles } Cl^-}{\text{ml. soln. C}}$$

$$= \frac{20 \text{ ml. soln. A} \times \left(\dfrac{0.60 \text{ mmole } Cl^-}{1 \text{ ml. soln. A}}\right) + 40 \text{ ml. soln. B} \left(\dfrac{0.80 \text{ mmole } Cl^-}{1 \text{ ml. soln. B}}\right)}{60 \text{ ml. soln. C}}$$

$$= \frac{(12 + 32) \text{ mmoles } Cl^-}{60 \text{ ml. soln. C}} = 0.73 \ M \ Cl^-$$

Note that since all solutions are electrically neutral, the sum of the concentration of positive ions times their individual charges must equal the same sum for the negative charges. In this case, 0.20 M Na^+ + 0.53 M K^+ = 0.73 M Cl^-.

Example: 15 ml. of 0.20 M $Al_2(SO_4)_3$ is mixed with 60 ml. of 0.30 M Na_2SO_4. What ions are present in the final solution and what are their concentrations?

Answer: The 0.20 M $Al_2(SO_4)_3$ soln. (A) contains 0.40 M Al^{3+} and 0.60 M $(SO_4^=)$, while the 0.30 M Na_2SO_4 soln. (B) has 0.60 M Na^+ and 0.30 M $(SO_4^=)$. The final volume of the mixture, soln. C, is approximately the sum of the initial volumes, or 75 ml.

$$\text{Conc. (Al}^{3+}\text{)} = \frac{15 \text{ ml. soln. A} \times \left(\dfrac{0.40 \text{ mmole Al}^{3+}}{1 \text{ ml. soln. A}}\right)}{75 \text{ ml. soln. C}} = 0.080 \ M \ \text{Al}^{3+}$$

$$\text{Conc. (Na}^{+}\text{)} = \frac{60 \text{ ml. soln. B} \times \left(\dfrac{0.60 \text{ mmole Na}^{+}}{1 \text{ ml. soln. B}}\right)}{75 \text{ ml. soln. C}} = 0.48 \ M \ \text{Na}^{+}$$

$$\text{Conc. (SO}_4^{=}\text{)} =$$
$$\frac{15 \text{ ml. soln. A} \times \left(\dfrac{0.60 \text{ mmole SO}_4^{=}}{1 \text{ ml. soln. A}}\right) + 60 \text{ ml. soln. B} \times \left(\dfrac{0.30 \text{ mmole SO}_4^{=}}{1 \text{ ml. soln. B}}\right)}{75 \text{ ml. soln. C}}$$
$$= \frac{(9 + 18) \text{ mmoles SO}_4^{=}}{75 \text{ ml. soln. C}} = 0.36 \ M \ \text{SO}_4^{=}$$

We can check equivalence of positive and negative charges in the final solution C (electroneutrality) by noting that:

$$\text{molarity of positive charge} = 3 \times (\text{Al}^{3+}) + 1 \times (\text{Na}^{+})$$
$$= 0.24 + 0.48 = 0.72 \ M$$
$$\text{molarity of negative charge} = 2 \times (\text{SO}_4^{=}) = 0.72 \ M$$

9. Reactions in Ionic Solutions

In dealing with reactions taking place in ionic solutions, we should write our balanced chemical equations to show only the ionic species involved. Thus, when NaCl solution is added to $AgNO_3$ solution, there is a reaction in which the insoluble salt, AgCl, precipitates. We would normally write this as a balanced equation, as follows:

$$AgNO_3 + NaCl \rightarrow AgCl \downarrow + NaNO_3$$

However, since all Ag^+ ion-containing solutions will react with all Cl^- ion containing solutions to produce the same insoluble AgCl, there seems to be no point in including the other ions which take no part in the reaction and, in fact, remain unchanged in the final solution. In the above case, the (NO_3) ions from the $AgNO_3$ solution and the (Na^+) ions from the NaCl solution are unaffected and are to be omitted from the balanced chemical equation. The ionic form of this chemical reaction is then:

$$Ag^+ + Cl^- \rightarrow AgCl$$

In similar fashion, mixing solutions of $CaCl_2$ and $Al_2(SO_4)_3$ leads to a precipitation of insoluble $CaSO_4$. The ionic equation is:

$$Ca^{+2} + SO_4^{=} \rightarrow CaSO_4$$

See Appendix III.6 for solubility rules to enable one to predict precipitation reactions for salts.

Aside from salts, strong acids are another species of compound which are in water solution, mixtures of ions. Thus, 0.20 M HCl contains 0.20 M H^+ ions and 0.20 M Cl^- ions. Also, 0.1 M H_2SO_4 contains 0.2 M H^+ ions and 0.1 M $SO_4^=$ ions. Weak acids, such as acetic acid (HAc) are essentially unionized and a 0.8 M HAc solution should be written as 0.8 M HAc. See Chapter 14, section 13 for lists of weak and strong acids.

A base is a substance capable of neutralizing an acid. Some bases are salts whose negative ion is the OH^- ion. The reaction of neutralization is just the reaction of the H^+ of the acid and the OH^- or other species of the base to form water:

$$\text{neutralization: } H^+ + OH^- \rightarrow HOH$$

For salts of the OH^{-1} the base may be considered the OH^{-1} ion. It is not the only base. Solutions of NH_3 are moderately basic and slightly ionic. The base NH_3 may be neutralized by H^+ from an acid to form the weak acid NH_4^+. Thus, NH_3 solutions react with HCl solutions to form NH_4Cl solutions. The ionic reaction is:

$$NH_3 + H^+ \rightarrow NH_4^+$$

As we shall see later, any substance capable of combining with H^+ ion will be considered a base.

Example: How many ml. of 0.20 M $CaCl_2$ are needed to completely react with 35 ml. of a 0.15 M $AgNO_3$ solution to precipitate AgCl? What species are present in the final solution and what are their concentrations?

Answer: The reaction is: $Ag^{+1} + Cl^{-1} \rightarrow AgCl \rightarrow$ Hence:

$$35 \text{ ml. AgNO}_3 \text{ soln.} = 35 \text{ ml. AgNO}_3 \text{ soln. } \left(\frac{0.15 \text{ mmole Ag}^+}{1 \text{ ml. AgNO}_3 \text{ soln.}} \right)$$

$$\left(\frac{1 \text{ mmole Cl}^-}{1 \text{ mmole Ag}^+} \right) \times \left(\frac{1 \text{ ml. CaCl}_2 \text{ soln.}}{0.40 \text{ mmole Cl}^-} \right)$$

$$= \frac{35 \times 0.15}{0.40} \text{ ml. CaCl}_2 \text{ soln.}$$

$$= 13 \text{ ml. CaCl}_2 \text{ soln.}$$

The final solution is $13 + 35 = 48$ ml. in volume. All Ag^+ ions and Cl^- ions have reacted so that only Ca^{2+} and NO_3^- ions are left. Their concentrations are:

$$\text{Conc. (Ca}^{+2}) = \frac{13 \text{ ml. CaCl}_2 \text{ soln.} \times \left(\frac{0.20 \text{ mmole Ca}^{2+}}{1 \text{ ml. CaCl}_2 \text{ soln.}} \right)}{48 \text{ ml. soln.}} = 0.054 \text{ } M$$

$$\text{Conc. (NO}_3^-) = \frac{35 \text{ ml. AgNO}_3 \text{ soln.} \times \left(\dfrac{0.15 \text{ mmole NO}_3^-}{1 \text{ ml. AgNO}_3 \text{ soln.}} \right)}{48 \text{ ml. soln.}} = 0.109 \ M$$

$$\rightarrow 0.11 \ M$$

Note once again that positive and negative charge have the same concentrations and electroneutrality is conserved.

Example: 40 ml. of 0.25 M $Al_2(SO_4)_3$ is mixed with 80 ml. of 0.30 M $BaCl_2$. $BaSO_4$ will precipitate out since it is insoluble. What species are left in the final solution and what are their concentrations?

Answer: The chemical reaction is: $Ba^{++} + SO_4^{=} \rightarrow BaSO_4 \downarrow$

Let us first see how much Ba^{++} and $SO_4^{=}$ were introduced into the solution:

$$80 \text{ ml. BaCl}_2 \text{ soln.} = 80 \text{ ml. BaCl}_2 \text{ soln.} \times \left(\frac{0.30 \text{ mmole Ba}^{++}}{1 \text{ ml. BaCl}_2 \text{ soln.}} \right)$$

$$= 24 \text{ mmoles Ba}^{++}$$

$$40 \text{ ml. Al}_2(SO_4)_3 \text{ soln.} = 40 \text{ ml. Al}_2(SO_4)_3 \text{ soln.} \times \left(\frac{0.75 \text{ mmole SO}_4^{=}}{1 \text{ ml. Al}_2(SO_4)_3 \text{ soln.}} \right)$$

$$= 30 \text{ mmoles SO}_4^{=}$$

Since 1 mole of $SO_4^{=}$ needs only 1 mole of Ba^{++}, we have more $SO_4^{=}$ than is needed for reaction, and 6 mmoles of $SO_4^{=}$ will be left in the final solution whose volume = 40 + 80 = 120 ml.

Its concentration is:

$$\text{Conc. (SO}_4^{=}) = \frac{6 \text{ mmoles SO}_4^{=}}{120 \text{ ml. soln.}} = 0.05 \ M \ (SO_4^{=})$$

The Al^{+3} and Cl^- are, of course, non-reactive and their concentrations are:

$$\text{Conc. (Al}^{+3}) = \frac{40 \text{ ml. Al}_2(SO_4)_3 \text{ soln.} \times \left(\dfrac{0.50 \text{ mmole Al}^{+3}}{1 \text{ ml. Al}_2(SO_4)_3 \text{ soln.}} \right)}{120 \text{ ml. soln.}}$$

$$= 0.17 \ M \ Al^{+3}$$

$$\text{Conc. (Cl}^-) = \frac{80 \text{ ml. BaCl}_2 \text{ soln.} \times \left(\dfrac{0.60 \text{ mmole Cl}^-}{1 \text{ ml. BaCl}_2 \text{ soln.}} \right)}{120 \text{ ml. soln.}}$$

$$= 0.40 \ M \ Cl^-$$

The reader should calculate the total amounts of positive and negative ions to see if electroneutrality is conserved.

---

10. Density and Specific Gravity

In industrial operation, composition of solutions is frequently measured by their density. The reason for this is that it is very simple to measure the density of a solution by means of the depth to which a hollow-stemmed, weighted bob will sink in it. Such an instrument is known as a hydrometer.

Thus the attendant in a gasoline station can tell very quickly and very accurately the condition of a car's storage battery by measuring with a hydrometer the density of the sulfuric acid in the battery. In order to make use of such a measurement, the relation between density and composition must be known. Such relations must be obtained by experiment. Many chemical handbooks contain tables giving the density of common solutions as functions of their composition. With such information available it is always possible to translate such data into the other concentration units discussed.

Example: A bottle of commercial sulfuric acid (H_2SO_4) is labeled 86% sulfuric acid: density, 1.787 g./cc. What is the molarity of this solution?

Answer: We want to find the number of moles of H_2SO_4 in 1 l. of this solution (i.e., molarity). The student should check the origin of the conversion factors.

$$1 \text{ l. soln.} = 1 \text{ l. soln.} \times \left(\frac{1000 \text{ cc.}}{1 \text{ l.}} \right) \quad \left(\frac{1.787 \text{ g. soln.}}{1 \text{ cc. soln.}} \right) \times \left(\frac{86 \text{ g. } H_2SO_4}{100 \text{ g. soln.}} \right)$$

$$\times \left(\frac{1 \text{ mole } H_2SO_4}{98 \text{ g. } H_2SO_4} \right) = \frac{1000 \times 1.787 \times 86}{100 \times 98} \text{ moles } H_2SO_4$$

$$= 15.7 \text{ moles } H_2SO_4$$

The concentration is thus 15.7 *M*.

Density, as we have used it, is always expressed in units of mass divided by volume. It is frequently valuable to have instead of an absolute density scale a relative density scale. In a relative density scale we would express the density of a substance by saying how many times more dense or less dense it was than another substance.

Such a relative density scale is in common use industrially, and the units are known as specific gravity. It is defined with respect to the density of water at 4°C.

$$\boxed{\text{Specific gravity of a substance} = \frac{\text{Density of the substance}}{\text{Density of water at } 4°C}.}$$

Water is thus the standard for such a scale. When we say that the specific gravity of a substance is 2, we mean that it is 2 times as dense as water at 4°C. If we wish to find the absolute density of the substance we then multiply its specific gravity by the density of water at 4°C.   Since the density of water at 4°C. is 1.000 g./cc., as far as the metric system is concerned the numerical value of the specific gravity will be the same as the density in grams per cubic centimeter. In other scales we do not have this simple relation.

Example: The specific gravity of concentrated sulfuric acid (96%) at 20°C. is 1.836.   What is its density in grams per cubic centimeter? in pounds per cubic foot?

Answer: From the above, the density must be also 1.836 g./cc.   To find the density in pounds per cubic foot we can make our usual conversion, or we can find out the density of water in pounds per cubic foot.   Since the latter is 62.4 lb./ft.^{3}:

$$\text{Density of sulfuric acid} = 1.836 \times 62.4 \text{ lb./ft.}^3 = 114.6 \text{ lb./ft.}^3$$

---

11. Problems

1. If the concentration of a solution is known in weight per cent, what property of the solution should be measured in dispensing it?   What if the concentration were expressed as molarity?

2. Of a solution whose normality is known, 100 g. are weighed out.   What additional property must be known to compute the amount of solute dispensed?

3. Make the following conversions:

    (a) 3 M H_2SO_4 to normality.

    (b) 0.1 N $Ca(OH)_2$ to molarity.

    (c) 5 g. Na_2SO_4/l. to molarity.

    (d) 20 mg. $CuSO_4$/ml. to molarity.

    (e) 3 N $Al_2(SO_4)_3$ to millimole $Al_2(SO_4)_3$/ml.

    (f) 0.46 mmole K_3PO_4/ml. to g. K_3PO_4/l.

    (g) 2.4 mg. $CaCl_2$/ml. to normality.

    (h) 5% $NaCl$ solution to moles $NaCl$/kg. solvent (molality).

    (i) 1 molal solution $NaCl$ to weight per cent.

4. How many moles of $Al_2(SO_4)_3$ are there in 20 cc. of a 3.0 M solution?

5. How many grams of $Ca(OH)_2$ are there in 800 cc. of a 0.12 N solution? How many millimoles?   How many milliequivalents?

6. How many equivalents of $Al_2(CrO_4)_3$ are there in 60 ml. of a solution having a concentration of 20 mg. $Al_2(CrO_4)_3$/ml.?   How many millimoles?   How many grams?

7. A reaction requires 12 g. of H_2SO_4.   How many cubic centimeters of a 3.0 M solution should you use?

8. A reaction requires 3.4 mmoles of Na_3PO_4.   How many milliliters of a 1.8 N solution should you use?

9. How would you prepare 150 ml. of a 3.5 N solution of $Ca(NO_3)_2$?

10. How would you prepare 240 ml. of a solution of Na_2CO_3 containing 3.8 g. Na_2CO_3/l.?

11. How would you prepare 25 ml. of a 1.2 M solution of KCl from a stock solution which is 3.0 M?

12. How would you prepare 15 cc. of a 0.45 N solution of $CuSO_4$ from a stock solution which is 2.4 M?

13. 150 cc. of 3.0 M K_2SO_4 is mixed with 80 cc. of 2.0 M $NaNO_3$. What is the concentration of each salt in the final solution?

14. A stock solution of $CuSO_4$ is 1.0 M. How would you make from this 80 cc. of a solution containing 20 mg. $CuSO_4$/ml.?

15. How many grams of Cu are there in 100 ml. of a 0.50 M solution of $CuSO_4$? How many millimoles?

16. How would you make a solution containing 5.0 mg. Cu/ml. from a stock solution which is 0.80 M $Cu(NO_3)_2$?

17. How many millimoles of Zn are there in 5.0 ml. of a solution which is 2.4 M $ZnCl_2$?

18. $2Al + 3Zn(NO_3)_2 \rightarrow 2Al(NO_3)_3 + 3Zn$. How many grams of Al will react with 50 ml. of a 0.40 M solution of $Zn(NO_3)_2$?

19. $BaCl_2 + H_2SO_4 \rightarrow 2HCl + BaSO_4\downarrow$. How many grams of $BaCl_2$ are needed to react with 90 ml. of a 0.48 N solution of H_2SO_4.? How many grams of $BaSO_4$ will be produced?

20. How many cubic centimeters STP of HCl *gas* are needed to neutralize 40 ml. of a 0.80 M solution of KOH. How many millimoles of KCl will be formed?

21. How many milliliters of 3.0 M H_2SO_4 are needed to neutralize 200 ml. of 0.34 N $Ca(OH)_2$? How many milliequivalents of $CaSO_4$ are formed?

22. How many milliliters of 1.4 N NaOH will completely neutralize 80 ml. of 0.72 M H_3PO_4? How many grams of Na_3PO_4 are formed?

23. 40 ml. of 0.56 N NaOH just neutralize 1.75 g. of an unknown acid. What is the equivalent weight of the acid?

24. 60 ml. of a 0.75 N solution of KOH neutralize 44 ml. of a solution of H_2SO_4. What is the molarity of the acid?

25. The concentrated ammonia solution in laboratories is a 26.0% solution of NH_3. If its density is 0.904 g./cc. what is its molarity? its normality? its specific gravity?

26. Concentrated hydrochloric acid is 37.1% by weight HCl and has a specific gravity of 1.184. What is its molarity? its normality? its molality?

27. Laboratory nitric acid is 68% by weight with a density of 1.405 g./cc. How many milliliters of this acid are needed to neutralize 800 ml. of 0.040 M $Ca(OH)_2$ solution?

28. What is the density and molarity of O_2 gas at STP?

29. What is the pressure of a 0.40 M concentration of Cl_2 gas at 40°C?

30. At 20°C. and 1 atm. total pressure, the partial pressure of O_2 in the atmosphere is about 150 mm. Hg. What is its molarity?

31. What species are present in 0.040 M $Al_2(SO_4)_3$?

32. What species are present in 0.050 M HCl?

33. What species are present in 0.065 M $ZnCl_2$?

34. 30 ml. of 0.12 M $CaCl_2$ is mixed with 60 ml. of 0.20 M $AlCl_3$. What species are present in the final solution and what are their concentrations?

35. 40 ml. of 0.15 M HCl are mixed with 60 ml. of 0.20 M NaOH. What species are present in the final solution and what are their concentrations?

36. Cu metal reacts with dilute HNO_3 according to the following ionic equation:
$$3Cu + 8H^+ + 2NO_3^- \rightarrow 3\,Cu^{++} + 2NO \uparrow + 4H_2O$$
140 mg. Cu is allowed to react to completion with 40 ml. of 0.15 M HNO_3. What species are present in the final solution and what are their concentrations?

CHAPTER IX

The Physical Properties of Solutions

1. Properties of Ideal Solutions

Pure substances may be characterized by their physical properties. They have very sharp melting points, very sharp boiling points (at 1 atm. pressure), definite densities, definite vapor pressures, etc. Thus pure water has a melting point of 0°C., a boiling point of 100°C. (at 1 atm. pressure), density of 1.000 g./cc. (at 4° C.), a vapor pressure of 17.36 mm. Hg (at 20°C.), etc.

These properties can provide a reliable means of identifying a pure substance. It is of some interest to know how such physical properties are affected when a solution of two pure substances is made. In general we might expect that the physical properties of the solution (mixture) will be intermediate between the properties of the two components of the mixture. Thus we might reasonably expect that a 50% by volume mixture of two liquids might have a density intermediate between the densities of the individual liquids, and similarly for other properties.

Investigations have shown that this is approximately true for large numbers of solutions. A solution for which it would be exactly borne out is called an *ideal solution*. In practice there are very few ideal solutions. But many solutions are almost ideal, especially when very dilute (i.e., containing only a small amount of solute).

We use the term *colligative* to describe those properties of a solution which may be calculated by taking simple arithmetic averages of the properties of solute and solvent. As we shall see, the colligative properties of solutions can give us much valuable information about the properties of the solute if the properties of the solvent are known.

2. Vapor Pressure of Solutions—Raoult's Law

When a non-volatile solute, (e.g., sugar) is added to a volatile solvent (e.g., water), it is found that the vapor pressure of the solvent is reduced. In dilute solutions it is found that the reduction is pro-

112

portional to the molecular concentration of the solvent. Or, conversely, the vapor pressure of the solution is proportional to the molecular fraction of the solvent present in it.

Thus if we take 10 molecules of sugar and 990 molecules of water to make a solution, the vapor pressure of the water in the solution will be $^{990}/_{1000}$ of what it was before the sugar was dissolved in it. This observation, which will hold true for other substances besides sugar and other solvents besides water, is known as Raoult's law. In terms of molar units:

Raoult's Law: The vapor pressure of a volatile solvent in a dilute solution is proportional to its mole fraction.

Definition:

Mole fraction of a substance A in a mixture of substances $= \dfrac{\text{Number of moles } A}{\text{Total number of moles of all substances}}$

Example: 1 mole of sugar ($C_6H_{12}O_6$) is added to 19 moles of water at 30°C. If the vapor pressure of pure water at 30°C. is 31.51 torr, what will the vapor pressure of the mixture be?

Answer: From Raoult's Law:

Vapor pressure of solution = Mole fraction of water × Vapor pressure of pure water

On substitution, since the mole fraction of H_2O is $^{19}/_{20}$:

$$\text{V.P. soln.} = \frac{19}{20} \times 31.51 \text{ torr} = 29.93 \text{ torr}$$

---

3. Molecular Weight of Solute—Vapor Pressure Lowering

It can be seen that Raoult's law provides a very useful method of measuring molecular weights of substances that are not capable of being vaporized. Thus, if we find that when we add 60 g. of an unknown compound to 9 moles of water the vapor pressure of the water is lowered by 10%, we can see that we must have added 1 mole of solute and we can conclude that its molecular weight is 60 g/mole.

Example: It is found that the addition of 20 g. of a solid to 160 g. of water lowers the vapor pressure at 25°C. from 23.52 torr to 22.80 torr. What is the molecular weight of the solid?

Answer: From Raoult's Law:

Vapor pressure of solution = Mole fraction of water × Vapor pressure of pure water

Solving this equation for the mole fraction of water we find:

$$\text{Mole fraction } H_2O = \frac{\text{V.P. soln.}}{\text{V.P. pure } H_2O} = \frac{22.80}{23.52}$$

$$= 0.969$$

But the mole fraction by definition $= \dfrac{\text{Moles of } H_2O}{\text{Moles of solute + Moles of } H_2O} = 0.969.$

Solving this equation for moles of solute:

$$\text{Moles solute} = \left(\frac{\text{Moles } H_2O}{0.969}\right) - (\text{Moles } H_2O) = \frac{0.031}{0.969} \times (\text{Moles } H_2O)$$

$$= \frac{0.031}{0.969} \times 160 \text{ g. } H_2O \times \left(\frac{1 \text{ mole } H_2O}{18 \text{ g. } H_2O}\right)$$

$$= 0.284$$

Then:

$$\text{Molecular weight solute} = \frac{20 \text{ g.}}{0.284 \text{ mole}} = 70 \text{ g./mole}$$

Do problems 1, 2, and 3 at the end of the chapter.

4. Molecular Weight of Solute—Boiling Point Elevation

It is a much simpler procedure to measure the elevation in boiling point of a solvent (at constant pressure) than to measure the change in vapor pressure (at constant temperature). It can be shown (by not too simple reasoning) that the addition of a non-volatile solute to a volatile solvent will raise its boiling point by an amount which is proportional to the number of moles of solute added. That is, 1 mole of any solute added to the same quantity of solvent will always produce the same elevation of the boiling point.

For each solvent, measurements have been made of the number of degrees C. increase in boiling point produced by adding 1 mole of non-volatile solute to 1000 g. of solvent (1 kg. of solvent). This number is known as the *molal boiling point constant* ($K_{B.P.}$) and can be used to calculate the molecular weights of solutes from laboratory experiments. This constant ($K_{B.P.}$) has the unit of °C.-kg. solvent/mole solute.

Example: The addition of 42.5 g. of a solid, X, to 800 g. of water produces an increase in boiling point of 0.34°C. If $K_{B.P.}$ for water is 0.52°C.-kg. H_2O/mole solute, what is the molecular weight of the solute, X?

Answer: We want to know the number of moles of X in 42.5 g. X. From the increase in boiling point we can calculate how many moles of X have been added:

$$0.34°C. = 0.34°\cancel{C.} \times \left(\frac{1\text{ mole }X}{0.52°\cancel{C.}\text{-}\cancel{kg.}\ \cancel{H_2O}}\right) \times \left(\frac{1\ \cancel{kg.}}{1000\ \cancel{g.}}\right) \times 800\ \cancel{g.}\ \cancel{H_2O}$$

$$= \frac{0.34 \times 800}{0.52 \times 1000}\text{ mole }X = 0.523\text{ mole }X$$

Then:

$$\text{Molecular weight }X = \frac{42.5\text{ g. }X}{0.523\text{ mole }X} = 81\text{ g./mole}$$

Do problems 4, 5, and 6 at the end of the chapter.

5. Molecular Weight of Solute—Freezing Point Depression

It is found that the addition of a solute to a solvent will lower its freezing point by an amount which depends only on the number of moles of solute which have been added and not on other properties of the solute. Thus we can use the observed lowering of the freezing point of a solvent to obtain the molecular weight of the solute dissolved in it. The tabulated value of the known depression of the freezing point in degrees C. of a solvent containing 1 mole of solute per kilogram of solvent is known as the *molal freezing point* constant ($K_{F.P.}$) and has the unit of °C.-kg. solvent/mole solute.

Example: The addition of 30.0 g. of a solute, X, to 650 g. of water lowers its freezing point by 0.82°C. If the $K_{F.P.}$ (H_2O) is 1.86°C.-kg. H_2O/mole solute, what is the molecular weight of X?

Answer: As before, we must determine the number of moles of X in 30.0 g. X. The ratio of the two will be the molecular weight of X. The following shows how all steps may be combined in one:

$$\text{Molecular weight }X = \frac{\text{Grams }X}{\text{Moles }X}$$

$$= \frac{30.0\text{ g. }X}{0.82°\cancel{C.} \times \left(\dfrac{1\text{ mole }X}{1.86°\cancel{C.}\text{-}\cancel{kg.}\ \cancel{H_2O}}\right) \times \left(\dfrac{1\ \cancel{kg.}}{1000\ \cancel{g.}}\right) \times 650\ \cancel{g.}\ \cancel{H_2O}}$$

$$= \frac{30.0\text{ g. }X}{\dfrac{0.82 \times 650}{1.86 \times 1000}\text{ moles }X} = \frac{30.0 \times 1.86 \times 1000}{0.82 \times 650}\text{ g./mole}$$

$$= 105\text{ g./mole}$$

Do problems 7 and 8 at the end of the chapter.

6. Molecular Weight of Solute—Osmotic Pressure

If a solution containing a solute is enclosed in a membrane (i.e., a cellophane bag) which is immersed in a pure solvent (see Figure 2), the natural tendency of the system will be for the solute to distribute itself equally between the solution inside the bag and the solvent outside.

If, however, the pores of the membrane are such that solute molecules cannot pass through the walls of the membrane, then the solution will exert a pressure against these walls which can be measured by a manometer. It is called *osmotic pressure*. Figure 2 illustrates a simple apparatus for measuring osmotic pressure.

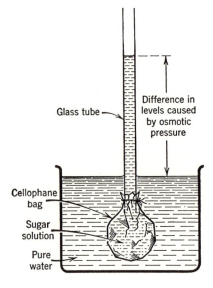

The osmotic pressure is independent of the solvent and, in dilute solutions, depends only on the temperature (T), the volume (V) enclosed by the membrane, and the number of moles of solute therein. It is found experimentally that the osmotic pressure (abbreviated by the Greek letter, π) is related to these other properties, just as though the solute were acting as a gas. The pressure, π, is given by the ideal gas law: $\pi \times V = nRT$. (See Chapter VI.) R is the ideal gas constant, 82.1 cc.-atm./mole-°K.; T is the absolute temperature; and n is the number of moles of solute.

Fig. 2.   Apparatus for demonstrating osmotic pressure.

Since osmotic pressure can be quite high (many atmospheres) and pressure is easily measured, the measurement of osmotic pressure provides a convenient method for obtaining the molecular weights of very large molecules, such as proteins.

Since $n/V =$ Moles/Liter = Concentration (c), we can express the osmotic pressure in units of concentration: $\pi = cRT$. In this equation c is the molarity of the solution.

Example: A solution containing 30 g. of protein per liter exerts an osmotic pressure of 9.4 torr at 25°C. What is the molecular weight of the protein?

Answer: Solving the osmotic pressure equation for the concentration:

$$c = \frac{\pi}{RT}$$

Substituting, we find:

$$c = \frac{9.4 \text{ torr} \times \left(\dfrac{1 \text{ atm.}}{760 \text{ torr}} \right)}{0.0821 \dfrac{\text{l. atm.}}{\text{mole-}^\circ\text{K.}} \times 298^\circ\text{K.}}$$

$$= \frac{9.4}{760 \times 0.0821 \times 298} \frac{\text{mole}}{\text{l.}} = 5.1 \times 10^{-4} \frac{\text{mole}}{\text{l.}}$$

Thus 30 g. protein/l. is equivalent to 5.1×10^{-4} mole/l.

$$\text{Molecular weight} = \frac{\text{Grams}}{\text{Moles}} = \frac{30 \text{ g.}}{5.1 \times 10^{-4} \text{ mole}} = 5.9 \times 10^4 \text{ g./mole}$$

$$= 59{,}000 \text{ g./mole}$$

7. Colligative Properties of Ionic Solutions

Many substances are not made of molecules but rather of ions, notably the class of compounds referred to as salts. When a formula weight of these compounds is placed in water, the change in freezing point, boiling point, and vapor pressure is usually close to some simple multiple of the expected change.

Thus, 1 mole of $NaCl$/kg. H_2O will show an increase in boiling point of $2 \times 0.52^\circ$C., and a lowering of freezing point of $2 \times 1.86^\circ$C.* Similarly, 1 mole of $CaCl_2$/kg. H_2O will have three times as great changes in properties (see section 8, Chap. IX).

The explanation for this behavior is quite simple. Sodium chloride ($NaCl$) is not made up of $NaCl$ molecules but rather of Na ions and Cl ions. When we dissolve 1 mole of $NaCl$ in water we are in reality dissolving a mixture of 1 mole of Na ions and 1 mole of Cl ions. Thus our solution contains not one solute ($NaCl$) but two solutes (Na ions and Cl ions). Consequently, we have two moles of solute and twice the expected change in properties. Similarly, 1 mole of $CaCl_2$ is actually 3 moles of solute—1 mole of Ca ion and 2 moles of Cl ion.

When dealing with salts which are in reality mixtures of ions, we must treat each ion as an individual and separate quantity.

* These values are approximate. The true values are somewhat lower, owing to the electrical forces of interaction between the ions (see section on activities).

8. Problems

1. Calculate the mole fraction and the molality of the solutes in each of the following solutions:

    (a) 50 g. of ethyl alcohol (C_2H_6O) in 400 g. of water (H_2O).

    (b) 70 g. of benzene (C_6H_6) in 180 g. of acetone (C_3H_6O).

    (c) 28 g. of NaCl in 280 g. of H_2O.

    (d) 13 g. of H_2SO_4 in 90 g. of H_2O.

2. The vapor pressure of ethyl alcohol at 20°C. is 43.9 mm. Hg. What is the vapor pressure of a solution at 20°C. containing 50 g. of ethyl alcohol (C_2H_6O) and 14 g. of phenol (C_6H_6O)?

3. The vapor pressure of water at 60°C. is 149.4 torr. What is the vapor pressure of a solution containing 25 g. of glucose ($C_6H_{12}O_6$) in 150 g. of water at 60°C.?

4. What concentration of glucose ($C_6H_{12}O_6$) in water is needed to raise its boiling point by 1.3°C.? $K_{B.P.}(H_2O)$ = 0.52°C.-kg. H_2O/mole solute.

5. If 25.0 g. of an unknown compound, X, dissolved in 340 g. of benzene raises its boiling point by 1.38°C., what is the molecular weight of X? $K_{B.P.}(C_6H_6)$ = 2.53°C.-kg. C_6H_6/mole solute.

6. It is found that, when 32.0 g. of a compound, X, are dissolved in 450 g. of CCl_4, the boiling point is raised by 6.21°C. Calculate the molecular weight of X. $K_{B.P.}(CCl_4)$ = 5.03°C.-kg. CCl_4/mole solute.

7. How many grams of diethylene glycol ($C_4H_{10}O_3$) should be added to 400 g. of water to give a solution which will not freeze at −10°C.? $K_{F.P.}(H_2O)$ = −1.86°C.-kg. H_2O/mole solute.

8. When 8.7 g. of a substance, X, are dissolved in 60 g. of benzene, the freezing point is depressed by 6.3°C. If $K_{F.P.}(C_6H_6)$ = 4.9°C.-kg. C_6H_6/mole solute, calculate the molecular weight of X.

9. What is the osmotic pressure of a 1.0 M solution of sucrose ($C_{12}H_{22}O_{11}$) at 20°C.?

10. A solution containing 4.5 g./l. of an unknown compound, X, exerts an osmotic pressure of 30 mm. Hg at 15°C. What is the molecular weight of X?

11. What is the freezing point lowering of a 0.3 molal solution of $CaCl_2$ in water?

12. What is the boiling point elevation of a 0.20 molal solution of $Al_2(SO_4)_3$?

13. When 40 g. of nitrobenzene ($C_6H_5NO_2$) are dissolved in 600 g. of naphthalene ($C_{10}H_8$), the freezing point of the solution is found to be 76.5°C. If pure naphthalene freezes at 80.2°C. calculate the molar freezing point constant of naphthalene.

14. The melting point of pure NaCl is 801°C., that of pure KCl is 776°C. NaCl and KCl when molten are completely miscible in all proportions. How can you account for the fact that a 50 mole per cent mixture of NaCl and KCl has a melting point close to 700°C.

15. $K_{F.P.}$ (Pb) = 8.5°C.-KgPb/Mole solute. How many grams of Sn must be dissolved in 400 g. of Pb to produce an alloy (solder) with a melting point of 220°C? The melting point of pure Pb is 327°C.

16. A eutectic mixture of two substances is that mixture of the two which has the lowest melting point. For Sn and Pb the eutectic mixture (soft solder) contains 67% by weight of Sn and melts at 183°C. If pure Sn melts at 232°C. calculate the $K_{F.P.}$ (Sn).

17. The melting point of pure Ag is 961°C. An alloy containing 25% of Cu by weight melts at 780°C. Calculate $K_{F.P.}$ (Ag).

PROBLEMS

18. In the commercial electrolysis of Mg metal, the electrolysis is conducted a molten solution of $MgCl_2$ in which appreciable quantities of NaCl and KCl have been added. Explain why this might be done. Why would KCl be used and not $CaCl_2$ or $ZnCl_2$.

19. $K_{F.P.}$ (C_6H_6) = 4.9°C.-Kg C_6H_6/mole solute. (a) How many grams of polystyrene of molecular weight 9000 would have to be dissolved in 100 g. of benzene (C_6H_6) to lower its freezing point by 1.05°C.?

   (b) If only 5 g. of the polymer were dissolved, by what amount would the freezing point of the benzene be lowered?

   (c) Do you think this lowering could be measured easily?

20. Nucleic acids, the materials in cells which contain genetic information, have molecular weights of about 2 to 6 million. Do you think that it is practical to measure these molecular weights by freezing point depressions? Explain.

CHAPTER X

Chemical Equilibrium

1. Reversible Reactions

A great many chemical reactions are reversible. That is, the products of the reaction may themselves react chemically to reform the original reactants. As a result of this ambidextrousness, such reactions do not generally go to completion. At 400°C., hydrogen (gas) and iodine (gas) will react to produce hydrogen iodide (gas). However, hydrogen iodide molecules simultaneously react with each other to form the initial reactants, hydrogen and iodine. Such reactions are designated by the use of a double arrow:

$$H_2\ (g) + I_2\ (g) \rightleftarrows 2HI\ (g)$$

If 1 mole of hydrogen and 1 mole of iodine are put into a 100-l. flask at a temperature of 400°C., the reaction will proceed until about 1.60 moles of HI have been produced and 0.20 mole each of H_2 and I_2 remain. Thereafter no further changes occur. If 2 moles of HI are placed in the same flask, it will decompose until 0.20 mole each of H_2 and I_2 are produced and 1.60 moles of HI remain. The reaction in each instance is said to have reached equilibrium, and the concentrations of the different species at the equilibrium point are known as the equilibrium concentrations. In this particular experiment they are 1.6×10^{-2} mole/l. of HI and 2.0×10^{-3} mole/l. of H_2 and I_2.

It is clear that at the equilibrium point no further changes occur because the rates of the forward and backward reactions are equal.

2. Le Châtelier's Principle

One of the most important laws governing the behavior of such reversible systems is known as Le Châtelier's principle. It enables us to predict qualitatively the way in which the equilibrium concentrations will change when external changes are made in the equilibrium system.

Le Châtelier's Principle: When a system is in equilibrium, a change in the properties of the system will cause the equilibrium concentration to shift in that direction which will tend to absorb the effect of the change.

The changes which we shall consider are temperature, total pressure, and addition or removal of one of the reacting species.

3. Effect of Temperature Changes on Equilibrium

Let us write the previous equilibrium to include the heat of the reaction:

$$H_2 + I_2 \rightleftarrows 2HI + 4 \text{ kcal.}$$

Consider the flask containing equilibrium concentrations of H_2, I_2, and HI. What will happen to the equilibrium concentrations if the temperature is raised? Note that the volume is constant.

Le Châtelier's principle tells us that the equilibrium point will shift in a direction to absorb the effect of the increase in temperature. To increase temperature, we must add heat energy. If some of the HI molecules decompose, they will absorb energy from their surroundings since this reverse reaction is endothermic.

Thus we can answer the question by saying that if the temperature is increased (at constant volume) the concentration of HI will decrease and the concentrations of H_2 and I_2 will increase.

Conversely, if the temperature is lowered the reverse will happen. However, we cannot from Le Châtelier's principle predict the extent of the change.

4. Effect of Changing Concentrations

Suppose some I_2 is added to our equilibrium mixture. How will this affect the concentrations?

Again, reasoning from Le Châtelier's principle, the system can absorb this stress if some hydrogen reacts with the iodine to produce more HI.

Or, in general, if we increase the concentration of any of the materials, the equilibrium point will be shifted in that direction that will use up this material. Again we cannot say how great the change will be from Le Châtelier's principle.

5. Effect of Pressure Changes

To illustrate the effect of pressure changes let us consider the Haber process for the production of ammonia (NH_3).

$$N_2 (g) + 3H_2 (g) \rightleftarrows 2NH_3 (g) + 22 \text{ kcal.}$$

Suppose we have an equilibrium mixture of N_2, H_2, and NH_3 in a cylinder fitted with a piston, and we compress the mixture by lowering the piston. How is the equilibrium affected? (*Note:* the temperature is to be kept constant.)

We are dealing here with a gas reaction. Gases are quite compressible. If all the N_2 and H_2 were to react, the system would occupy only 2 gas volumes. If all the NH_3 were decomposed, the system would occupy 4 gas volumes. Thus, as the N_2 and H_2 react to produce NH_3, there is a reduction in volume occupied by the gases. This gives us our answer. If the pressure is increased the system will accommodate itself by moving in that direction which will occupy a smaller volume. By occupying a smaller volume it will reduce its pressure. Thus an increase in pressure will increase the concentration of NH_3 and decrease that of N_2 and H_2.

Not all systems behave in this manner. In the previous illustration of $H_2 (g) + I_2 (g) \rightleftarrows 2HI (g)$, both products and reactants occupy the same gas volume and changing the pressure will have no affect on the equilibrium point.

In reactions involving solids or liquids only, since these are almost incompressible, changing pressure has practically no effect on the equilibrium point.

In systems involving solids or liquids and gases we need pay attention only to the gases involved.

Example: Consider the equilibrium:

$$PtO_2 (s) + 2Cl_2 (g) \rightleftarrows PtCl_4 (s) + O_2 (g) + 1 \text{ kcal.}$$

If we increase the pressure, how is the equilibrium affected?

Answer: Since the molar volumes of the solids are very small compared to the gases, consider only the gas volumes. Two moles of Cl_2 occupies twice the volume of 1 mole of O_2 at the same temperature and volume.

Thus, increasing the pressure will increase the equilibrium concentration of O_2 and decrease Cl_2.

Do problem 1 at the end of the chapter.

6. Law of Mass Action—Equilibrium Constants

It has been found that there is a quantitative relation between the equilibrium concentrations present in a system at equilibrium. This relation is known as the law of mass action.

Law of Mass Action: If the product of the concentrations of *all* of the products are divided by the product of the concentrations of *all* the reactants, the ratio formed is a constant under all changes except that of temperature.*

Consider the equilibrium,

$$H_2\ (g) + I_2\ (g) \rightleftarrows 2HI\ (g)$$

Then by the law of mass action the ratio:

$$\frac{C_{HI} \times C_{HI}}{C_{H_2} \times C_{I_2}} = \frac{C^2_{HI}}{C_{H_2} \times C_{I_2}} \text{ is a constant } = K_{eq.}(HI)$$

These equilibrium constants can be measured at each temperature and can be changed only by changing the temperature. They will have complex units depending on the units used to measure the concentrations. In the above case the units for $K_{eq.}(HI)$ are (moles HI)^2 per (moles H_2) $\times$ (moles I_2), if the units representing the concentrations are moles per liter. Generally this is so and values are listed for $K_{eq.}$ without stating units, it being understood that concentrations are in moles per liter.

Example: Write the equilibrium constant for:

$$N_2 + 3H_2 \rightleftarrows 2NH_3$$

Answer:

$$K_{eq.} = \frac{C^2_{NH_3}}{C_{N_2} \times C^3_{H_2}}$$

Units are $\dfrac{(\text{Moles NH}_3)^2 - \text{Liter}^2}{(\text{Moles N}_2) \times (\text{Moles H}_2)^3}$

The equilibrium constant is a defined quantity, and the student should observe the rules of the definition. If this is accepted, then the equations given become regular algebraic equations wherein we simply substitute numbers when we want an answer.

The utility of the law of mass action is that it gives us a method of calculating precisely how the equilibrium concentrations change when there is a change in the external conditions (except for temperature).

Do problem 2 at the end of the chapter.

* This must be clarified by the understanding that, if a certain substance appears in the balanced equation with a coefficient 3, then the ratio must contain its concentration multiplied threefold (i.e., the third power).

7. Calculation of Equilibrium Constants

We can calculate $K_{eq.}$ for a given equilibrium if we know the equilibrium concentrations. Thus if we return to section 1 we will find that at 400°C. the concentrations of H_2, I_2, and HI in equilibrium are 2×10^{-3} mole $H_2/l.$, 2×10^{-3} mole $I_2/l.$, and 1.6×10^{-2} mole HI/l. From these we can calculate the equilibrium constant:

$$K_{eq.} = \frac{C^2_{HI}}{C_{H_2} \times C_{I_2}} = \frac{(1.60)^2 \times 10^{-4}}{(0.20)(0.20) \times 10^{-4}} = \frac{2.56}{0.040} = 64 \quad \text{(at 400°C.)}$$

We have followed general practice in omitting mention of units although they may be inserted if desired.

Similarly, it is found that for the reaction

$$2SO_2\ (g) + O_2\ (g) \rightleftarrows 2SO_3\ (g) + 47 \text{ kcal}$$

a set of equilibrium concentrations at 527°C. are found to be

$$C_{SO_3} = 4 \text{ moles/l.}; \quad C_{O_2} = 0.5 \text{ mole/l.}; \quad C_{SO_2} = 0.2 \text{ mole/l.}$$

We can write

$$K_{eq.} = \frac{C'^2_{SO_3}}{C'^2_{SO_2} \times C_{O_2}} = \frac{4^2}{(0.2)^2 \times (0.5)} = \frac{16}{0.04 \times 0.5}$$

or

$$K_{eq.} = 800 \qquad \text{(at 527°C.)}$$

The concentration of gases can be expressed in terms of molar units as we have done, or equally well at a fixed temperature in terms of partial pressures. In either case, the resulting equilibrium constant will be a constant, but the numerical values and units will differ if the number of moles of gas changes during the reaction.

From the ideal gas law, pressure P and molar concentration ($C = n/v$) are related by the expression:

$$P = CRT \text{ or } C = RT/P$$

Hence, for the equilibria $N_2O_4 \rightleftarrows 2NO_2$

$$K_{eq.(C)} = \frac{C^2_{NO_2}}{C_{N_2O_4}} \qquad\qquad K_{eq.(P)} = \frac{P^2_{NO_2}}{P_{N_2O_4}}$$

But, $P_{NO_2} = C_{NO_2}(RT)$ and $P_{N_2O_4} = C_{N_2O_4}(RT)$

Hence, $K_{eq.(P)} = \dfrac{C^2_{NO_2}(RT)^2}{C_{N_2O_4}(RT)} = K_{eq.(C)} \times RT$

In general, $K_{eq.(P)} = K_{eq.(C)} \times (RT)^{\Delta n}$

where Δn is the change in numbers of moles of gases during the reaction.

Do problem 3 at the end of the chapter.

8. Calculation of Equilibrium Concentrations

If we know $K_{eq.}$ and have sufficient additional information, we can always calculate the equilibrium concentrations at a given set of conditions.

Example:

$$2SO_2 + O_2 \rightleftarrows 2SO_3$$

$K_{eq.} = 800$ at $527°C$.   If the concentrations of SO_2 and SO_3 are 2.0 moles $SO_2/l.$ and 10 moles $SO_3/l.$, what is the C_{O_2}?

Answer: Solving the K_{eq} for C_{O_2} we have

$$C_{O_2} = \frac{C^2_{SO_3}}{C^2_{SO_2} \times K_{eq.}}$$

and on substitution:

$$C_{O_2} = \frac{(10)^2}{(2)^2 \times 800} = \frac{100}{4 \times 800} = 0.031 \text{ mole } O_2/l.$$

Example:

$$H_2 + I_2 \rightleftarrows 2HI$$

If $K_{eq.} = 64$ at $400°C.$, calculate the equilibrium concentration when 2 moles of H_2 and 2 moles of I_2 are put into a 10-l. flask at $400°C$.

Answer: Let us make a table of starting concentrations and equilibrium concentrations.   Let X be the number of moles per liter of H_2 used up.

	Starting Concentration	Equilibrium Concentration
H_2	0.2 mole/l.	$0.2 - X$
I_2	0.2 mole/l.	$0.2 - X$
HI	0	$2X$

(*Note:* From X moles of H_2 we make $2X$ moles of HI and use up X moles of I_2.)
Then, by definition:

$$K_{eq} = \frac{C^2_{HI}}{C_{H_2} \times C_{I_2}}$$

Now substitute from the table:

$$64 = \frac{(2X)^2}{(0.2 - X)(0.2 - X)} = \frac{(2X)^2}{(0.2 - X)^2}$$

This equation may be solved by taking the square roots of both sides:

$$8 = \frac{2X}{0.2 - X}$$

Solving for X we find:

$$1.6 - 8X = 2X \quad \text{or} \quad 10X = 1.6 \quad \text{or} \quad X = 0.16 \text{ mole/l.}$$

The equilibrium concentrations are thus:

$$C_{H_2} = 0.04 \text{ mole/l.}; \quad C_{I_2} = 0.04 \text{ mole/l.}; \quad C_{HI} = 0.32 \text{ mole/l.}$$

Note that the equation would be more difficult to solve if the starting concentrations of H_2 and I_2 had been different. We shall tackle these more complicated equations when we discuss ionic equilibria.

9. Heterogeneous Equilibria—Concentrations of Solids and Liquids

We have thus far discussed equilibria in which all the reacting substances were distributed uniformly throughout the volume of the reacting systems. These are called *homogeneous equilibria*. How shall we treat systems in which this is not the case?

If we put 1 g. of $BaCO_3$ into a quartz bulb and heat it to 900°C. we shall find the following equilibrium:

$$BaCO_3 \ (s) \rightleftarrows BaO \ (s) + CO_2 \ (g) - 64 \text{ kcal}$$

But the $BaCO_3$ and BaO are solids and will occupy only a small part of the total volume of the quartz bulb. The CO_2 gas which is dispersed throughout will react with BaO only when the two are in contact. Such systems in which there is more than one phase present are called *heterogeneous systems*.

The answer to the problem of calculating concentrations in such systems is as follows. The concentration of a substance is equal to the mass (in grams or moles) divided by the *volume occupied just by that substance!*

Thus if we put 2 g. of ice into a 12-l. bulb along with some gases, the concentration of water (in the form of ice) is not 2 g. $H_2O/12$ l. but 2 g. $H_2O/2.2$ cc. $= 0.9$ g. $H_2O/cc.$ (the density of ice being 0.9 g./cc.). In fact, the reader will observe that if we put in 20 g. of ice its concentration is still 0.9 g. $H_2O/cc.$ (or 50 moles $H_2O/l.$ by conversion).

This consideration leads us to an important conclusion. The concentration of *pure* solids taking part in an equilibrium can be calculated from the density of the solid and do not depend on the volume of the vessel used! Furthermore, their concentrations are not affected by other substances we may add or remove or by changes in pressure.

We can thus combine these concentrations with the number representing the equilibrium constant, which is also a constant. The result will be a new equilibrium constant which expresses a simple relation between the concentrations of those materials which do change their concentrations with changes in external conditions.

Example:

$$BaCO_3 \ (s) \rightleftarrows BaO \ (s) + CO_2 \ (g)$$

Write the equilibrium constant.
 Answer:

$$K_{eq} = \frac{C_{BaO} \times C_{CO_2}}{C_{BaCO_3}}$$

But $BaCO_3$ and BaO are pure solids, and so their concentrations are constant and will not change. Rearranging the equation we find:

$$\frac{K_{eq.} \times C_{BaCO_3}}{C_{BaO}} = \boxed{C_{CO_2} = K'_{eq.}} \quad \text{(the new equilibrium constant)}$$

$$\text{since } \frac{K_{eq.} \times C_{BaCO_3}}{C_{BaO}} \text{ is itself a constant.}$$

We can conclude in this case that at equilibrium the concentration of CO_2 is constant and does not depend on the amounts of BaO or $BaCO_3$ in the system!

Example:

$$CuO \ (s) + CO \ (g) \rightleftarrows Cu \ (s) + CO_2 \ (g) + 30.5 \ \text{kcal.}$$

What is the equilibrium constant?
 Answer:

$$K_{eq.} = \frac{C_{Cu} \times C_{CO_2}}{C_{CuO} \times C_{CO}}$$

But again, since Cu and CuO are pure solids with constant concentrations, we simplify:

$$K'_{eq.} = \frac{K_{eq.} \times C_{CuO}}{C_{Cu}} = \frac{C_{CO_2}}{C_{CO}}$$

Whenever we have heterogeneous equilibria we shall omit the concentrations of pure solids and write the simpler expression involving only the concentrations that can be changed.

10. Different Ways of Writing Equilibria

There is something very arbitrary about our use of the words reactants and products in relation to reversible reactions. Thus in the case of NO_2 gas forming N_2O_4:

$$2NO_2 \ (g) \rightleftarrows N_2O_4 \ (g) + 14 \ \text{kcal.} \qquad (F)$$

we find that at 1 atm. total pressure and 60°C. the equilibrium system consists largely of N_2O_4, whereas at 0.1 atm. total pressure and 100°C., it is largely NO_2. In the latter case we would do well to write the reaction in reverse fashion:

$$14 \text{ kcal} + N_2O_4 \ (g) \rightleftarrows 2NO_2 \ (g) \qquad (R)$$

How does such a practice affect our use of equilibrium constants? For the first equation (F) we can write K_F as

$$K_F = \frac{C_{N_2O_4}}{C^2_{NO_2}}$$

and for the reverse equation (R), we can write an equilibrium constant (let us label it K_R):

$$K_R = \frac{C^2_{NO_2}}{C_{N_2O_4}}$$

We see that K_R is the reciprocal of K_F. This leads to a very important rule. *When we reverse the writing of a reaction we invert the equilibrium constant* (i.e., reactants become products and vice versa). However, we see that the basic mass law expression is not affected since if at equilibrium $C_{N_2O_4}/C^2_{NO_2}$ is a constant so is the reciprocal $C^2_{NO_2}/C_{N_2O_4}$. The constants are merely different in the two cases and reciprocals of each other.

Another problem which arises comes from our convention of writing balanced equations with small, whole numbers as coefficients. We can just as well double or triple or half all the coefficients and obtain just as valid an equation! For example,

$$4NO_2 \ (g) \rightleftarrows 2N_2O_4 \ (g) + 28 \text{ kcal.} \qquad (F_2)$$

$$6NO_2 \ (g) \rightleftarrows 3N_2O_4 \ (g) + 42 \text{ kcal.} \qquad (F_3)$$

$$NO_2 \ (g) \rightleftarrows \tfrac{1}{2}N_2O_4 \ (g) + 7 \text{ kcal.} \qquad (F_{1/2})$$

For each of these three ways of writing the balanced equation we can write an appropriate equilibrium constant.

$$K_{F_2} = \frac{C^2_{N_2O_4}}{C^4_{NO_2}} \qquad K_{F_3} = \frac{C^3_{N_2O_4}}{C^6_{NO_2}} \qquad K_{F_{1/2}} = \frac{C^{1/2}_{N_2O_4}}{C_{NO_2}}$$

We note that $K_{F_2} = K_F{}^2$, $K_{F_3} = K_F{}^3$, and $K_{F_{1/2}} = K_F{}^{1/2}$. Each of these equilibrium constants is very simply related to the original one, K_F, being a simple power or root of K_F.

This leads us to a second important rule, that when we multiply the coefficients in a balanced equation by some number n, the new equation has an equilibrium constant K_n which is just the nth power of our original equilibrium constant K (i.e., $K_n = K^n$). Note that if we represent reversing the equation by the process of multiplication by -1, then this new rule includes the previous rule as well (e.g., $K_{-1} = K^{-1}$).

Do problems 10 to 13 at the end of the chapter.

Once again we see that the essential equilibrium relation is unaffected since if a certain ratio of concentrations is constant, the same ratio raised to any power will also be constant.

11. Adding and Subtracting Equilibria

Very often in the laboratory we shall be concerned with more than one equilibrium occurring in the same system. In these instances we will want to add or subtract equations together to obtain a new equation which may involve substances of more interest to us. As an example let us consider the successive ionization of a weak acid such as H_2S. We can write:

$$H_2S \rightleftarrows H^+ + HS^- \tag{I}$$

$$HS^- \rightleftarrows H^+ + S^= \tag{II}$$

If we are interested only in the $S^=$ ion and not in the intermediate HS^- ion, we can add these two equilibria together in a way which causes HS^- to cancel, then:

$$H_2S \rightleftarrows 2H^+ + S^= \tag{III}$$

For these three equations we can write the three equilibrium constants.

$$K_I = \frac{C_{HS^-} \times C_{H^+}}{C_{H_2S}} \qquad K_{II} = \frac{C_{H^+} \times C_{S^=}}{C_{HS^-}} \qquad K_{III} = \frac{C_{H^+}^2 \times C_{S^=}}{C_{H_2S}}$$

We note that $K_{III} = K_I \times K_{II}$. This gives us another very important rule:

> When we add two equilibria to obtain a third equilibrium equation, the equilibrium constants are multiplied together to obtain the constant for the new equation.

Example: A saturated solution of $CaCO_3$ (s) contains Ca^{++} and $CO_3^=$ ions in equilibrium with the solid salt. When some H^+ ions are added to the saturated solution, there is an equilibrium between H^+, $CO_3^=$, and the weak acid HCO_3^-. Write equations for each of the equilibria in the system and the over-all equilibria. Write the equilibrium constants for each of these and the relation between them.

Answer:

$$CaCO_3 (s) \rightleftarrows Ca^{++} + CO_3^{=} \tag{I}$$

$$CO_3^{=} + H^+ \rightleftarrows HCO_3^- \tag{II}$$

Adding we have

$$CaCO_3 (s) + H^+ \rightleftarrows Ca^{++} + HCO_3^- \tag{III}$$

$$K_I = C_{Ca^{++}} \times C_{CO_3^-} \quad \text{(by convention we omit } C_{CaCO_3} (s) \text{)}$$

$$K_{II} = \frac{C_{HCO_3^-}}{C_{H^+} \times C_{CO_3^-}} \qquad K_{III} = \frac{C_{Ca^{++}} \times C_{HCO_3^-}}{C_{H^+}} = K_I \times K_{II}$$

Do Problems 14 to 19 at end of the chapter.

12. Problems

1. Predict from Le Châtelier's principle the effect of temperature increase and pressure increase, for each of the following equilibria:

(a) $4HCl (g) + O_2 (g) \rightleftarrows 2Cl_2 (g) + 2H_2O (g) +$ Heat

(b) $CO (g) + H_2O (g) \rightleftarrows CO_2 (g) + H_2 (g) +$ Heat

(c) $CO (g) + H_2 (g) \rightleftarrows C (s) + H_2O (g) +$ Heat

(d) $N_2O_4 (g) \rightleftarrows 2NO_2 (g) -$ Heat

(e) $Fe_3O_4 (s) + 4H_2 (g) \rightleftarrows 3Fe (s) + 4H_2O (g) +$ Heat

(f) $NH_4HS (s) \rightleftarrows NH_3 (g) + H_2S (g) -$ Heat

(g) $2O_3 \rightleftarrows 3O_2 +$ Heat

2. Write expressions for the equilibrium constants of all the equilibria in problem 1.

3. The equilibrium constant for reaction 1(b) above at 800°C. is 1.2. Calculate the concentrations of all substances when 2.0 moles each of CO and H_2O are put into a 5-l. flask at 800°C.

4. At 21°C., K_{eq}. for the dissociation of N_2O_4 [reaction 1(d) above] is 0.48 (mole NO_2)^2-l./moles N_2O_4. Calculate the concentration of NO_2 in equilibrium with 0.36 mole N_2O_4/l. at 21°C.

5. $PCl_3 (g) + Cl_2 (g) \rightleftarrows PCl_5 (g) +$ Heat. What is the equilibrium constant for this reaction, given that an equilibrium mixture in a 12-l. flask contains 0.42 mole of PCl_5, 1.76 moles of Cl_2, and 0.10 mole of PCl_3?

6. At 400°C., calculate the concentrations of H_2, I_2, and HI that will be in equilibrium [given K_{eq}.(HI) $= 64$ at 400°C.] if 1.4 moles of HI is put in a 20-l. flask and allowed to come to equilibrium.

7. At any temperature ice may be in equilibrium with water vapor:

$$H_2O (s) \rightleftarrows H_2O (g) -$$ Heat

Write the equilibrium constant. What can you conclude about the concentration of water vapor if more ice is added to the system?

8. What is the concentration of solid $CaCO_3$ (density $= 2.711$ g./cc.)? Will this be very different at 20 atm. pressure?

9. Given the following equilibrium between carbon tetrachloride liquid and its vapor:

$$CCl_4 (l) \rightleftarrows CCl_4 (g) -$$ Heat

PROBLEMS

(a) Write the equilibrium constant.

(b) What is the concentration of CCl_4 (l) (density = 1.59 g./cc.)?

(c) What happens to the equilibrium concentration of CCl_4 as the temperature is lowered? (d) if some of the liquid CCl_4 is removed?

10. Write K_{eq} for each of the following:

(a) $\frac{1}{2}H_2$ (g) + $\frac{1}{2}I_2$ (g) $\rightleftarrows$ HI (g) + 2 kcal.

(b) $\frac{1}{3}Al_2O_3$ (s) $\rightleftarrows$ $\frac{2}{3}Al$ (l) + $\frac{1}{2}O_2$ (g)

(c) $4H_2O_2$ (g) $\rightleftarrows 4H_2O$ (g) + $2O_2$ (g)

(d) H_2O_2 (g) $\rightleftarrows H_2O$ (g) + $\frac{1}{2}O_2$ (g)

(e) Cu_2S (s) + $2O_2$ (g) $\rightleftarrows 2CuO$ (s) + SO_2 (g)

(f) $\frac{1}{2}Cu_2S$ (s) + O_2 (g) $\rightleftarrows CuO$ (s) + $\frac{1}{2}SO_2$ (g)

11. (a) If the value of K_{eq} for reaction 10(d) has the value 6×10^6 at 800°C., what is the value of K_{eq} for reaction 10(c) at this same temperature? (b) What are the units of these two constants?

12. (a) At 600°C. the value of K_{eq} for reaction 10(f) is 70. Calculate K_{eq} for reaction 10(e) at 600°C. (b) What are the units of these two constants?

13. Write the equilibrium expressions for the reverse of all the reactions in problem 1.

14. Combine equation 1(b) and 1(c) in problem 1 so as to eliminate H_2 (g). How is the equilibrium constant for this new equation related to K_b and K_c?

15. Combine equations 1(b) and 1(c) in problem 1 so as to eliminate CO (g). How is the equilibrium constant for this new equation related to K_b and K_c?

16. Combine equations 1(c) and 1(e) in problem 1 so as to eliminate H_2O. How is the equilibrium constant for this new equation related to K_c and K_e?

17. (a) Write equilibrium equations for the three successive ionizations of H_3PO_4 in water solution. (b) How is the equilibrium constant for the following equilibrium related to these three constants?

$$H_2PO_4^- \rightleftarrows 2H^+ + PO_4^\equiv$$

18. HCN and HAc (acetic acid) are both weak acids for which we can write equilibrium equations expressing their incomplete ionization in water solution:

$$HCN \rightleftarrows H^+ + CN^-$$

$$HAc \rightleftarrows H^+ + Ac^-$$

(a) Write an equilibrium expression for the equilibrium which obtains when HAc is added to NaCN solution

$$HAc + CN^- \rightleftarrows HCN + Ac^-$$

(b) How is the equilibrium constant for this reaction related to K(HCN) and K(HAc?)

19. If K_I is the equilibrium constant for the reaction:

$$O_2 \text{ (g)} + 2NO \text{ (g)} \rightleftarrows 2NO_2 \text{ (g)}$$

and K_{II} is for

$$4NO \text{ (g)} + 2Cl_2 \text{ (g)} \rightleftarrows 4NOCl \text{ (g)}$$

calculate K_{III} for the reaction

$$NO_2 \text{ (g)} + \frac{1}{2}Cl_2 \text{ (g)} \rightleftarrows NOCl \text{ (g)} + \frac{1}{2}O_2 \text{ (g)}$$

in terms of K_I and K_{II}.

CHAPTER XI

Electrolysis of Ionic Solutions

1. Classification of Compounds by Electrical Properties

We can classify compounds into two categories, depending on whether or not they will conduct an electric current when placed in water solution.

1. Electrolytes: Those compounds whose water solutions will conduct an electric current (e.g., NaCl, Na_2SO_4, HCl, H_2SO_4).
2. Non-Electrolytes: Those compounds whose water solutions will not conduct an electric current (e.g., sugar, alcohol, acetone).

We can further divide class 1, the electrolytes, into two categories:

True Electrolytes: Electrolytes that will conduct an electric current in the pure, molten state (e.g., NaCl, $CuSO_4$).
Pseudo-Electrolytes: Electrolytes that will not conduct an electric current in the pure, molten state (e.g., all acids, $SnCl_2$, $AlCl_3$).

The present chapter is concerned with the chemical reactions that occur when an electric current is passed through a solution of an electrolyte, a phenomenon referred to as electrolysis.

2. Reactions Occurring during Electrolysis

When the terminals of a battery or any other source of direct current are connected to metal wires and the wires inserted into a water solution of an electrolyte, current will flow through the solution. As a result of this passage of current, chemical reactions take place at the surfaces of the electrodes dipping into the solution.

Negatively charged electrons move from the battery to its negative terminal (cathode). This cathode becomes electrically charged (negative), and the electrons on it react with the materials surrounding it.

At the positively charged terminal (anode), the materials in the solution react with it, giving up their electrons to it. These electrons can now return to the battery, completing the electric circuit.

Electrolysis is thus a set of reactions at the anode (+) and cathode (−) which involve the transfer of electrons between the materials in solution and the electrodes. Such electron-transfer reactions are referred to as oxidations (loss of electrons) or reductions (gain of electrons), and the two always occur together.

During electrolysis three possible types of reactions may occur at the electrodes:

The electrode reacts: The material of which the electrode is made (e.g., zinc, copper, carbon) may react with the solution.

The solvent reacts: The water which is always present may gain or lose electrons, itself being broken up into products.

The solute reacts: One or more of the ions present in the solute material will gain or lose electrons at the electrode and be deposited or transformed.

It is possible to predict which of these three possible reactions takes place under given conditions if we know the relative tendencies of the different alternative reactions. These may be predicted from the electromotive series (see Chapter XV).

Under certain conditions two or more of the possible reactions may occur simultaneously.

3. Reactions at the Cathode

The cathode is negatively charged. Positive ions in the solution will migrate towards it, and negative ions will be repelled. If the positive ion of a fairly inactive metal (e.g., copper, silver, lead, zinc) is present, then this ion will combine with the electrons present on the cathode and be deposited as free metal.

Case I. Inactive Metals. If we electrolyze a solution of $CuCl_2$, the reaction at the cathode may be written:

$$Cu^{+2} + 2e^{-1} \rightarrow Cu^0 \quad \text{(reduction at cathode)}$$

These equations must be electrically balanced as well as chemically balanced. The exponents refer to the units of charge present in the particle.

The copper ion (Cu^{+2}) with 2 plus charges combines with 2 electrons (e^{-1}), each bearing 1 negative charge, and is deposited as neutral copper metal (Cu^0). This reaction representing a gain in electrons by Cu^{+2} ions is called reduction (since the positive charge of the Cu^{+2} ions is reduced).

Case II.   Active Metals.   If, on the other hand, we pass a current through a solution of NaCl, the Na^{+1} ions will not readily combine with an electron and, instead, water molecules are preferentially decomposed, liberating hydrogen gas.

$$2H_2O + 2e^{-1} \rightarrow 2OH^{-1} + \underline{H_2\uparrow} \qquad \text{(reduction)}$$

It is seldom that the cathode ever reacts with the solution.

4. Reactions at the Anode

The anode is positively charged and attracts negative ions and repels positive ions.   If the negative ion has little attraction for electrons, it will lose them to the anode and be deposited or transformed.   If the negative ion has a strong attraction for electrons, then water will be decomposed, losing electrons and producing oxygen gas.   If the anode itself is composed of an active metal, then it will go into solution in the form of positive metal ions, leaving the electrons behind.

Case I.   Deposition of Anions.   If an electric current is passed through a solution of $CuCl_2$, the Cl^{-1} ions will react:

$$2Cl^{-1} \rightarrow \underline{Cl_2{}^0\uparrow} + 2e^{-1} \qquad \text{(oxidation at anode)}$$

This process of loss of electrons (by Cl^{-1} ions) is called oxidation. Oxidation always takes place at the anode.

Case II.   Reaction of the Anode Metal.   If the anode is made of a metal which is more easily oxidized than the anion, then it will react.   Thus, if a copper anode had been used in the electrolysis of a solution such as $ZnSO_4$, the Cu is more easily oxidized than the $SO_4{}^{-2}$ ion and the reaction that occurs is

$$Cu^0 \rightarrow Cu^{+2} + 2e^{-1} \qquad \text{(oxidation at anode)}$$

Case III.   Decomposition of Water at Anode.   If the metal anode is inert (e.g., platinum) and the anion more difficult to oxidize than water, then the water is preferentially decomposed as follows:

$$2H_2O \rightarrow \underline{O_2{}^0\uparrow} + 4H^{+1} + 4e^{-1} \qquad \text{(oxidation at anode)}$$

We will defer a discussion of which of these types of reactions occur until Chapter XV.   However, the student should practice writing these ion-electron equations and should familiarize himself with their balancing.

5. Quantitative Relations—Faraday's Law

We can state Faraday's law as follows:

Faraday's Law:

1. The amount of chemical reaction that takes place at each electrode is directly proportional to the total amount of electricity that has passed through the solution.
2. The number of equivalents of reaction that takes place at the cathode is exactly equal to the number of equivalents of reaction at the anode. (This is understandable from our knowledge of electricity. For every electron that enters the solution at the cathode, an equal number must leave at the anode to complete the circuit.)

To employ these laws for quantitative work we must first define some practical units.

Definition: One equivalent of electricity is that amount of charge (electrons) needed to release 1 equivalent of cation at the cathode (and simultaneously 1 equivalent of anion at the anode). It is called the *faraday*.

By experiment we find that 1 equivalent of electricity (i.e., 1 mole of electrons) is equal to 96,500 coulombs:

$$1 \text{ eq. electricity} = 1 \text{ faraday} = 1 \text{ mole electron} \quad \text{(definition)}$$

$$1 \text{ faraday} = 96,500 \text{ coulombs} \quad \text{(experiment)}$$

Since 1 ampere of current is defined as 1 coulomb/sec., we can calculate the total number of coulombs of electricity passing through a solution by multiplying the current by the time:

$$\text{Coulombs} = \text{Amperes} \times \text{Seconds}$$

Example: One equivalent of electricity is passed through a solution of sulfuric acid. What are the products of the reaction?
Answer: At the cathode:

$$1 \text{ eq. hydrogen} = 1 \text{ g. } H_2 = 11.2 \text{ l. STP } H_2$$

At the anode:

$$1 \text{ eq. oxygen} = 8 \text{ g. } O_2 = 5.6 \text{ l. STP } O_2$$

Example: 20 amp. are passed through a concentrated solution of sodium chloride for 50 sec. What are the products of the reaction?

Answer:

20 amp. $\times$ 50 sec. = 1000 coulombs = 1000 ~~coulombs~~ $\times \left(\dfrac{1 \text{ eq. elec.}}{96,500 \text{ ~~coulombs~~}} \right)$

$$= 0.0104 \text{ eq.}$$

At the cathode:

$$0.0104 \text{ ~~eq.~~ } \text{~~H}_2\text{~~} \times \left(\frac{11.2 \text{ l. STP H}_2}{1 \text{ ~~eq. H}_2\text{~~}} \right) = 0.116 \text{ l. STP H}_2$$

At the anode:

$$0.0104 \text{ eq. Cl}_2 = 0.369 \text{ g. Cl}_2 = 0.0104 \text{ ~~eq. Cl}_2\text{~~} \times \left(\frac{11.2 \text{ l. STP Cl}_2}{1 \text{ ~~eq. Cl}_2\text{~~}} \right)$$

$$= 0.116 \text{ l. STP Cl}_2$$

Note: It will turn out that the valence of an ion is equal numerically to the number of charges on it.   This is helpful in calculating the equivalent weights of ions.

The concept of equivalent is frequently a confusing one, and so it is desirable to be able to do calculations on electron reactions without using them.   In such cases, one must write balanced equations for the individual electrode reactions and use the stoichiometric coefficients to obtain chemical conversion factors.

Example: How many coulombs of electricity are required to produce 100 cc. STP of O_2 gas at the anode during the electrolysis of water?

Answer: First, write the balanced anode equation:

$$2 H_2O \rightarrow 4H^+ + O_2 \uparrow + 4e^-$$

We see that the chemical conversion factor is:

$$4 \text{ moles electrons} \equiv 1 \text{ mole } O_2 \text{ gas}$$

Hence:

$$100 \text{ cc. STP } O_2 = 100 \text{ cc. STP } O_2 \left(\frac{1 \text{ mole } O_2}{22,400 \text{ cc. STP } O_2} \right)$$

$$\left(\frac{4 \text{ moles } e^-}{1 \text{ mole } O_2} \right) \left(\frac{96,500 \text{ coulombs}}{1 \text{ mole } e^-} \right)$$

$$= \frac{100 \times 4 \times 96,500}{22,400} \text{ coulombs}$$

$$= 1720 \text{ coulombs}$$

---

6. Problems

1. Write balanced ion-electron equations for each of the following:
    (a) The deposition of Cr^{+3} ions at the cathode.
    (b) The reaction of Zn metal anode to give Zn^{+2} ions.

(c) The anode and cathode reactions occurring during the electrolysis of $AgNO_3$ with inert electrodes.

(d) The anode and cathode reactions occurring during the electrolysis of NH_4Cl with inert electrodes.

2. Make the following conversions:

(a) 480 coulombs to faradays.

(b) 5×10^{10} electrons to coulombs.

(c) 1.4 moles of electrons to coulombs.

(d) 0.680 faraday to coulombs.

(e) 0.20 equivalent of electricity to coulombs.

(f) 80 amp.-min. to coulombs.

(g) 8500 coulombs to ampere-hours.

3. A current of 12 amp. is passed through a solution of $CrCl_3$ for 40 min. How many grams of Cr and how many liters STP of Cl_2 gas are produced?

4. A current of 2 amp. is passed through HNO_3 solution for 40 sec. How many millimoles of H_2 and O_2 gas, respectively, are formed?

5. How long must a silver plater allow a current of 20 amp. to run through a $AgNO_3$ bath to deposit 40 g. of Ag? How many moles of O_2 gas will be formed?

6. (a) How many amperes of current will be required to deposit 40 g. of Cu per hour at the cathode of an electrolysis cell containing $CuSO_4$? (b) How many liters STP of O_2 will be liberated per hour at the anode?

7. (a) How many grams of Zn will be deposited at the cathode of an electrolysis cell containing $ZnSO_4$ by the passage of 43,000 coulombs of electricity? (b) If a current of 15 amp. is used, how long a time is needed?

8. At low voltages Fe^{+++} ions will be reduced to Fe^{++} in the electrolysis of $FeCl_3$ solutions. How long will it take a current of 18 amp. to reduce 10 g. of $FeCl_3$ to the lower valence state? How many liters STP of Cl_2 will be liberated at the anode?

9. At low voltages V^{+5} ions (actually $VO_2{}^+$) will be reduced to V^{++} on electrolysis of vanadyl salts. How many grams of vanadium will be reduced from the upper to the lower valence state by the passage of 80,000 coulombs?

10. In the electrolysis of molten Al_2O_3 solutions in cryolite, Al is liberated at the cathode and the graphite anode reacts with the oxide to form CO gas. Write balanced equations for the anode and cathode reactions. How many liters STP of CO will be formed during an electrolysis in which 160,000 coulombs are passed through a cell? How many moles of Al will be deposited?

11. The electroplating of Au (gold) is done from solution of KCN containing $Au(CN)_4{}^-$ complex ions. Write an equation for the balanced cathode reaction. If reactive Au metal is used for the anode, write the balanced equation for the anode reaction. How many hours will be required for a 60 amp. current to deposit 16 g. of Au?

12. When hot solutions of NaCl are electrolyzed at high voltages, the Cl^- is oxidized at the anode to $ClO_4{}^-$ (the oxygen comes from the H_2O). Write a balanced equation for the anode reaction. How many moles of $ClO_4{}^-$ will be formed at the anode in 3 hr. by an 18 amp. current?

CHAPTER XII

Simple Equilibria in Ionic Solutions

1. What Is Present in an Ionic Solution?

Before discussing equilibria in ionic solutions, let us first restate some of the properties of ionic substances.

An ionic compound (such as NaCl) is different from non-ionic substances in that it is in reality a mixture of two species, the Na^{+1} ion and the Cl^{-1} ion. When placed in a water solution, these two ions exist separately and have their own individual properties. They may also react independently. A 1 M solution of NaCl should thus be looked upon as a solution containing two solutes, 1 mole of Na^{+1} ions and 1 mole of Cl^{-1} ions. Similarly, a 1 M solution of $Al_2(SO_4)_3$ contains 2 moles of Al^{+3} ions and 3 moles of SO_4^{-2} ions.

When we mix a solution of NaCl with a solution of KNO_3, the resulting solution does not contain either NaCl or KNO_3 molecules. Instead it contains Na^{+1} ions, K^{+1} ions, Cl^{-1} ions, and NO_3^{-1} ions. The same result may be obtained by mixing $NaNO_3$ solution with KCl solution.

2. The Solubility of Ionic Compounds

If we add some sodium chloride to water it will dissolve. If we keep adding the solid salt, we eventually reach a point at which no more will dissolve. The solution is then said to be *saturated* with sodium chloride. Such a system is also in equilibrium; in fact, it represents a *heterogeneous equilibrium* between the solid sodium chloride and the Na^{+1} ions and the Cl^{-1} ions present in the solution. We can represent the equilibrium by an equation:

$$NaCl\ (s) \rightleftarrows Na^{+1}\ (aq) + Cl^{-1}\ (aq)$$

The symbol (aq) indicates that the ions exist in the aqueous solution, as distinct from the NaCl (s) which is present as pure solid. We shall generally omit the symbol (aq), it being understood from the equation.

For such a heterogeneous equilibrium we can write an equilibrium constant:

$$K_{eq.} = \frac{C_{Na^{+1}} \times C_{Cl^{-1}}}{C_{NaCl}}$$

However, since the NaCl is present as pure solid its concentration is constant, and this expression may be simplified by combining $K_{eq.}$ with C_{NaCl} (see Chapter X, section 9):

$$K_{eq.} \times C_{NaCl} = C_{Na^{+1}} \times C_{Cl^{-1}} = K_{S.P.}$$

The simplified relation thus obtained states that in an *equilibrium system* containing pure, *solid NaCl* and *dissolved Na^{+1} and Cl^{-1} ions*, the product of the concentrations of the Na^{+1} ions and Cl^{-1} ions is equal to a constant which is called the *solubility product constant* ($K_{S.P.}$).

3. The Solubility Product Constant

The significance of the solubility product relation is of some interest. In the case of NaCl, mentioned above, the relation predicts that, if we add to the saturated solution some substance containing Na^{+1} ions (e.g., $NaNO_3$), then in order to maintain the product of $C_{Na^{+1}}$ and $C_{Cl^{-1}}$ constant the Cl^{-1} ions must decrease. This occurs when some of the Cl^{-1} ions in the solution react with some of the Na^{+1} ions to form more pure, solid NaCl. From the numerical value of the solubility product constant we can predict exactly how much solid will form and what the final concentration will be.

Before discussing such calculations in detail, let us see how the constants should be written for more complex salts. The following list gives a representative set of such salts and the algebraic expressions for their solubility product constants:

Salt	Solubility Product*
Ag_2CrO_4	$K_{S.P.}(Ag_2CrO_4) = C^2_{Ag^{+1}} \times C_{CrO_4^{-2}}$
$CaSO_4$	$K_{S.P.}(CaSO_4) = C_{Ca^{+2}} \times C_{SO_4^{-2}}$
CaF_2	$K_{S.P.}(CaF_2) = C_{Ca^{+2}} \times C^2_{F^{-2}}$
Bi_2S_3	$K_{S.P.}(Bi_2S_3) = C^2_{Bi^{+2}} \times C^3_{S^{-2}}$

* The name solubility product is misleading. Although it is true that a small value of $K_{S.P.}$ for a particular salt means that the salt is not very soluble, and vice versa, nevertheless the $K_{S.P.}$ *is not proportional* to the solubility of the salt. This is true even though we shall see that it is possible to calculate the solubility of a salt if we know its $K_{S.P.}$ and, conversely, we may calculate $K_{S.P.}$ for a salt if we know its solubility.

By common agreement the units for the concentrations in these expressions are always moles per liter.

Do problem 1 at the end of the chapter.

4. Measurement of the Solubility Product Constant

From the definition of $K_{\text{S.P.}}$ we see that it may be calculated if we are given the values of the concentration of the ions of the salt that are present in an equilibrium solution.

The most straightforward method of obtaining such information is to take a solution containing the ions in equilibrium with the solid salt and analyze the *solution* to see what the ionic concentrations are.

The simplest of such procedures is to take the salt itself and shake it up with a large volume of water until a saturated solution is formed. This can then be filtered, 1 l. of the clear solution taken, the water evaporated, and the total amount of salt measured by weighing the solid residue.

Example: When 1 l. of a saturated solution of $CaCO_3$ is evaporated to dryness, the residue is found to weigh 7.0 mg. What is the $K_{\text{S.P.}}(CaCO_3)$?

Answer: The saturated solution contained 7.0 mg. $CaCO_3$ dissolved in it. This is

$$7.0 \times 10^{-3}\,\text{g. } CaCO_3 \times \left(\frac{1 \text{ mole } CaCO_3}{100\,\text{g. } CaCO_3}\right) = 7.0 \times 10^{-5} \text{ mole } CaCO_3$$

Since there were no other substances present, this means that the saturated solution contained 7.0×10^{-5} mole Ca^{+2} ions/l. and 7.0×10^{-5} mole CO_3^{-2} ions/l. By definition, the solubility product is

$$K_{\text{S.P.}}(CaCO_3) = C_{Ca^{+2}} \times C_{CO_3^{-2}}$$

On substitution we have

$$K_{\text{S.P.}}(CaCO_3) = (7.0 \times 10^{-5})(7.0 \times 10^{-5})$$
$$= 49 \times 10^{-10} = 4.9 \times 10^{-9}$$

Example: At 20°C. a saturated solution of $PbCl_2$ is found to contain 4.50 g. of $PbCl_2$ per liter dissolved in it. What is $K_{\text{S.P.}}(PbCl_2)$?

Answer:

$$4.50 \text{ g. } PbCl_2/\text{l.} = \left(\frac{4.50 \text{ g. } PbCl_2}{\text{l.}}\right) \times \left(\frac{1 \text{ mole } PbCl_2}{278 \text{ g. } PbCl_2}\right)$$

$$= 1.62 \times 10^{-2} \text{ mole } PbCl_2/\text{l.}$$

But 1.62×10^{-2} mole $PbCl_2$/l. in solution will produce 1.62×10^{-2} mole Pb^{+2} ions/l. and $2 \times 1.62 \times 10^{-2}$ mole Cl^{-1} ions/l. $= 3.24 \times 10^{-2}$ moles Cl^{-1}/l.

By definition:

$$K_{\text{S.P.}}(PbCl_2) = C_{Pb^{+2}} \times C^2_{Cl^{-1}}$$

and on substitution:

$$K_{\text{S.P.}} = (1.62 \times 10^{-2}) \times (3.24 \times 10^{-2})^2$$
$$= 1.62 \times 10^{-2} \times 10.5 \times 10^{-4}$$
$$= 1.70 \times 10^{-5}$$

If from other measurements we know the value of the $K_{\text{S.P.}}$ for a salt at a given temperature, then it is possible to calculate the solubility of the salt in water at that temperature.*

Example: The $K_{\text{S.P.}}$(AgCl) is known to be 1.6×10^{-10} at 20°C.   Calculate from this the solubility of AgCl in water at 20°C.

Answer: By definition:

$$K_{\text{S.P.}} = C_{\text{Ag}^{+1}} \times C_{\text{Cl}^{-1}}$$

If we have a saturated solution of AgCl, then let us call the solubility of AgCl X moles/l.   However, if X moles/l. of AgCl is dissolved, the solution will contain X moles/l. of Ag^{+1} ions and X moles/l. of Cl^{-1} ions.   We can now substitute these values and solve for X.

Then

$$K_{\text{S.P.}} = (X)(X) = 1.6 \times 10^{-10}$$

$$X^2 = 1.6 \times 10^{-10}$$

$$X = 1.26 \times 10^{-5} \text{ mole/l.} \rightarrow 1.3 \ M$$

Example: The $K_{\text{S.P.}}$ for PbBr$_2$ is 6.3×10^{-6} at 20°C.   What is the solubility of PbBr$_2$ at 20°C.?

Answer: By definition:

$$K_{\text{S.P.}} = C_{\text{Pb}^{+2}} \times (C_{\text{Br}^{-1}})^2$$

Let us set X = the solubility of PbBr$_2$ in moles per liter.   But X moles PbBr$_2$/l. will produce X moles Pb^{+2}/l. and $2X$ moles Br^{-1}/l.   On substitution and solving for X we find:

$$K_{\text{S.P.}}(\text{PbBr}_2) = (X)(2X)^2 = 6.3 \times 10^{-6}$$

or

$$4X^3 = 6.3 \times 10^{-6}$$

Then

$$X^3 = 1.58 \times 10^{-6}$$

$$X = 1.17 \times 10^{-2} \text{ mole/l.} \rightarrow 1.2 \ M$$

Do problems 2–5 at the end of the chapter.

* The algebraic expression for $K_{\text{S.P.}}$ is given by definition as a certain product of concentrations of ions raised to certain powers.   In problems dealing with this relation, the student must decide what the concentrations of the ions are and then substitute as required into the equation.   Once the concentrations are decided on, the rest is pure arithmetic.

5. Use of the Solubility Product Constant—Common Ion Effect

We have thus far discussed the solubility product relation for solutions containing only the ions of the pure salt. However, in most laboratory experiments a salt is made to precipitate in the presence of other ions as well, and the simple relations between the concentrations of the positive and negative ions which we obtain from the formula no longer hold.

Thus we can prepare a saturated solution of AgCl (containing Ag^{+1} ions and Cl^{-1} ions) by dissolving solid AgCl in water. In this solution, because all the Ag^{+1} and Cl^{-1} ions come from AgCl, their concentrations are equal. However, it is possible to make a saturated solution of AgCl by adding a small amount of $AgNO_3$ to a large amount of a solution of NaCl. Ag^{+1} ions will unite with Cl^{-1} ions (if enough of each are present), and a white precipitate of AgCl will form. The solution will contain Na^{+1} ions, NO_3^{-1} ions, a large excess of Cl^{-1} ions, and a small amount of Ag^{+1} ions. Because of the method of preparation, the concentrations of Ag^{+1} and Cl^{-1} are no longer equal. Nevertheless, despite all these complications, the system is an equilibrium system and the solubility product relation holds true: $K_{S.P.} = C_{Ag^{+1}} \times C_{Cl^{-1}}$. If we know $K_{S.P.}$ and either the $C_{Ag^{+1}}$ or $C_{Cl^{-1}}$, this equation permits us to solve for the unknown concentration.

Example: A few drops of a solution of $AgNO_3$ are added to a large amount of a 0.01 M solution of NaCl and a ppt. of AgCl is formed. On analysis it is found that the solution contains 0.010 moles/l. of Cl^{-1} ions. If $K_{S.P.}$ (AgCl) = 1.6 $\times$ 10^{-10}, calculate the $C_{Ag^{+1}}$ left in the solution.

Answer: By definition:

$$K_{S.P.} \text{ (AgCl)} = C_{Ag^{+1}} \times C_{Cl^{-1}}$$

But we are given:

$$K_{S.P.} = 1.6 \times 10^{-10}; \quad C_{Cl^{-1}} = 1.0 \times 10^{-2}$$

Then on solving for $C_{Ag^{+1}}$ ions and substituting:

$$C_{Ag^{+1}} = \frac{K_{S.P.}(AgCl)}{C_{Cl^{-1}}} = \frac{1.6 \times 10^{-10}}{1.0 \times 10^{-2}} = 1.6 \times 10^{-8} \text{ mole/l.}$$

Example: What concentration of Ag^{+1} ions will be in equilibrium with a saturated solution containing a ppt. of Ag_2CrO_4 and a CrO_4^{-2} ion concentration of 0.40 mole/l.?

$$K_{S.P.} \text{ (Ag}_2\text{CrO}_4) = 1.1 \times 10^{-11}$$

Answer: By definition:

$$K_{S.P.} \text{ (Ag}_2\text{CrO}_4) = C^2_{Ag^{+1}} \times C_{CrO_4^{-2}}$$

We are given

$$C_{CrO_4^{-2}} = 0.40; \quad K_{S.P.} = 1.1 \times 10^{-11}$$

Solving for $C_{Ag^{+1}}$ and substituting:

$$C^2_{Ag^{+1}} = \frac{K_{S.P.}}{C_{CrO_4^{-2}}} = \frac{1.1 \times 10^{-11}}{0.40} = 2.75 \times 10^{-11}$$

$$C^2_{Ag^{+2}} = 27.5 \times 10^{-12}$$

and taking square roots:

$$C_{Ag^{+1}} = 5.2 \times 10^{-6} \text{ mole/l.}$$

The solubility product relation may be used to answer the following types of questions:

1. When will precipitates form?
2. How much excess of one reagent is needed to reduce the concentration of a certain ion to a given value? (Or, in another form: How far towards completion can we drive ionic reactions?)

Example: A solution contains a concentration of Pb^{+2} ions of 2×10^{-3} mole/l. To this is added enough solid NaCl to bring the Cl^{-1} ion concentration to 3×10^{-2} mole/l. Will a ppt. of $PbCl_2$ form, given that $K_{S.P.}(PbCl_2) = 1.7 \times 10^{-5}$? (Ignore volume change.)

Answer: By definition:

$$K_{S.P.}(PbCl_2) = C_{Pb^{+2}} \times C^2_{Cl^{-1}}$$

This relation holds true only in a solution saturated with $PbCl_2$. If the solution is not saturated (i.e., more $PbCl_2$ can be dissolved in it) then the product of the $C_{Pb^{+2}} \times C^2_{Cl^{-1}}$ will be less than the $K_{S.P.}(PbCl_2)$. (No solution may contain so many Pb^{+2} ions and Cl^{-1} ions that $C_{Pb^{+2}} \times C^2_{Cl^{-1}}$ is greater than $K_{S.P.}(PbCl_2)$ unless the solution is *supersaturated*.) In the problem given:

$$C_{Pb^{+2}} \times C^2_{Cl^{-1}} = (2 \times 10^{-3})(3 \times 10^{-2})^2$$

$$= 2 \times 10^{-3} \times 9 \times 10^{-4}$$

$$= 18 \times 10^{-7} = 1.8 \times 10^{-6}$$

But 1.8×10^{-6} is clearly a smaller number than 1.7×10^{-5} which is $K_{S.P.}(PbCl_2)$, and so the solution is not saturated, and no ppt. of $PbCl_2$ will form.

Example: A solution is contaminated with Pb^{+2} ions. By adding Na_2SO_4, $PbSO_4$ may be made to ppt. What concentration of SO_4^{-2} ions is needed to reduce the $C_{Pb^{+2}}$ ions to 2×10^{-6} mole/l.? Given:

$$K_{S.P.}(PbSO_4) = 1.8 \times 10^{-8}$$

Answer:

$$K_{S.P.}(PbSO_4) = C_{Pb^{+2}} \times C_{SO_4^{-2}}$$

Solving for $C_{SO_4^{-2}}$ ions:

$$C_{SO_4^{-2}} = \frac{K_{S.P.}}{C_{Pb^{+2}}}$$

Substituting:

$$C_{SO_4^{-2}} = \frac{1.8 \times 10^{-8}}{2 \times 10^{-6}} = 0.9 \times 10^{-2} = 9 \times 10^{-3} \text{ mole/l.}$$

Do problems 6–14 at the end of the chapter.

6. Limitations on the Use of the Solubility Product Relation

A. Temperature. The solubility of salts usually increases with increasing temperature. Thus the $K_{S.P.}$ for a salt will also increase with increasing temperature. Therefore values of $K_{S.P.}$ must be known for each individual temperature for which it is proposed to use the relation. This is illustrated in the table for $PbCl_2$.

Temperature	20°C.	60°C.	90°C.
Solubility (moles $PbCl_2$/l.)	0.033	0.065	0.100
$K_{S.P.}(PbCl_2)$	1.5×10^{-4}	12×10^{-4}	40×10^{-4}

B. Salt Effect—Activities. As the concentration of ions in the solution increases, the ions are hampered in their motion; that is, they move less freely. This is due mostly to the fact that positive ions will attract to themselves negative ions in preference to other positive ions. This crowd of ions of negative charge will hamper the mobility of the positive ions, and the same is true for the effect of positive ions on the mobility of negative ions.

Thus in a 1.00 M solution of NaCl both the Na^{+1} and Cl^{-1} ions will be considerably less mobile than the Na^{+1} and Cl^{-1} ions in a solution of NaCl which is less concentrated, for example, 0.001 M NaCl. The result of this crowding on the properties of the Na^{+1} and Cl^{-1} ions is to make it appear as though their concentrations were lower. That is, they will react more slowly in the concentrated solutions. This seemingly smaller concentration can be observed experimentally and is called the "effective" or "active" concentration of ions.

When the effective or active concentrations of the ions are used to calculate the $K_{S.P.}$ for a solution, this corrected solubility product constant is called the *activity product constant*. The effect of this crowding can be as much as 10% or 20% in more concentrated solutions.

As an example, the solubility of AgCl is increased by some 10% by adding 1 mole/l. of KNO_3 to water. The K^{+1} and NO_3^{-1} ions impede the motions of the Ag^{+1} and Cl^{-1} ions and thus decrease their tendency to form solid AgCl.

For these reasons the $K_{S.P.}$ relation may be in error by as much as 20% or even higher if the solution contains large concentrations of salts. The relation is likewise very inaccurate when applied to salts like NaCl which are quite soluble, because a saturated solution of NaCl will contain almost 1 mole/l. of NaCl.

As a rough rule of thumb, the $K_{S.P.}$ relation is accurate only for salts whose solubilities are 1×10^{-2} mole/l. or less. Also, it is not very accurate even for these salts if the solution contains large concentrations of other dissolved salts.

It should be noted, however, that, if the concentration of other salts is known and the "activities" of the ions are known, then the corrected ion activity product may be used instead of the $K_{S.P.}$. This is, however, quite involved, and we shall not discuss it further.

7. Incomplete Ionization—Weak Electrolytes

True salts are solids whose crystal lattices are made up of positive and negative ions. These are completely ionized in the solid state. When dissolved in water they remain 100% ionized. There are no molecules of such salts (e.g., all salts of the alkali metals, etc.)!

There is a large class of compounds which are not true salts although they display salt-like properties. They are characterized by the property of not conducting the electric current in the molten state but conducting the electric current when dissolved in water solution. Such substances are called pseudo-electrolytes. They are molecular, not ionic, in structure. The ability of their water solutions to conduct the electric current is due to a reversible, chemical reaction of their molecules with molecules of water to produce ions. *All acids* and most salts of the amphoteric metals belong to this category of pseudo-electrolytes.

In a water solution of a pseudo-electrolyte there will exist positive ions, negative ions, and undissociated molecules of the pseudo-electrolyte. These will all be in equilibrium, and the equilibrium may be represented by balanced chemical equations:

$$HCl + H_2O \rightleftarrows H_3O^{+1} + Cl^{-1}$$

$$HC_2H_3O_2 + H_2O \rightleftarrows H_3O^{+1} + C_2H_3O_2^{-1}$$

$$FeCl_3 + 6H_2O \rightleftarrows Fe(H_2O)_6^{+3} + 3Cl^{-1}$$

We can divide these pseudo-electrolytes into two classes. Class I are known as strong electrolytes. Class II are known as weak electrolytes. Strong electrolytes are those which are almost 100% ionized in water solution. (HCl; HNO_3; H_2SO_4; HBr; HI; $HClO_4$ are the only common acids which are strong electrolytes.) Weak electrolytes are those which are only ionized to a small extent in water solution (acetic acid, $HgCl_2$, HClO, H_2S, H_2CO_3).

8. Ionization Constants

The reversible reaction of water with the molecules of weak electrolytes to produce ions is a *homogeneous equilibrium* and, as such, we can apply to it the law of mass action and write an equilibrium constant expression.

Thus, for the ionization of acetic acid (abbreviated, HAc; Ac represents the acetate group $C_2H_3O_2$):

$$HAc + H_2O \rightleftarrows H_3O^{+1} + Ac^{-1}$$

$$K_{eq.} = \frac{C_{H_3O^{+1}} \times C_{Ac^{-1}}}{C_{H_2O} \times C_{HAc}} \quad \text{(by definition of equilibrium constant)}$$

Now for the reasons discussed previously this relation is fairly accurate only if the concentrations of the ions are low and the solution is not too concentrated with respect to other ions. But in solutions which are more dilute than 1 M, the main part of the solution is made up of water, and the water concentration of these solutions is not very different from pure water (55.5 moles $H_2O/l.$). To the extent of accuracy of our expression then, we can combine the almost constant concentration of water C_{H_2O} with the constant $K_{eq.}$ and write a simplified expression:

$$K_{eq.} \times C_{H_2O} = \frac{C_{H_3O^{+1}} \times C_{Ac^{-1}}}{C_{HAc}} = K_{ion} \text{ (HAc)}$$

This new constant is known as the ionization constant and is abbreviated K_{ion}.

For weak electrolytes that may ionize in several steps there will be a different chemical equation for each step, and a K_{ion} can be written for each step   Or, if desired, the steps may be combined into a single step representing the overall ionization and a third K_{ion} written for this equilibrium. The relations used in any given problem will depend on which species we are interested in. This is illustrated for the weak electrolyte H_2S as follows:

Step I: $H_2S + H_2O \rightleftarrows H_3O^{+1} + HS^{-1}$

Step II: $HS^{-1} + H_2O \rightleftarrows H_3O^{+1} + S^{-2}$

If our interest is only in the amount of sulfide ion (S^{-2}) that is present in the solution, we can add both equations together and obtain the direct equilibrium between this ion and the H_2S.

Overall Equilibrium: $H_2S + 2H_2O \rightleftarrows 2H_3O^{+1} + S^{-2}$

The ionization constants for these different steps can each be written separately. (*Note:* In a given problem we would use the ionization constant which relates those ions and molecules in which we are interested. See Chapter X, sections 10, 11.)

Step I: $\dfrac{(C_{H_3O^{+1}}) \cdot (C_{HS^{-1}})}{(C_{H_2S}})} = K_{ion}(H_2S)$ (first ionization)

$= 1 \times 10^{-7}$ at 20°C.   (by experiment)

Step II: $\dfrac{(C_{H_3O^{+1}}) \cdot (C_{S^{-2}})}{(C_{HS^{-1}})} = K_{ion}(HS^{-1})$ (second ionization)

$= 1 \times 10^{-15}$ at 20°C.   (by experiment)

The ionization constant for the overall ionization into sulfide ion may be obtained from the third equation above (overall equilibrium):

$$\dfrac{(C_{H_3O^{+1}})^2 \cdot (C_{S^{-2}})}{(C_{H_2S})} = K_{ion}(H_2S) \qquad \text{(overall ionization)}$$

$= 1 \times 10^{-22}$ at 20°C.   (by experiment)

Note that mathematically the products of the ionization constants for steps I and II are equal to the overall ionization constant.

Problem: Write the three equilibrium equations for the step-wise ionization of phosphoric acid (H_3PO_4). Write also the overall equation for the second ionization, and the overall equation for the third ionization. Write the ionization constant equations for each of these five equilibria.

The student should work this out by himself, following the scheme outlined above. Look up the experimental values for these steps in any textbook or in the *Handbook of Chemistry and Physics*.

The smaller the ionization constant of a weak electrolyte, the fewer ions will it give in solution and thus the weaker will it be as an electrolyte. $K_{ion}(HCN) = 4.0 \times 10^{-10}$ at 20°C.; $K_{ion}(HC_2H_3O_2) = 1.8 \times 10^{-5}$ at 20°C. Comparing these values we see that K_{ion} is greater for acetic acid, and so we can conclude that acetic acid is a stronger acid than hydrocyanic acid.

Do problem 15 at the end of the chapter.

9. Relation between Per Cent Ionization and Ionization Constant

The ionization constant expression relates the concentrations of ions of a weak electrolyte and the concentrations of undissociated molecules. If all these concentrations are known, the value of K_{ion} may be calculated. Conversely, if K_{ion} is known, then the concentrations of the ions and the molecules may be calculated.

Example: A 0.40 M solution of HAc is found to be 0.67% ionized. Calculate the concentrations of all ions and molecules present in the solution and the $K_{ion}(\text{HAc})$.

Answer: Since 0.67% of the HAc is ionized, 99.33% is left in the form of HAc. The concentrations are:

$$C_{\text{HAc}} = 99.33\% \text{ of } 0.40 \text{ mole HAc/l.} = 0.9933 \times 0.4 \ M$$

$$= 0.397 \ M \quad \rightarrow 0.40 \ M$$

Every molecule of HAc that ionizes produces 1 H_3O^{+1} ion and 1 Ac^{-1} ion. Thus their concentrations are equal and given by

$$C_{\text{Ac}^{-1}} = C_{\text{H}_2\text{O}^{+1}} = 0.67\% \text{ of } 0.40 \text{ mole HAc/l.} = 0.0067 \times 0.4 \ M$$

$$= 0.00268 \text{ mole/l.} = 2.68 \times 10^{-3} \ M$$

By definition:

$$K_{ion}(\text{HAc}) = \frac{(C_{\text{H}_2\text{O}^{+1}}) \cdot (C_{\text{Ac}^{-1}})}{(C_{\text{HAc}})}$$

Then, on substitution:

$$K_{ion}(\text{HAc}) = \frac{(2.68 \times 10^{-3})(2.68 \times 10^{-3})}{0.397} = \frac{7.20 \times 10^{-6}}{0.397}$$

$$= 1.8 \times 10^{-5}$$

Example:

$$K_{ion}(\text{HF}) \text{ at } 20°\text{C.} = 7.2 \times 10^{-4}$$

Calculate the concentrations of H_3O^{+1}, F^{-1}, and HF in a 0.60 M solution of HF. Calculate the per cent ionization of HF.

Answer: Let X = the number of moles HF/l. that have dissociated. Then the concentration left will be:

$$C_{\text{HF}} = 0.60 - X$$

But since every HF that ionizes produces $1H_3O^{+1}$ and $1F^{-1}$:

$$C_{\text{F}^{-1}} = C_{\text{H}_2\text{O}^{+1}} = X$$

By definition:

$$K_{ion}(\text{HF}) = \frac{C_{\text{H}_2\text{O}^{+1}} \times C_{\text{F}^{-1}}}{C_{\text{HF}}}$$

On substituting the given information and concentrations:

$$7.2 \times 10^{-4} = \frac{(X)(X)}{0.60 - X} = \frac{X^2}{0.6 - X}$$

This relation, if solved for X, would give a quadratic equation which we could solve by means of a formula. However, we can simplify our task by making an approximation.

HF is a weak acid since K_{ion} is small (7.2×10^{-4}). This means that X is small. If this is so, then it will not be making much of an error to set $C_{HF} = 0.60$ instead of $(0.60 - X)$. That is, we ignore the X compared to 0.60. The equation now becomes:

$$7.2 \times 10^{-4} = \frac{X^2}{0.6}$$

or

$$X^2 = 0.6 \times 7.2 \times 10^{-4} = 4.32 \times 10^{-4}$$

and

$$X = 2.1 \times 10^{-2} \text{ mole/l.}$$

Thus our answer is:

$$C_{F^{-1}} = C_{H_3O^{+1}} = 0.021 \text{ mole/l.}$$

$$C_{HF} = 0.579 \text{ mole/l.} \rightarrow 0.58 \ M$$

and

$$\text{Per cent ionization} = \frac{\text{Amount of HF ionized}}{\text{Original amount of HF}} \times 100 = \frac{0.021}{0.60} \times 100 = 3.5\%$$

Note: The approximation is justified since X did turn out to be a small number. If an exact calculation is made, the answer still turns out to be $X = 0.021$. Such approximations will be used constantly.

Do problems 16 and 17 at the end of the chapter.

10. Use of the Ionization Constant—Common Ion Effect

Although we have thus far illustrated the ionization constant relation by using weak acids, the same methods apply to the dissociation of weak bases and weak salts. This is illustrated in the following table:

Substance	Equilibria	K_{ion}
Ammonia (NH_3)	$NH_3 + H_2O \rightleftarrows NH_4^{+1} + OH^{-1}$	$\dfrac{C_{NH_4^{+1}} \times C_{OH^{-1}}}{C_{NH_3}}$
Methylamine (CH_3NH_2)	$CH_3NH_2 + H_2O \rightleftarrows CH_3NH_3^{+1} + OH^{-1}$	$\dfrac{C_{CH_3NH_3^{+1}} \times C_{OH^{-1}}}{C_{CH_3NH_2}}$
Mercury (II) chloride $(HgCl_2)$	$HgCl_2 \rightleftarrows HgCl^{+1} + Cl^{-1}$	$\dfrac{C_{HgCl^{+1}} \times C_{Cl^{-1}}}{C_{HgCl_2}} = K_I$
	$HgCl^{+1} \rightleftarrows Hg^{+2} + Cl^{-1}$	$\dfrac{C_{Hg^{+2}} \times C_{Cl^{-1}}}{C_{HgCl^{+1}}} = K_{II}$

In the discussion thus far we have limited ourselves to examples in which both ions came only from the weak electrolyte. However, it is far more frequent in the laboratory to have other salts present in the solution which may have ions in common with the weak electrolyte. Then we no longer have the simple relationship between the concentrations of the ions and the un-ionized weak electrolyte.

Example: To a solution of 0.2 M HAc, some HCl (g) is added until the concentration of HCl is 0.3 M. What substances are now present in the solution, and what are their concentrations?

$$K_{ion}(HAc) = 1.8 \times 10^{-5}$$

Answer: HCl is a strong electrolyte. We may consider it completely dissociated into 0.3 M H_3O^{+1} ions and 0.3 M Cl^{-1} ions. On adding it to the weakly dissociated acetic acid (HAc) and applying Le Châtelier's principle we will see that the H_3O^{+1} ions from the HCl will cause the HAc equilibrium to be shifted towards lesser dissociation.

Let X = number of moles of HAc/l. that are still dissociated. Then there are present in the solution:

$$C_{Cl^{-1}} = 0.3 \; M$$

$$C_{H_3O^{+1}} = 0.3 + X$$

$$C_{Ac^{-1}} = X$$

$$C_{HAc} = 0.2 - X$$

By definition:

$$K_{ion}(HAc) = \frac{C_{H_3O^{+1}} \times C_{Ac^{-1}}}{C_{HAc}}$$

On substitution:

$$1.8 \times 10^{-5} = \frac{(0.3 + X)(X)}{0.2 - X}$$

This equation can be solved exactly for X. However, to simplify our work, let us make an approximation. X will be very small, since HAc is a weak acid. We can thus replace $0.3 + X$ by 0.3 and, similarly, $0.2 - X$ by 0.2. The equation now becomes:

$$1.8 \times 10^{-5} = \frac{(0.3)(X)}{(0.2)} = 1.5X$$

Then

$$X = \frac{1.8 \times 10^{-5}}{1.5} = 1.2 \times 10^{-5} \text{ mole/l.}$$

Note: This justifies our approximation, since $X = 1.2 \times 10^{-5}$, which is indeed small compared to 0.3 or 0.2.

Our answer is:

$$C_{Cl^{-1}} = 0.3 \; M$$

$$C_{H_3O^{+1}} = 0.3 \; M$$

$$C_{Ac^{-1}} = 1.2 \times 10^{-5} \; M$$

$$C_{HAc} = 0.2 \; M$$

Example: A solution contains 0.4 M HAc and 0.2 M NaAc. What ions and molecules are present in it, and what are their concentrations?

$$K_{ion}(HAc) = 1.8 \times 10^{-5}$$

Answer: From the completely ionized NaAc we get 0.2 M Na^{+1} ions and 0.2 M Ac^{-1} ions. Let X = moles HAc/l. that dissociate. Then we have

$$C_{Na^{+1}} = 0.2 \ M$$
$$C_{Ac^{-1}} = 0.2 + X$$
$$C_{H_3O^{+1}} = X$$
$$C_{HAc} = 0.4 - X$$

Again:

$$K_{ion}(HAc) = \frac{C_{Ac^{-1}} \times C_{H_3O^{+1}}}{C_{HAc}}$$

and on substitution:

$$1.8 \times 10^{-5} = \frac{(0.2 + X)(X)}{(0.4 - X)}$$

and, making the approximation that X is small compared to 0.2 or 0.4:

$$1.8 \times 10^{-5} = \frac{(0.2)(X)}{0.4} = \frac{X}{2}$$

Thus

$$X = 2 \times 1.8 \times 10^{-5} = 3.6 \times 10^{-5} \ mole/l.$$

and

$$C_{Ac^{-1}} = 0.2 \ M$$
$$C_{H_3O^{+1}} = 3.6 \times 10^{-5} \ M$$
$$C_{HAc} = 0.4 \ M$$

Do problems 18–20 at the end of the chapter.

11. Buffer Solutions

When an ion is present in a solution in small concentrations, a small amount of another substance that will react with it may almost completely remove it. Quite frequently it is important to have an ion present in small concentrations and yet have reserves of other compounds present such that, if a small amount of it is removed, these others will act to replenish it, that is, bring back the original concentration. An ion whose concentration may be low but stabilized in this way is said to be buffered, and the solution containing such an ion is called a *buffer solution*.

A buffer solution of any ion may be made by mixing a weak electrolyte containing that ion together with a large amount of a salt containing the other ion of the weak electrolyte.

Thus a mixture of HAc and NaAc constitute a buffer solution for the H^+ ion (actually H_3O^{+1}). Similarly a mixture of HAc and HCl are a buffer solution for the Ac^{-1} ion.

The behavior of these solutions is indicated by the following:

A solution of HAc and NaAc $\xleftarrow{\text{contains}}$
$\begin{cases} 1. \text{ A large amount of } Ac^{-1} \text{ ions} \\ 2. \text{ A large amount of HAc molecules} \\ 3. \text{ A small amount of } H_3O^{+1} \text{ ions} \end{cases}$

(*Note:* We omit the Na^{+1} ions which do not concern us. KAc would work just as well.)

If we add to this solution something that contains H_3O^{+1} ions (e.g., HCl), part of the reservoir of Ac^{-1} ions will react with it to form more HAc, thus tending to restore the original concentration of H_3O^{+1}.

If, on the other hand, we add to the solution something that removes H_3O^{+1} (such as NaOH which neutralizes it), part of the large reservoir of HAc molecules will ionize to restore the original concentration.

These restorative mechanisms are indicated by the equilibrium between H_3O^{+1} ions and the reservoirs:

$$\boxed{\text{HAc} + \text{H}_2\text{O}} \rightleftarrows \boxed{\text{Ac}^{-1}} + \boxed{\text{H}_3\text{O}^{+1}}$$

By varying the concentrations of weak acid and salt we can fix the concentration of H_3O^{+1} within broad limits. The concentration of the buffered ion, H_3O^{+1}, can be calculated by the methods outlined in the preceding paragraph on the common ion effect.

Example: What concentration of NaAc and HAc would you use to make a buffered solution in which $C_{H_3O^{+1}} = 1 \times 10^{-4}$?

Answer: Let $C_{NaAc} = X$ and $C_{HAc} = Y$; 1×10^{-4} = concentration of HAc that dissociates. Then we will have in the solution:

$$C_{Na^{+1}} = X$$

$$C_{Ac^{-1}} = X + 1 \times 10^{-4}$$

$$C_{H_3O^{+1}} = 1 \times 10^{-4}$$

$$C_{HAc} = Y - 1 \times 10^{-4}$$

$$K_{ion}(HAc) = 1.8 \times 10^{-5} = \frac{C_{H_3O^{+1}} \times C_{Ac^{-1}}}{C_{HAc}}$$

On substitution:

$$1.8 \times 10^{-5} = \frac{(1 \times 10^{-4})(X + 1 \times 10^{-4})}{(Y - 1 \times 10^{-4})}$$

But note again that 1×10^{-4} will be small compared to X and Y, so we can approximate:

$$1.8 \times 10^{-5} = \frac{(1 \times 10^{-4})(X)}{(Y)}$$

Solving for the ratio:

$$\frac{X}{Y} = \frac{1.8 \times 10^{-5}}{1 \times 10^{-4}} = 1.8 \times 10^{-1} = 0.18$$

Thus any concentrations of NaAc and HAc that are in the ratio $C_{\text{NaAc}}/C_{\text{HAc}} = 0.18$ will provide a buffer solution in which $C_{\text{H}_3\text{O}^{+1}} = 1 \times 10^{-4}$. The following combinations are taken as illustrations:

$$C_{\text{NaAc}} = 0.18 \ M; \quad C_{\text{HAc}} = 1 \ M$$

$$C_{\text{NaAc}} = 0.018 \ M; \quad C_{\text{HAc}} = 0.1 \ M$$

Will $C_{\text{NaAc}} = 1.8 \times 10^{-4}$; $C_{\text{HAc}} = 10 \times 10^{-4}$ work? Why not?

---

12. Brönsted Acids and Bases. Conjugate Pairs

Historically chemists have pictured acids as substances which contain displaceable hydrogen and bases as substances which contain OH^- ions. However, with the discovery that acids may vary enormously in the amount of their dissociation it became necessary to extend the definition of acids.

One of the most fruitful definitions was provided by Brönsted who defined an acid as a substance capable of dissociating a proton in solution. Thus (ignoring the question of hydration) we may write for the dissociation of acetic acid in water solution:

$$\text{HAc} \rightleftarrows \text{H}^+ + \text{Ac}^-$$

Now to the extent that the acid is weak or strong the equilibrium point will lie more or less to the right in this equation. There are two ways of discussing such an equilibrium, one from the point of view of the reactants (acid), the other from the point of view of the products (H^+ and Ac^-). Thus we may say that acetic acid is a weak acid because HAc has little tendency to dissociate, or conversely we may say that acetic acid is a weak acid because Ac^- ion has a large affinity for H^+. These two points of view are equivalent and not independent.

Brönsted thus chose to define a base as any substance capable of associating with a proton. The combination of the base and proton (e.g., $\text{Ac}^- + \text{H}^+$) is obviously an acid by virtue of the first definition, and the acid and the base which differ in structure by only a proton were dubbed a conjugate, acid-base pair.

Thus HAc is an acid and Ac^- is its conjugate base. Similarly the following form acid-base, conjugate pairs: H_3O^+/H_2O; HCl/Cl^-; HNO_2/NO_2^-; H_2SO_4/HSO_4^-; $HSO_4^-/SO_4^=$; H_2CO_3/HCO_3^-; H_2S/HS^-; NH_4^+/NH_3; $HPO_4^=/PO_4^{\equiv}$; NH_3/NH_2^-; $OH^-/O^=$.

We now note that if an acid is very strong, such as HCl, then its conjugate base must be very weak. Conversely if an acid is very weak, such as HCN, then its conjugate base CN^- must be very strong. There is thus a complete symmetry between acids and bases. As we shall see in the next chapter this provides a much more satisfactory point of view of such reactions as hydrolysis than is provided by the older definition of acids and bases.

We can define the association constant of a base $(K_{assoc.})$ as the equilibrium constant for the reaction:

$$Base + H^+ \rightleftarrows (H\ Base)^+$$

or for acetate ion:

$$Ac^- + H^+ \rightleftarrows HAc \qquad K_{assoc.}(Ac^-) = \frac{C_{HAc}}{C_{H^+} \times C_{Ac^-}}$$

Note that:

$$K_{assoc.}(Ac^-) = \frac{1}{K_{ion}(HAc)}$$

or:

$$K_{ion}(HAc) \times K_{assoc.}(Ac^-) = 1$$

13. Displacement Reactions of Weak Acids

It is convenient to establish 5 categories of acids and their conjugate bases. These are listed as follows in terms of their ionization constants.

TABLE XVI

RELATIVE STRENGTH OF SOME CONJUGATE ACID-BASE PAIRS

Acid Strength	Acid K_{ion}	Base Strength	Base K_{assoc}	Example
Very Strong	10^4 to 1	Very Weak	10^{-4} to 1	HCl/Cl^-
Strong	1 to 10^{-4}	Weak	1 to 10^4	$HSO_4^-/SO_4^=$
Moderate	10^{-4} to 10^{-8}	Moderate	10^4 to 10^8	HAc/Ac^-
Weak	10^{-8} to 10^{-12}	Strong	10^8 to 10^{12}	$HCO_3^-/CO_3^=$
Very Weak	10^{-12} to 10^{-16}	Very Strong	10^{12} to 10^{16}	$HS^-/S^=$

Here we have chosen every 4 powers of 10 in K_{ion} as the dividing line for a new category. Note the complete symmetry in the designation of the strengths of acids and their conjugate bases.

An immediate use of such categories is that it enables us to talk qualitatively and quantitatively about the reactions of weak acids and bases. For example, $CO_3^=$ by the Table XVI is a strong base, whereas HAc is a moderate acid. We should then expect the two to react:

$$HAc + CO_3^= \rightleftarrows Ac^- + HCO_3^- \qquad K_{eq.}$$

In this equation the equilibrium should be well to the right, i.e., in favor of the formation of the weaker acid. We can describe this behavior by saying that HCO_3^- ($K_{ion} = 4 \times 10^{-11}$) is a weaker acid than HAc ($K_{ion} = 1.8 \times 10^{-5}$) or equivalently by saying that $CO_3^=$ ($K_{assoc.} = 2.5 \times 10^{+12}$) is a stronger base than Ac^- ($K_{assoc.} = 5.5 \times 10^6$).

Such reactions in which the ion of a weak acid displaces the ion of a stronger acid are referred to as displacement reactions. Quantitatively we can observe that the equation is a combination of two equilibria:

$$HAc \rightleftarrows H^+ + Ac^- \qquad K_{ion}(HAc)$$

$$HCO_3^- \rightleftarrows H^+ + CO_3^= \qquad K_{ion}(HCO_3^-)$$

If we subtract the $HCO_3^-/CO_3^=$ equation from the HAc/Ac^- equation we obtain the displacement equation and so by our rule (Chapter X, section 11) the equilibrium constant for the displacement reaction is given by the ratio of ionization constants. $K_{eq.} = K_{ion}(HAc)/K_{ion}(HCO_3^-)$. From the data above we see that $K_{eq.} = 4.5 \times 10^5$. The equilibrium lies well over to the right.

Example: Given $K_{ion}(NH_4^+) = 5.5 \times 10^{-10}$; $K_{ion}(HNO_2) = 4.0 \times 10^{-4}$, calculate $K_{eq.}$ for the reaction:

$$NH_3 + HNO_2 \rightleftarrows NH_4^+ + NO_2^-$$

Answer: By our usual rule for combining equilibria we see that

$$K_{eq.} = \frac{K_{ion}(HNO_2)}{K_{ion}(NH_4^+)} = \frac{4.0 \times 10^{-4}}{5.5 \times 10^{-10}} = 7.2 \times 10^5$$

We note again that 7.2×10^5 is a very large number, so that the equilibrium lies very far over in favor of the products.

Example: Will H_2CO_3 react appreciably with CN^-?

$$K_{ion}(H_2CO_3) = 4.3 \times 10^{-7}; \qquad K_{ion}(HCN) = 4 \times 10^{-10}$$

Answer: HCN is a weaker acid than H_2CO_3 (see K_{ion}). Thus CN^- will displace HCO_3^- from H_2CO_3 (or, CN^- is a stronger base than HCO_3^-). The reaction is:

$$H_2CO_3 + CN^- \rightleftarrows HCO_3^- + HCN$$

and its equilibrium constant is:

$$K_{eq.} = \frac{K_{ion}(H_2CO_3)}{K_{ion}(HCN)} = \frac{4.3 \times 10^{-7}}{4 \times 10^{-10}} = 1.1 \times 10^3$$

Note: Appreciable amounts of CN^- and H_2CO_3 *will be present at equilibrium.*

TABLE XVII
DISSOCIATION CONSTANTS OF SOME COMMON ACIDS
(FIRST DISSOCIATION ONLY)

Acid	K_{ion}	Acid	K_{ion}
H_2SO_4	$>10^2$	HSO_3^-	1×10^{-7}
HNO_3	$>10^2$	H_2S	1×10^{-7}
HCl	$>10^2$	$H_2PO_4^-$	6.2×10^{-8}
HBr	$>10^2$	NH_4^+	5.5×10^{-10}
$HClO_4$	$>10^2$	HCN	4.0×10^{-10}
HSO_4^-	1.2×10^{-2}	HCO_3^-	4.7×10^{-11}
HF	7×10^{-4}	$HPO_4^=$	1×10^{-12}
HNO_2	5×10^{-4}	HOH	1×10^{-14}
HAc	1.8×10^{-5}	HS^-	1×10^{-15}
H_2CO_3	4.3×10^{-7}		

Note: Because of the symmetry between acids and bases we shall here deal only with the K_{ion}(acids), relying on the relations between acid and base to derive $K_{assoc.}$(base). Thus for the base NH_3 we list its properties in terms of the conjugate acid NH_4^+.

14. Problems

1. Write $K_{S.P.}$ for each of the following salts:

(a) AgI.

(b) CuS.

(c) $Ca_3(PO_4)_2$

(d) $Mg(OH)_2$.

(e) Sb_2S_3.

(f) $Fe_4[Fe(CN)_6]_3$.

2. $K_{S.P.}(BaCrO_4) = 2 \times 10^{-10}$ at 20°C. Calculate the molar solubility of $BaCrO_4$ in water at 20°C.

3. $K_{S.P.}(PbF_2) = 3.7 \times 10^{-8}$ at 20°C. Calculate the molar solubility of PbF_2 at 20°C.

4. $K_{S.P.}[Al(OH)_3] = 1.9 \times 10^{-33}$ at 20°C. Calculate the molar solubility of $Al(OH)_3$ at 20°C. *Note:* This neglects ionization of H_2O.

5. For each of the following salts, whose solubility is given at 20°C., calculate the value of $K_{S.P.}$:

(a) Solubility of $AgBr$ is 5.8×10^{-7} mole/l.

(b) Solubility of $BaCO_3$ is 7×10^{-5} mole/l.

(c) Solubility of $Pb(OH)_2$ is 4.2×10^{-6} mole/l.

(d) Solubility of Bi_2S_3 is 1.7×10^{-15} mole/l.

6. $K_{S.P.}(BaCrO_4) = 2 \times 10^{-10}$. (a) What is the solubility of $BaCrO_4$ in 0.040 M $BaCl_2$? (b) What is its solubility in 0.25 M K_2CrO_4?

7. $K_{S.P.}(PbF_2) = 3.7 \times 10^{-8}$. (a) What is the solubility of PbF_2 in 0.050 M $Pb(NO_3)_2$? (b) What is its solubility in 0.020 M NaF?

8. $K_{\text{S.P.}}[\text{Mg(OH)}_2] = 5.5 \times 10^{-12}$. (*a*) What is the solubility of Mg(OH)_2 in 0.0010 M NaOH? (*b*) What concentration of OH^{-1} ions in a solution will ensure that $C_{\text{Mg}^{+2}\text{ ion}}$ is less than $4 \times 10^{-8} M$? Neglect ionization of H_2O.

9. $K_{\text{S.P.}}(\text{CuCl}) = 1.8 \times 10^{-7}$. A solution contains a concentration of Cu^{+1} ions $= 2 \times 10^{-4} M$. Solid NaCl is added until $C_{\text{Cl}^{-1}}$ is $1 \times 10^{-2} M$. Was this high enough to cause a ppt. of CuCl to form? (Neglect volume change.)

10. $K_{\text{S.P.}}(\text{PbI}_2) = 8.7 \times 10^{-9}$. To a solution in which $C_{\text{Pb}^{+2}} = 3 \times 10^{-3} M$, solid NaI is added until $C_{\text{I}^{-1}}$ is $1 \times 10^{-4} M$. Will a ppt. of PbI_2 form?

11. If 50 cc. of solution in which $C_{\text{Ag}^{+1}} = 3 \times 10^{-4} M$ is added to 100 cc. of a solution in which $C_{\text{Cl}^{-1}} = 2 \times 10^{-6} M$, will a ppt. of AgCl form? $K_{\text{S.P.}}(\text{AgCl}) = 1.7 \times 10^{-10}$. (*Note:* volume change is important.)

12. A solution contains $1 \times 10^{-2} M$ Cl^{-1} ions and $1 \times 10^{-3} M$ I^{-1} ions. $K_{\text{S.P.}}(\text{AgCl}) = 1.7 \times 10^{-10}$. $K_{\text{S.P.}}(\text{AgI}) = 8.5 \times 10^{-17}$.

    (*a*) If a salt of Ag^{+1} ion is added dropwise to this, which ppt. forms first, AgCl or AgI?

    (*b*) What is the largest $C_{\text{Ag}^{+1}}$ which may exist in this solution without pptg. AgCl?

    (*c*) When this concentration is reached, what will the $C_{\text{I}^{-1}}$ be?

13. To a solution containing 0.04 M Cd^{+2} ions and 0.3 M Zn^{+2} ions some S^{-2} ions are added. $K_{\text{S.P.}}(\text{ZnS}) = 4.5 \times 10^{-24}$. $K_{\text{S.P.}}(\text{CdS}) = 1.4 \times 10^{-28}$.

    (*a*) What is the largest $C_{\text{S}^{-2}}$ ion that can exist in this solution without pptg. ZnS?

    (*b*) What will the $C_{\text{Cd}^{+2}}$ ion be at this point?

From your answer, what can you conclude regarding the possibility of using S^{-2} ions to make a quantitative separation of Zn^{+2} and Cd^{+2} ions?

14. A solution contains 0.02 M Ca^{+2} ions and 0.003 M Ba^{+2} ions. Could you make a separation of Ba^{+2} ions and Ca^{+2} ions by using SO_4^{-2} to ppt. one of them? Explain. $K_{\text{S.P.}}(\text{CaSO}_4) = 2.4 \times 10^{-5}$; $K_{\text{S.P.}}(\text{BaSO}_4) = 9.9 \times 10^{-11}$.

15. Write K_{ion} expressions for each of the following weak electrolytes:

    (*a*) HCN.

    (*b*) HClO.

    (*c*) H_2SO_3 (first ionization).

    (*d*) HCO_3^{-1} ions.

    (*e*) PbAc_2 (first ionization).

    (*f*) PbAc^{+1} ions.

    (*g*) SnCl_2 (overall ionization).

    (*h*) Fe(OH)^{+1} ions.

16. Given the percentage of ionization of each of the following solutions, calculate the concentration of ions and molecules present in the solution and the K_{ion} of the weak electrolyte.

    (*a*) 0.5 M HNO_2 is 3.0% ionized.

    (*b*) 0.04 M HF is 13.4% ionized.

    (*c*) 0.08 M NH_3 solution is 1.5% ionized.

    (*d*) 0.003 M H_2S is 0.6% ionized (consider first ionization only).

    (*e*) 0.060 M H_2CO_3 solution is 0.27% ionized (consider first ionization only).

17. From the known ionization constants of the following weak electrolytes, calculate the concentrations of all ions and molecules present in the solutions given and the percentage of ionization:

    (*a*) 0.20 M HClO; $K_{\text{ion}}(\text{HClO}) = 5.6 \times 10^{-8}$.

    (*b*) 0.020 M H_2CO_3; $K_{\text{ion}}(\text{H}_2\text{CO}_3) = 4.3 \times 10^{-7}$ (first ionization).

    (*c*) 0.004 M HAc; $K_{\text{ion}}(\text{HAc}) = 1.8 \times 10^{-5}$.

    (*d*) 0.45 M NH_3 solution; $K_{\text{ion}}(\text{NH}_3) = 1.8 \times 10^{-5}$.

    (*e*) 0.65 M $\text{H}_2\text{PO}_4^{-1}$ ions; $K_{\text{ion}}(\text{H}_2\text{PO}_4^{-1}) = 6.2 \times 10^{-8}$ (only one ionization).

18. What concentration of Ac^{-1} ions will reduce $C_{H_3O^{+1}}$ ion to 2×10^{-4} M in a 0.40 M solution of HAc?

19. What concentration of H_3O^{+1} ions will reduce $C_{S^{-2}}$ ion to 2×10^{-18} M in a 0.10 M solution of H_2S?

20. What concentration of H_3O^{+1} ions will reduce $C_{HS^{-1}}$ ion to 2×10^{-6} M in a 0.10 M solution of H_2S?

21. What mixtures would provide buffer solutions for each of the following ions?

(a) H_3O^{+1}.

(b) OH^{-1}.

(c) F^{-1}.

(d) NO_2^{-1}.

(e) Pb^{+2}.

(f) HS^{-1}.

(g) HCO_3^{-1}.

(h) CO_3^{-2}.

22. How would you make a buffer solution from NaAc and HAc in which the $C_{H_3O^{+1}}$ was 2×10^{-5} M?

23. How would you make a buffer solution from HCl and HAc in which the $C_{Ac^{-1}\,ion}$ was 4×10^{-4} M?

24. What are the units of $K_{S.P.}(AgCl)$? of $K_{S.P.}(PbCl_2)$?

25. What are the units of $K_{ion}(HAc)$? of $K_{ion}(HF)$?

26. Convert $K_{S.P.}(AgCl)$ in which the units of concentration are expressed in moles per liter to the $K_{S.P.}$ when the units of concentration are expressed in grams per liter. $K_{S.P.}(AgCl) = 1.7 \times 10^{-10}$.

27. Convert $K_{ion}(HAc)$ to the K_{ion} in which all concentrations are expressed in grams per liter.

28. How would you make a buffer solution from NH_4^+ and NH_3 in which $C_{H_3O^+} = 1 \times 10^{-9}$ M. $K_{ion}(NH_4^+) = 5.5 \times 10^{-10}$.

29. A solution is made to contain initially 0.4 M HCN and 0.3 M NaCN. Calculate C_{H^+}.

30. How would you make a buffer solution in which $C_{H_3O^+} = 2 \times 10^{-2}$ M using $SO_4^=$ and HSO_4^-.

31. How would you make a buffer in which C_{CN^-} was 3×10^{-9} M using HCl and HCN?

32. How would you make a buffer in which $C_{NO_2^-}$ was 2×10^{-7} M?

33. Write the formulae for the conjugate bases of each of the following acids: HF; NH_2^-; H_2O_2; C_2H_2; HO^-; $H_2SiO_4^=$.

34. Write the formulae for the conjugate acids of each of the following bases: $S^=$; $O_2^=$; HF; H_2SO_4; CH_3OH; NH_3; $PO_4^\equiv$.

35. $K_{ion}(H_2PO_4^-) = 6.2 \times 10^{-8}$.   Calculate the association constant for the conjugate base.   Classify the base strength according to section 13.

36. $K_{assoc.}(OH^-) = 10^{14}$.   Calculate the acid ionization constant of the conjugate acid.   Classify the acid strength according to section 13.

37. $K_{assoc.}(Cl^-) = 10^{-2}$.   Calculate the ionization constant of the conjugate acid.   Classify both acid and base strengths according to section 13.

38. Calculate $K_{eq.}$ for the reaction of $CO_3^=$ with HCN.   Does it go to completion?

39. $K_{ion}(HS^-) = 1 \times 10^{-15}$; $K_{ion}(HPO_4^-) = 1 \times 10^{-12}$.   Calculate $K_{eq.}$ for the reaction of $S^=$ with HPO_4^-.   Does it go to completion?

40. $K_{ion}(HOH) = 1 \times 10^{-14}$; $K_{ion}(HPO_4^-) = 1 \times 10^{-12}$.   Calculate $K_{eq.}$ for the reaction of $PO_4^\equiv$ with HOH.   Does it go to completion?   Note this particular displacement reaction is termed hydrolysis.

CHAPTER XIII

The Ionization of Water—Hydrolysis

1. Ionization of Water

To the processes going on in water solutions of ions, which we have already discussed, we must now add the following complication. Water is itself a weak electrolyte, owing to the process of self-ionization. The equilibrium of this weak electrolyte may be represented by the following equation:

$$H_2O + H_2O \rightleftarrows H_3O^{+1} + OH^{-1} - 14 \text{ kcal.}$$

This self-ionization occurs in every solution containing water! Thus all water solutions always contain some H_3O^{+1} ions and some OH^{-1} ions. Only when these two concentrations are equal, as they are in pure water, can we say that the solution is neutral. In acid solution we will still have some OH^{-1} ions. The solution is said to be acid only because $C_{H_3O^{+1}}$ ion is greater than $C_{OH^{-1}}$, and conversely for basic solutions.

We can apply the law of mass action to this equilibrium:

$$K_{\text{eq.}} = \frac{C_{H_3O^{+1}} \times C_{OH^{-1}}}{(C_{H_2O})^2}$$

But since the C_{H_2O} is very nearly the same, either in pure water or dilute solutions (namely, about 55 moles $H_2O/l.$), we can combine this concentration with the $K_{\text{eq.}}$ and write a simplified expression:

$$K_{\text{eq.}} \times (C_{H_2O})^2 = C_{H_3O^{+1}} \times C_{OH^{-1}} = K_{\text{ion}}(H_2O)$$

This new constant $K_{\text{ion}}(H_2O)$ has been measured experimentally and found to be equal to 1×10^{-14} at 25°C. It is found to increase as the temperature increases, being 56×10^{-14} at 100°C. We shall generally use the value at 25°C. unless otherwise indicated.

This expression for K_{ion} of water tells us that in every water solution at 25°C. the product of the concentrations of OH^{-1} and H_3O^{+1} ions is equal to 1×10^{-14}. If we know one of these, we can always calculate the other.

159

Example: What substances are present in a 0.020 M solution of NaOH?   What are their concentrations?

Answer: NaOH is an ionic salt (base).   A 0.020 M solution of NaOH will have 0.020 M Na^{+1} ions and 0.020 M OH^{-1} ions.   However, we will also get some OH^{-1} ions and H_3O^{+1} ions from the ionization of water itself.   These amounts will be small, since water is such a weak electrolyte.   Since $C_{OH^{-1}}$ from NaOH is large (2×10^{-2} M) we can neglect the added OH^{-1} ions from the water.   To calculate the H_3O^{+1} ions:

$$K_{ion}(H_2O) = C_{OH^{-1}} \times C_{H_3O^{+1}}$$

Solving for $C_{H_3O^{+1}}$:

$$C_{H_3O^{+1}} = \frac{K_{ion}(H_2O)}{C_{OH^{-1}}}$$

and substituting:

$$C_{H_3O^{+1}} = \frac{1 \times 10^{-14}}{2 \times 10^{-2}} = 5 \times 10^{-13} \ M$$

Note: Our approximation that the amounts of OH^{-1} and H_3O^{+1} ions from the ionization of water are small is certainly justified.

Summarizing:

$$C_{Na^{+1}} = 2 \times 10^{-2} \ M$$

$$C_{OH^{-1}} = 2 \times 10^{-2} \ M$$

$$C_{H_3O^{+1}} = 5 \times 10^{-13} \ M$$

$$C_{H_2O} \simeq 55 \ M$$

The approximation adopted above will be generally applicable Almost any solute that contributes OH^{-1} ions (like a base) or H_3O^{+1} (like an acid) will suppress the ionization of water immensely, and we can neglect the additional ions contributed by this ionization.

Do problems 1 and 2 at the end of the chapter.

2. Neutrality of Solutions

The simultaneous existence of H_3O^{+1} and OH^{-1} ions in all water solutions forces us to adopt a new definition of neutrality.

Definition: A *neutral solution* is one in which

$$C_{OH^{-1}} = C_{H_3O^{+1}}$$

Since the product of these concentrations is 1×10^{-14} $[K_{ion} (H_2O)]$ at 25°C., their concentrations in a neutral solution at 25°C. are

Neutral solution at 25°C.:

$$C_{OH^{-1}} = C_{H_3O^{+1}} = 1 \times 10^{-7} \ M$$

Thus, in an acid solution $C_{H_3O^{+1}}$ is greater than $1 \times 10^{-7} M$. Similarly, in a basic solution $C_{OH^{-1}}$ is greater than $1 \times 10^{-7} M$ (at 25°C.).

3. Logarithmic Units—The p Scale—pH

Science is always seeking abbreviations and shorthand to enable it to express briefly and compactly its facts and its theories. To avoid the inconvenience of writing exponentials (powers of 10), it has been found useful to express very large and small numbers in logarithmic units. For this purpose the following definition is employed:

Definition:

$$p\text{H} = - \log C_{H_3O^{+1}}$$

In general, a negative logarithm or p scale has been adopted with the following interpretation. When a quantity is preceded by the letter p the combination means -1 times the logarithm of the quantity following the letter p, e.g.:

$$pK_{ion} \text{ means } (- \log K_{ion})$$

$$p\text{OH} \text{ means } (- \log C_{OH^{-1}})$$

$$pK_{S.P.} \text{ means } (- \log K_{S.P.})$$

A pH of 2 means that $- \log C_{H_3O^{+1}} = 2$ or, changing the minus sign from left to right, $\log C_{H_3O^{+1}} = -2$. Since the logarithm of 10^{-2} is -2, we see that a pH of 2 signifies a concentration of hydrogen ion equal to 10^{-2} mole/l.

The following table shows the relation between the p scale and the normal concentration units in the case of pH.

RELATION BETWEEN pH AND CONCENTRATION OF HYDROGEN IONS

pH	-2	-1	0	1	2	3	4	...	11	...	16
$C_{H_3O^{+1}}$	100	10	1	0.1	0.01	1×10^{-3}	1×10^{-4}	...	1×10^{-11}	...	1×10^{-16}

Example: The concentration of hydrogen ion in a certain solution is $3 \times 10^{-4} M$. What is the pH of the solution?

Answer:

$$p\text{H} = - \log C_{H_3O^{+1}} = - \log (3 \times 10^{-4})$$

$$= -(\log 3 + \log 10^{-4})$$

$$= -[0.48 + (-4)] = -(0.48 - 4)$$

$$= -(-3.52) = 3.52$$

Note: It is always possible to check the answer to such a problem, at least approximately. Note that 3×10^{-4} is larger than 10^{-4} and smaller than 10^{-3}. Since the pH of the former is 4 and the pH of the latter is 3, we see that the correct pH must be between 3 and 4.

Example: The pH of a solution is 4.65. What is the concentration of hydrogen ion in the solution?

Answer: (Note first that, since 4.65 is between 4 and 5, the hydrogen ion concentration must lie between 10^{-4} and 10^{-5} M.)

$$pH = -\log C_{H_3O^{+1}} = 4.65$$

Therefore

$$\log C_{H_3O^{+1}} = -4.65; \quad C_{H_3O^+} = 10^{-4.65} \; M$$

In looking up antilogarithms we must break up the number into two parts; a positive decimal (the mantissa) which can then be found in logarithm tables or on a slide rule, and a whole number (the characteristic), either positive or negative, which will appear as a power of 10 in the answer.

$$\log C_{H_3O^{+1}} = -4.65 = -5.00 + 0.35; \quad C_{H_3O^+} = 10^{-5} \times 10^{+0.35} \; M$$

Therefore

$$C_{H_3O^{+1}} = (10^{-5}) \cdot (2.24)$$

$$= 2.24 \times 10^{-5} \; M$$

Check: 2.24×10^{-5} is between 10^{-4} and 10^{-5}.

Note: If the student lacks familiarity with logarithms he should consult the Appendix or a suitable textbook.

Do problems 3 and 4 at the end of the chapter.

4. A Paradox—When Is the Ionization of Water Significant?

We have thus far used the $K_{ion}(H_2O)$ relation to compute the concentration of either OH^{-1} or H_3O^{+1} if one of these is known. In doing this we ignored the contribution of ions from water to the given concentration. Is such a procedure always justified?

Let us take an example of when it is not. Let us compute the concentration of ions in a 1×10^{-8} M solution of HCl. Proceeding as usual we write $C_{Cl^{-1}} = 1 \times 10^{-8}$ M; $C_{H_3O^{+1}} = 1 \times 10^{-8}$ M, and, calculating $C_{OH^{-1}ion}$,

$$C_{OH^{-1}ion} = \frac{K_{ion}(H_2O)}{C_{H_3O^{+1}}} = \frac{1 \times 10^{-14}}{1 \times 10^{-8}} = 1 \times 10^{-6}$$

Thus we find that in an acid solution of 1×10^{-8} M HCl $C_{OH^{-1}ion}$ is greater than $C_{H_3O^{+1}ion}$! That is, the solution is basic! The error is easy to detect. In pure water $C_{H_3O^{+1}} = 1 \times 10^{-7}$ M. By adding

HCl (1×10^{-8} mole/l.) we can only increase this. But in the example chosen the amount of added H_3O^{+1} was less than the amount already present from the ionization of water.

Whenever the amount of added OH^{-1} ions or H_3O^{+1} ions is about the same or less than 10^{-7} M, then we cannot ignore the amounts arising from the ionization of water! The given problem must be solved as follows:

Let X = concentration of H_3O^{+1} and OH^{-1} from ionization of water. Then

$$C_{OH^{-1}} = X; \quad C_{H_3O^{+1}} = X + 1 \times 10^{-8}$$

$$K_{ion}(H_2O) = 1 \times 10^{-14} = (X)(X + 1 \times 10^{-8}) = X^2 + X \cdot 10^{-8}$$

This quadratic equation can be solved accurately to give X. However, a quick way to get an approximate answer is as follows: Since neutral water contains $C_{H_3O^{+1}} = 1 \times 10^{-7}$, let us add to this the amount obtained from the acid ($1 \times 10^{-8} = 0.1 \times 10^{-7}$). Their sum is $1 \times 10^{-7} + 0.1 \times 10^{-7} = 1.1 \times 10^{-7}$. But this is too large, since the addition of the acid suppresses the ionization of water. If we take the average of these two estimates, 1×10^{-7} and 1.1×10^{-7}, namely, 1.05×10^{-7}, we will be very close to the correct answer. The reader can check the accuracy of this answer by solving the quadratic for X. It gives $X = 9.5 \times 10^{-8}$. But $C_{H_3O^{+1}} = X + 1 \times 10^{-8} = 10.5 \times 10^{-8} = 1.05 \times 10^{-7}$, which checks well!

Do problem 5 at the end of the chapter.

5. Hydrolysis

Acids and bases are not the only substances that will upset the balance of OH^{-1} and H_3O^{+1} ions in water. When certain salts are placed in water it is found that the solution may become acidic or basic. This process is termed *hydrolysis*.

Etymologically, hydrolysis means "to dissolve or split with water." Technically, with special regard to ionic equilibria, the term represents the following process:

If a salt contains ions of a weak electrolyte, when it is placed in water solution those ions will react reversibly with the water molecules to form the *undissociated weak electrolyte* and either hydrogen or hydroxide ions.

If the ion is the ion of a weak base, then the product of hydrolysis will be the unionized weak base plus hydrogen ions and the solution will be acidic. If it is the ion of a weak acid, then the products will be the molecules of the weak acid plus hydroxide ions and the solution will be basic.   In this sense, hydrolysis may be thought of as the reverse reaction to neutralization.

Example: What will happen when sodium acetate is dissolved in water?

Answer: The acetate ion, being the ion of a weak acid, will react with water molecules to form un-ionized molecules of acetic acid and hydroxide ions.   Thus the solution will be basic.   The sodium ion is unaffected:

$$C_2H_3O_2^{-1} + H_2O \rightleftarrows HC_2H_3O_2 + OH^{-1}$$

Example: What ions and molecules will be present in a 0.2 M solution of sodium cyanide (NaCN)?   If the salt is 2% hydrolyzed, what are their concentrations?

Answer: *Note:* 2% hydrolyzed does not mean 2% ionized!   NaCN is a true salt and as such is 100% ionized in solution.   However, the cyanide ion (CN^{-1}) is the ion of a weak acid, HCN, and so the hydrolysis refers to the reaction of this ion with water.   Because of the hydrolysis of this ion, the solution will contain HCN molecules and excess OH^{-1} and therefore will be basic.   The sodium ion is not the ion of a weak base and will be totally unaffected.

$C_{Na^{+1}} = 0.2\ M$

$C_{CN^{-1}} = 98\%$ of $0.2\ M = 0.98 \times 0.2 = 0.196\ M$

$C_{HCN} = C_{OH^{-1}}$

    (since these are produced in equal amounts by the hydrolysis)

    $= 2\%$ of $0.2\ M = 0.02 \times 0.2 = 0.004\ M$

    $= 4 \times 10^{-3}\ M$

$C_{H_3O^{+1}} = \dfrac{1 \times 10^{-14}}{4 \times 10^{-3}} = 2.5 \times 10^{-12}\ M$

$C_{H_2O} = 55\ M$

To determine whether or not a given salt will undergo hydrolysis, we must look at the individual ions of the salt.   Only ions of either weak acids or weak bases will undergo hydrolysis.

If we look at the hydrolysis reaction of a base such as CN^- with HOH we note that it is just a special case of the displacement reactions we discussed in Chapter 12, section 13:

$$CN^- + HOH \rightleftarrows HCN + OH^-$$

Here the base CN^- is competing with the very strong base OH^-. In these systems HOH is acting as an acid.   Since it is a very weak acid ($K_{ion} = 10^{-14}$), we shall find that the equilibrium constants for hydrolyses are generally smaller than unity.   In a few special instances where we are dealing with even weaker acids than HOH, hydrolysis tends to be nearly complete. Thus NH_2^-, $O^=$, CH_3O^-, $O_2^=$ and $SiO_4^{\equiv}$ are more than 99% hydrolyzed into their conjugate acids in water solution.

Do problems 7–9 at the end of the chapter.

6. Calculation of the Extent of Hydrolysis—Hydrolysis Constants

We can always determine qualitatively which ions will undergo the greatest hydrolysis. The weaker the acid or base, the greater the extent of hydrolysis of its ions. Thus CN^{-1} ions will hydrolyze more completely than Ac^{-1} ions since $K_{ion}(HCN) = 4 \times 10^{-10}$ whereas $K_{ion}(HAc) = 1.8 \times 10^{-5}$.

The extent of hydrolysis may be calculated quantitatively if the K_{ion} of the weak acids and bases are known. Let us do this for the hydrolysis of the Ac^{-1} ion in a 0.04 M solution of NaAc. The equation for the hydrolysis is

$$Ac^{-1} + H_2O \rightleftarrows HAc + OH^{-1}$$

We can apply the law of mass action to this equilibrium:

$$K_{eq.} = \frac{C_{HAc} \times C_{OH^{-1}}}{C_{H_2O} \times C_{Ac^{-1}}}$$

Following the usual approximation we combine $K_{eq.}$ with the practically constant C_{H_2O} and obtain the simplified expression:

$$K_{eq.} \times C_{H_2O} = \frac{C_{HAc} \times C_{OH^{-1}}}{C_{Ac^{-1}}} = K_{hyd.}(Ac^{-1})$$

This $K_{hyd.}(Ac^{-1})$ is, however, related to $K_{ion}(HAc)$ and $K_{ion}(H_2O)$.

$$K_{ion}(H_2O) = C_{H_3O^{+1}} \times C_{OH^{-1}}$$

$$K_{ion}(HAc) = \frac{C_{Ac^{-1}} \times C_{H_3O^{+1}}}{C_{HAc}}$$

If we divide the first equation by the second we obtain

$$\frac{K_{ion}(H_2O)}{K_{ion}(HAc)} = \frac{C_{H_3O^{+1}} \times C_{OH^{-1}}}{C_{Ac^{-1}} \times C_{H_3O^{+1}}} \times C_{HAc} = \frac{C_{HAc} \times C_{OH^{-1}}}{C_{Ac^{-1}}}$$

We have thus derived a very important relation:

$$K_{hyd.}(Ac^{-1}) = \frac{K_{ion}(H_2O)}{K_{ion}(HAc)}$$

Thus the $K_{hyd.}$ for any given ion can be obtained by dividing $K_{ion}(H_2O)$ by the ionization constant of the acid. The same formula can be derived for the hydrolysis of basic ions.

We can now return to our original problem of the hydrolysis of 0.04 M NaAc. Let X = moles Ac^{-1} per liter that hydrolyze. Then $0.04 - X = C_{Ac^{-1}}$ that is left. Since each mole of Ac^{-1} hydrolyzing produces 1 mole of OH^{-1} ions and 1 mole of HAc, these concentrations must be equal to each other.

$$C_{HAc} = C_{OH^{-1}} = X$$

We can now substitute into the hydrolysis constant relation since $K_{ion}(HAc) = 1.8 \times 10^{-5}$ and $K_{ion}(H_2O) = 1 \times 10^{-14}$.

$$K_{hyd.}(Ac^{-1}) = \frac{K_{ion}(H_2O)}{K_{ion}(HAc)} = \frac{1 \times 10^{-14}}{1.8 \times 10^{-5}} = \frac{C_{OH^{-1}} \times C_{HAc}}{C_{Ac^{-1}}}$$

or

$$5.5 \times 10^{-10} = \frac{(X)(X)}{0.04 - X} = \frac{X^2}{0.04 - X}$$

But X will be small since $K_{hyd.}$ is small, and we can neglect X compared to 0.04. This approximation results in

$$5.5 \times 10^{-10} = \frac{X^2}{0.04}$$

Solving for X we find

$$X^2 = 0.04 \times 5.5 \times 10^{-10} = 22 \times 10^{-12}$$

Thus

$$X = 4.7 \times 10^{-6} \text{ mole/l.} = C_{OH^{-1}} = C_{HAc}$$

To calculate the percentage of hydrolysis:

$$\% \text{ hydrolysis} = \frac{\text{Amount of } Ac^{-1} \text{ hydrolyzed}}{\text{Original amount of } Ac^{-1}} \times 100$$

$$= \frac{4.7 \times 10^{-6} M}{4.0 \times 10^{-2}} \times 100$$

$$= 1.18 \times 10^{-2}\%$$

If we make other calculations we will find quite generally that the percentage of hydrolysis will be small.

Do problem 10 at the end of the chapter.

7. Salts of Weak Acids and Weak Bases

An interesting case in hydrolysis arises when we consider the behavior of a salt which contains ions of both a weak acid and a weak base. NH_4Ac, NH_4,CN, $AlAc_3$, and MgF_2 are examples of such combinations.

It should be pointed out here that the higher valence states metal ions form solutions which tend to be acidic. Thus Fe^{+3}, Al^{+3} and Mg^{+2} form acidic solutions. This comes about by the dissociation of the hydration shell of H_2O molecules which are always firmly attached to the metal ions in aqueous solutions. Actually just as we write H_3O^+ instead of H^+ we should write $Al(HOH)_6^{+++}$ instead of Al^{+++}, $Mg(HOH)_6^{++}$ instead of Mg^{++}, etc. The acidity then comes from the ionization reaction:

$$Al(HOH)_6^{+++} \rightleftarrows Al(HOH)_5(OH)^{++} + H^+$$

The acid is the hydrated metal ion, $Al(HOH)_6^{+++}$ and the conjugate base is the $Al(HOH)_5(OH)^{++}$ ion. Most aqueous metal ions have 6 HOH molecules attached to them, e.g., $Na(HOH)_6^+$; $Ca(HOH)_6^{++}$; $Fe(HOH)_6^{+++}$; $Ni(HOH)_6^{++}$; etc.

For salts such as NH_4CN in whose solutions both weak acids (NH_4^+) and weak bases (CN^-) are present we must first determine whether the solution will be predominantly acid or base by comparing the two reactions involving production of H^+ and OH^- respectively:

$(A) \quad NH_4^+ \rightleftarrows H^+ + NH_3 \qquad K_{ion}(NH_4^+) = 5.5 \times 10^{-10}$

$(B) \quad CN^- + HOH \rightleftarrows HCN + OH^- \qquad K_{hyd.} = \dfrac{K_{ion}(HOH)}{K_{ion}(HCN)}$

$$= 2.5 \times 10^{-5}$$

In this case we see that the hydrolysis of CN^- proceeds further to the right than does the ionization of the weak acid NH_4^+. Hence the solution of NH_4CN will be basic.

However, even when the acid A and base B reactions differ considerably, there is another mechanism which tends to bring them into approximate balance. In the above case, the excess OH^- ions produced by reaction B are very strong bases and will react with the normally weak acid NH_4^+ to produce the even weaker acid HOH:

$(C) \qquad OH^- + NH_4^+ \rightleftarrows NH_3 + HOH$

By removing the OH^{-1} ions, this second reaction tends to cause still more of the CN^{-1} ions to hydrolyze, and the final result is to bring about a great deal more hydrolysis of both the CN^{-1} and the NH_4^{+1} than would normally occur if each existed in a separate solution. The net result of these two reactions can be seen by adding them.

(D) $CN^{-1} + NH_4^{+1} \rightleftarrows NH_3 + HCN$ (H_2O and OH^{-1} cancel)

That is, the anions and cations tend to reverse the normal neutralization process and form undissociated acid and base. This tendency so far outweighs the differences in acid and base strength that, as a first approximation, we can always consider that the anions and cations hydrolyze to an equal extent.

In making calculations, no great error will be made if equation D is used as the basis of the calculation rather than the individual hydrolyses A and B. The slight differences in hydrolysis of anions and cations which do exist and are responsible for the solution being either acidic or basic may then be calculated rather simply.

Example: Compute the hydrolysis of a 0.3 M solution of NH_4CN.

$$K_{ion}(NH_4^+) = 5.5 \times 10^{-10}; \qquad K_{ion}(HCN) = 4 \times 10^{-10}$$

Answer: We first compute

$$K_{hyd.}(CN^-) = \frac{K_{ion}(HOH)}{K_{ion}(HCN)} = 2.5 \times 10^{-5}$$

and note that it will produce more OH^- than is produced by the feebler ionization of NH_4^+ ($K_{ion} = 5.5 \times 10^{-10}$). Hence $C_{OH^-} > C_{H^+}$ and the solution is basic. However, the difference cannot be large since the acid (NH_4^+) and base (CN^-) tend to react predominantly with each other. Thus to a first approximation, the reaction is: (letting X = concentration of CN^- reacting):

$$CN^- + NH_4^+ \rightleftarrows NH_3 + HCN$$

Concentrations: $(0.3 - X)$ $(0.3 - X)$ (X) (X)

This is a displacement reaction (Chapter XII, section 13) for which

$$K_{eq.} = \frac{K_{ion}(NH_4^+)}{K_{ion}(HCN)} = \frac{5.5 \times 10^{-10}}{4 \times 10^{-10}} = 1.4$$

$$= \frac{C_{NH_3} \times C_{HCN}}{C_{NH_4^+} \times C_{CN^-}} = \frac{X^2}{(0.3 - X)^2}$$

Taking square roots of both sides:

$$\frac{X}{0.3 - X} = 1.4^{1/2} = 1.18$$

Solving for X: $X = 0.162\ M \sim 0.16\ M$
Thus the salt is $0.16/0.300 \times 100 = 53\%$ hydrolyzed.

We can now calculate $C_{H_3O^+}$ from either acid:

$$K_{ion}(HCN) = 4 \times 10^{-10} = \frac{C_{H_3O^+} \times C_{CN^-}}{C_{HCN}}$$

or:
$$C_{H_3O^+} = \frac{C_{HCN}}{C_{CN^-}} \times 4 \times 10^{-10}$$

But $\quad C_{HCN} = X = 0.16\ M; \quad C_{CN^-} = 0.300 - X = 0.14\ M$

Therefore $\quad C_{H_3O^+} = \frac{0.16}{0.14} \times 4 \times 10^{-10} = 4.6 \times 10^{-10}\ M$

and $\quad C_{OH^-} = \frac{K_{ion}(HOH)}{C_{H_3O^+}} = \frac{1 \times 10^{-14}}{4.6 \times 10^{-10}} = 2.2 \times 10^{-5}\ M$

We would obtain the same result if we had used the $K_{ion}(NH_4^+)$ to calculate C_{OH^-}.

8. Buffer Solutions—Some Practical Considerations

We can obtain buffered solutions of H_3O^+ (or for simplicity H^+) by using a weak acid and the salt of the same weak acid (see Chapter 12, section 11). It is of interest to be able to know in terms of the desired $C_{H_3O^+}$ which weak acid to select. Let us see if there is some simple way of deciding this.

In a buffered solution, the pH is calculated from the equation for the dissociation of the weak acid. If we represent our acid by HB we can write:

$$HB \rightleftarrows H^+ + B^-$$

where B^- is now the conjugate base of HB. The ionization constant for HB now relates the concentrations:

$$K_{ion}(HB) = \frac{C_{H^+} \times C_{B^-}}{C_{HB}}$$

Solving this for C_{H^+} we find:

$$C_{H^+} = K_{ion}(HB) \times \left(\frac{C_{HB}}{C_{B^-}}\right)$$

Now it is not generally practical to use concentrations of acid (HB) or base (B^-) which are in excess of 1 M or less than about 0.1 M. If we deliberately restrict ourselves to these values as a practically useful range, then the ratio (C_{HB}/C_{B^-}) will never exceed 10 (e.g., when $C_{HB} = 1\ M$ and $C_{B^-} = 0.1\ M$), nor will it be less than 0.1 (e.g., when $C_{HB} = 0.1\ M$ and $C_{B^-} = 1\ M$). But since this ratio multiplied by $K_{ion}(HB)$ gives the C_{H^+}, we see that for any acid-base conjugate pair, chosen to buffer H^+ ion, the usable range of C_{H^+} will lie within one power of 10 of $K_{ion}(HB)$.

Thus if we want $C_{H^+} = 1 \times 10^{-5}$ we need a weak acid whose K_{ion} is within 1 power of 10, e.g., 10^{-4} to 10^{-6}. Similarly to buffer C_{H^+} at 1×10^{-9} we need a weak acid whose K_{ion} is in the range 1×10^{-8} to 1×10^{-10}.

Conversely if we know the K_{ion} of a weak acid, we know that it can be used to buffer H^+ in a range of C_{H^+}, which is within 1 power of 10 of this K_{ion}.

Example: $K_{ion}(HCO_3^-) = 4.7 \times 10^{-11}$. What are the practical ranges of H^+ over which HCO_3^- and $CO_3^=$ can be used as buffers?

Answer: From the foregoing we see that we can use $HCO_3^-/CO_3^=$ buffers to give C_{H^+} from 4.7×10^{-10} to 4.7×10^{-12}. As an example 0.3 M HCO_3^- and 0.6 M $CO_3^=$ will give:

$$C_{H^+} = K_{ion}(HCO_3^=) \times \frac{C_{HCO_3^-}}{C_{CO_3^-}} = 4.7 \times 10^{-11} \times \frac{0.3}{0.6}$$

$$= 2.4 \times 10^{-11}\ M \quad \text{or} \quad C_{OH^-} = \frac{1 \times 10^{-14}}{C_{H^+}}$$

$$= 4.2 \times 10^{-4}\ M$$

The reason for this practical limitation is that many acids and salts are not sufficiently soluble to give concentrations in excess of 1 M. Even for the ones that are sufficiently soluble it is preferable to work in more dilute solutions to avoid complications due to complex ion formation and abnormal activity effects (Chapter XII, section 6B).

The lower limitation is set by the fact that the reservoirs which provide the buffering action should be present in sufficiently large concentration so that they will not be consumed by any reaction that occurs in the buffer solution.

9. Titration of Weak Acids and Bases

When we titrate an acid or base we add equal numbers of equivalents of acid and base and form a salt and water. If the acid and base are both strong, the salt formed does not hydrolyze and the final solution will be neutral. That is, its pH will be 7.

If, however, either the acid of the base is weak, the final salt solution is hydrolyzed and not neutral. (A titration cannot be performed successfully if both acid and base are weak.)

The pH of the final salt solution may be estimated from the methods developed in the previous paragraphs on hydrolysis.

However, a good rule of thumb is: the pH of the final solution can be obtained by adding or subtracting from $7,\frac{1}{2} pK_{ion}$ of the acid or base.

Titration of Weak Acid: pH of salt = $7 + \frac{1}{2}pK_{ion}$(acid)

Titration of Weak Base: pH of salt = $7 - \frac{1}{2}pK_{ion}$ (base)

Thus, if we are titrating HAc with NaOH, the pH at the end-point will be the pH of NaAc (the salt formed) and is roughly $7 + \frac{1}{2}[pK_{ion}(\text{HAc})] = 7 + \frac{1}{2}(4.74) = 9.37$. This result depends slightly on the concentration of the solutions used, and the accurate answer may be obtained from the equations developed previously (see section on hydrolysis).

For the titration of NH_3 with HCl, the pH at the endpoint will be 4.63.

These results are important since we shall want to choose indicators for these titrations that change color at the proper pH values, not at 7.

10. Changes in pH During Titration—Titration Curves

In the laboratory operation of titrating an ordinary strong acid solution, such as 0.1 M HCl, with a strong base solution, such as 0.1 M NaOH, the accuracy of the titration depends on the sharpness of the end point of the titration. If we are using phenolphthalein as an indicator for example, then one drop of excess acid should turn the solution from red to colorless and conversely, one drop of excess base should turn the solution from colorless to pink.

Let us consider in detail how the pH of the final solution is affected by these additions of one drop in excess of either acid or base.

Let us suppose that the solution is neutral (pH = 7), and it has a total volume of about 50 ml. That is, we may have used 25 ml. each of acid and base. If we now add one drop further of the 0.1 M base, NaOH, we are adding about 1/20 ml. = 0.05 ml. This contains:

$$0.05 \text{ ml. base} = 0.05 \text{ ml. base} \left(\frac{0.1 \text{ mmole NaOH}}{1.0 \text{ ml. base}} \right)$$

$$= 0.005 \text{ mmole NaOH}$$

This 0.005 mmole NaOH in 50 ml. of solution gives an OH^- concentration of:

$$C_{OH^-} = \frac{0.005 \text{ mmole}}{50 \text{ ml. soln.}} = 1 \times 10^{-4} \ M$$

This is very much larger than the $C_{OH^-} = 1 \times 10^{-7} \ M$, which is present at equilibrium and represents a 1000-fold change in this concentration just by the addition of one drop of excess base. We can calculate the pH from the C_{H^+} given by:

$$C_{H^+} = \frac{K_{ion}(HOH)}{C_{OH^-}} = \frac{1 \times 10^{-14}}{1 \times 10^{-4}} = 1 \times 10^{-10} \, M$$

Hence, pH $= -\log C_{H^+} = 10$

A similar calculation will show that one drop (0.05 ml.) of excess acid will shift the pH by three units, but in the opposite direction, i.e., from 7 at neutrality, to 4.

It is this extreme sensitivity of the pH of the final neutral solution to the addition of very small amounts of excess acid or base that makes it possible to determine the end point of the titration with great sensitivity and thus perform titrations with great accuracy.   Note that during most of the titration procedure, there is relatively little change in the pH of the solution.   Very large changes in pH occur only near the end point.

Example: 25.0 ml. of 0.100 M NaOH are titrated with 0.100 M HCl. Calculate the pH when (*a*) 20.0 ml., (*b*) 24.0 ml., and (*c*) 24.5 ml. of acid have been added.

Answer: (*a*) When 20.0 ml. of acid have been added, 5.0 ml. of base are left unreacted in 45.0 ml. of solution, hence:

$$C_{OH^-} = \frac{5.0 \text{ ml. base} \times \left(\dfrac{0.100 \text{ mmole OH}^-}{1 \text{ ml. base}} \right)}{45.0 \text{ ml. soln.}}$$

$$= \frac{5.0 \times 0.100 \text{ mmole OH}^-}{45.0 \text{ ml. base}} = 0.011 \, M$$

$$C_{H^+} = \frac{1 \times 10^{-14}}{0.011} \approx 1 \times 10^{-12} \, M; \text{ pH} = 12$$

(*b*) When 24.0 ml. have been added, 1.0 ml. of base is left unreacted in 49 ml. of solution, and:

$$C_{OH^-} = \frac{1.0 \text{ ml. base} \times \left(\dfrac{0.100 \text{ mmole OH}^-}{1 \text{ ml. base}} \right)}{49.0 \text{ ml. soln.}} = 0.0020 \, M$$

$$C_{H^+} = 5 \times 10^{-12} \, M; \text{ pH} = 11.3$$

(*c*) When 24.5 ml. of acid have been added, there are 0.5 ml. of unreacted base in about 50 ml. of soln., and:

$$C_{OH^-} = \frac{0.5 \text{ ml. base} \times \left(\dfrac{0.100 \text{ mmole OH}^-}{1 \text{ ml. base}} \right)}{50 \text{ ml. soln.}} = 0.001 \, M$$

$$C_{H^+} = 1 \times 10^{-11} \, M; \text{ pH} = 11.0$$

The following diagram shows how the pH of the solution changes as 25 ml. of 0.100 M NaOH is neutralized by 25 ml. of 0.100 HCl, and then some excess acid is added.

The titration of a weak acid with a strong base (or vice versa) does not give so sharp a change in pH with excess volume of acid or base at the end point.   The ion of the weak acid tends to buffer the solution and make its pH vary slowly when small amounts of acid are added.

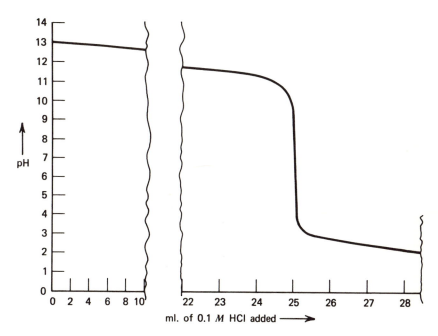

Figure 1. Titration Curve Showing pH Changes as 0.100 M HCl is added to 25.0 ml of 0.100 M NaOH.

11. Problems

In the following use K_{ion} as needed from Table XVII.

1. (*a*) What are the units of $K_{ion}(H_2O)$?
   (*b*) What would the value of $K_{ion}(H_2O)$ be if concentrations were expressed in grams per liter?

2. What ions are present in each of the following solutions, and what are their concentrations?

   (*a*) 0.0050 M HCl.   (*d*) 0.75 M NaOH.
   (*b*) 0.34 M NaCl.   (*e*) $3 \times 10^{-5} M$ KOH.
   (*c*) 0.60 M K$_2$SO$_4$.   (*f*) $2 \times 10^{-4} M$ ZnCl$_2$ (ignore hydrolysis of Zn^{+2}).

3. What is the pH and the pOH of each of the following solutions:

(a) $0.0010\ M$ HCl.

(b) $0.0010\ M$ H$_2$SO$_4$.

(c) $1.0 \times 10^{-4}\ M$ NaOH.

(d) $3.4 \times 10^{-4}\ M$ HCl.

(e) $0.06\ M$ H$_2$SO$_4$.

(f) $0.53\ M$ KOH.

(g) $0.034\ M$ Ca(OH)$_2$.

(h) $0.03\ M$ HAc (if it is 2.4% ionized).

4. Make the following conversions:

(a) Express $K_{ion} = 2.5 \times 10^{-6}$ as pK_{ion}.

(b) Express $C_{Ag^{+1}} = 3 \times 10^{-12}$ as $pC_{Ag^{+1}}$.

(c) Express $K_{eq.} = 8 \times 10^{-16}$ as $pK_{eq.}$.

(d) Express $pK_{S.P.} = 9.70$ as $K_{S.P.}$.

(e) Express $pOH = 6.30$ as $C_{OH^{-1}}$.

(f) Express $pK_{ion} = 8.64$ as K_{ion}.

5. What is the pH of a $2 \times 10^{-8}\ M$ solution of H$_2$SO$_4$?

6. For each of the following salts, write balanced equations representing the hydrolysis of those ions that do hydrolyze:

(a) NaAc.

(b) (NH$_4$)$_2$SO$_4$.

(c) ZnCl$_2$.

(d) FeCl$_3$.

(e) KCN.

(f) Fe(Ac)$_3$.

(g) Al$_2$(SO$_4$)$_3$.

(h) MgCO$_3$.

7. A $0.4\ M$ solution of NaClO is 0.22% hydrolyzed. What molecules and ions are present in the solution? What are their concentrations?

8. A $0.07\ M$ solution of KCN is 2.0% hydrolyzed. What molecules and ions are present in the solution? What are their concentrations?

9. A $0.25\ M$ solution of NaHCO$_3$ is 0.03% hydrolyzed. What molecules and ions are present in the solution? What are their concentrations? (Ignore dissociation of HCO$_3^{-1}$.)

10. Calculate hydrolysis constants for each of the following salt solutions. Compute also the pH of the solution and the percentage of hydrolysis.

(a) $0.05\ M$ NaAc.

(b) $0.008\ M$ NH$_4$Cl.

(c) $0.32\ M$ Na$_2$HPO$_4$ (ignore second step).

(d) $0.5\ M$ Na$_2$S (ignore second step).

(e) $0.64\ M$ KCN.

(f) $0.06\ M$ MgCl$_2$; $K_{ion}[Mg(HOH)^{++}] = 5 \times 10^{-11}$ (ignore second step).

(g) $0.40\ M$ NH$_4$Ac.

(h) $0.003\ M$ NH$_4$CN.

11. Calculate the pH to be expected at the endpoint in the following titrations:

(a) NH$_3$ with HCl.

(b) NaOH with HCN.

(c) KOH with H$_2$PO$_4^{-1}$.

(d) KOH with HS^{-1}.

12. To 50 ml. of a solution of $0.20\ M$ HAc is added 50 ml. of $0.20\ M$ NaOH. Ignoring any contractions in volume, compute the concentrations of Na$^+$ and HAc and H$^+$ in the final solution.

13. To 100 ml. of a $0.40\ M$ solution of HF is added 50 ml. of $0.50\ M$ NaOH.

(a) Compute C_{Na^+}, C_{F^-}, and C_{HF} approximately, ignoring any effects of hydrolysis.

(b) Compute the pH of the solution.

14. 60 ml. of $0.45\ M$ HCl are mixed with 40 ml. of $0.30\ M$ NaOH. Calculate the concentration of Cl$^-$, Na$^+$, H$^+$, and OH$^-$ in the final solution. What is its pH?

15. Calculate the pH of a solution made up to contain initially $0.12\ M$ Na$_2$CO$_3$ and $0.08\ M$ NaHCO$_3$.

16. What is the extent of hydrolysis and the pH of a solution of $0.2\ M$ FeCl$_3$? $K_{ion}[Fe(HOH)_6^{+3}] = 2 \times 10^{-4}$. Ignore any secondary ionization.

17. What is the extent of hydrolysis and the pH of a solution of $0.02M$ $Al_2(SO_4)_3$? $K_{ion}[Al(HOH)_6^{+++}] = 6 \times 10^{-6}$. Ignore any secondary ionization.

18. 1.0 ml. of 0.5 M HCl are added to 100 cc. of 0.003 M NaOH. What is the pH of the initial and final solutions? (Neglect volume change due to added acid.)

19. What effective pH's could be obtained from the following buffers?

(a) $HPO_4^=/PO_4^\equiv$ $K_{ion}(HPO_4^-) = 1 \times 10^{-12}$.
(b) $HSO_4^-/SO_4^=$ $K_{ion}(HSO_4^-) = 1.2 \times 10^{-2}$.
(c) $H_3BO_3/H_2BO_3^-$ $K_{ion}(H_3BO_3) = 6 \times 10^{-10}$.
(d) HF/F^- $K_{ion}(HF) = 7 \times 10^{-4}$.
(e) $HS^-/S^=$ $K_{ion}(HS^-) = 1 \times 10^{-15}$.

20. The solubility of very insoluble salts may be affected by hydrolysis if they contain the ions of weak acids. Thus $CaCO_3$ is more soluble than one would expect from considerations of its $K_{S.P.}$ alone. The reaction is:

$$CaCO_3\ (s) + HOH \rightleftarrows Ca^{++} + HCO_3^- + OH^-$$

For the following insoluble salts write similar balanced chemical reactions in the cases where you feel that hydrolysis might be significant:

(a) AgCl.   (c) $Fe(OH)_3$.   (e) CaF_2.
(b) ZnS.   (d) AgAc.   (f) $BaCrO_4$.

21. Given $K_{ion}(HCO_3^-) = 4.7 \times 10^{-11}$. $K_{S.P.}(CaCO_3) = 4.8 \times 10^{-9}$. Calculate $K_{eq.}$ for the hydrolysis reaction of $CaCO_3\ (s)$ given in the preceding problem.

22. $K_{S.P.}[Co(OH)_2] = 2 \times 10^{-16}$. Calculate the solubility of the salt in water taking into consideration the C_{OH^-} already present in water.

23. Given $K_{S.P.}(BaCO_3) = 4.9 \times 10^{-9}$, $K_{ion}(HCO_3^-) = 4.7 \times 10^{-11}$ and $K_{ion}(HAc) = 1.8 \times 10^{-5}$. (a) Write a balanced equation for the reaction of HAc and solid $BaCO_3$. (b) Calculate the equilibrium constant for this reaction. (c) Would you expect HAc to dissolve $BaCO_3$? Explain.

24. Suggest conjugate acid-base pairs to give the following buffered pH's:

(a) pH = 3.   (d) pH = 11.7.
(b) pH = 6.   (e) pH = 14.
(c) pH = 8.5.

25. Calculate the per cent of hydrolysis and the pH for a 0.2 M solution of $FeAc_3$. (Use K_{ion} from other problems.)

26. Calculate the per cent of hydrolysis and the pH of a 0.05 M solution of NH_4Ac.

27. Calculate the per cent of hydrolysis and the pH of a 0.04 M solution of $Al(CN)_3$. Use K_{ion} from preceding problems.

28. Calculate the per cent of hydrolysis and the pH of 0.2 M NH_4F.

29. Prove that if the titration of a weak acid with a strong base is stopped precisely at the half-way point that the C_{H^+} of the resulting solution is equal to K_{ion} (acid).

30. Show that if the weak acid $HCN(K_{ion} = 4 \times 10^{-10})$ is titrated with 0.1 M NaOH, the end point of the titration should occur at a pH of about 11.2. How will the pH of this solution change on the addition of one drop ($\sim$0.05 ml.) of the base to 50 ml. of solution?

31. Extremely weak acids whose K_{ion} are very much smaller than $K_{ion}(H_2O)$ have conjugate bases whose properties are very difficult to study in water solution. Can you explain why?

32. $K_{ion}(NH_3)$ is estimated at about 1×10^{-28}. What is the conjugate base of NH_3? Estimate its concentration in 1 M NH_3.

CHAPTER XIV

Additional Equilibria in Ionic Solutions

1. Ionization of Polyvalent Electrolytes

In Chapter XII we emphasized problems dealing with the first step in the ionization of a weak electrolyte. Many electrolytes, however, are polyvalent and ionize in a sequence of steps:

First step: $\qquad H_2O + H_2S \rightleftarrows HS^{-1} + H_3O^{+1}$ $\qquad\qquad K_{ion}(I)$

Second step: $\qquad H_2O + HS^{-1} \rightleftarrows S^{-2} + H_3O^{+1}$ $\qquad\qquad K_{ion}(II)$

or by addition of these two steps we have the overall equation:

$$2H_2O + H_2S \rightleftarrows S^{-2} + 2H_3O^{+1} \qquad K_{ion}(\text{overall})$$

If we are given C_{H_2S} and want to find $C_{HS^{-1}}$, we would use the first equation together with its $K_{ion}(I)$.

If we are given $C_{HS^{-1}}$ and want to find $C_{S^{-2}}$, we would use the second equation together with its $K_{ion}(II)$.

If we are given C_{H_2S} and want to find $C_{S^{-2}}$, we would use the overall equation which is equal to $K_{ion}(I)$ times $K_{ion}(II)$.

If we are given C_{H_2S} and want to find both $C_{HS^{-1}}$ and $C_{S^{-2}}$, then we proceed stepwise.

Example: What are the $C_{HS^{-1}}$, $C_{H_3O^{+1}}$ and $C_{S^{-2}}$ in a 0.03 M solution of H_2S. $K_{ion}(H_2S) = 1 \times 10^{-7}$; $K_{ion}(HS^{-1}) = 1 \times 10^{-15}$.

Answer: Proceeding as usual, let X = moles/l. of H_2S that ionize. Then we have in solution:

$$C_{H_2S} = 0.03 - X$$

$$C_{H_3O^{+1}} = X$$

$$C_{HS^{-1}} = X$$

But it will be objected that some of the HS^{-1} ionizes to produce more H_3O^{+1}, and so $C_{H_3O^{+1}}$ is more than X moles/l. and $C_{HS^{-1}}$ is less than X moles/l.

The answer to these objections is this: H_2S ionizes only weakly in the first step $[K_{ion}(I) = 1 \times 10^{-7}]$. This permits us to ignore X with respect to 0.03. Similarly, HS^{-1} ionizes even less in the second step, and so we can ignore the changes that this second ionization produces on $C_{H_3O^{+1}}$ and on $C_{HS^{-1}}$. Then

$$K_{ion}(H_2S) = \frac{C_{H_3O^{+1}} \times C_{HS^{-1}}}{C_{H_2S}}$$

On substitution:

$$1 \times 10^{-7} = \frac{(X)(X)}{0.03 - X} \quad \text{(approximating } 0.03 - X \cong 0.03\text{)}$$

Then

$$1 \times 10^{-7} = \frac{X^2}{0.03} \quad \text{or} \quad X^2 = 0.3 \times 10^{-8} = 30 \times 10^{-10}$$

Thus

$$X = 5.5 \times 10^{-5} \text{ moles/l.} = C_{H_3O^{+1}} = C_{HS^{-1}}$$

To find $C_{S^{-2}}$ we now use $K_{ion}(II)$:

$$K_{ion}(II) = \frac{\cancel{C_{H_3O^{+1}}} \times C_{S^{-2}}}{\cancel{C_{HS^{-1}}}}$$

But we have just agreed that $C_{H_3O^{+1}} = C_{HS^{-1}}$, so we can cancel them. Therefore

$$C_{S^{-2}} = K_{ion}(II) = 1 \times 10^{-15} \text{ mole/l.}$$

If there is a common ion added, then all calculations are simplified, as indicated in Chapter XIII.

2. Reactions of Amphoteric Ions

Some substances may react both as weak acids or as weak bases. Thus HCO_3^{-1} may do either:

As acid: $HCO_3^{-1} + H_2O \rightleftarrows H_3O^{+1} + CO_3^{-2}$

As base: $HCO_3^{-1} + H_2O \rightleftarrows OH^{-1} + H_2CO_3$

In general we see that if we have a polyvalent acid all of its intermediate ions can behave as acids or bases. This will be true of $H_2PO_4^-$, $HPO_4^=$, $H_2BO_3^-$, $HBO_3^=$, etc. The relative acidity or pH of the solution of such ions will depend on whether or not the acid-producing reaction proceeds more or less than the base-producing reaction (hydrolysis). But this can be quickly decided (as in the case of hydrolysis of salts of weak acids and bases) by comparing the equilibrium constants for these two reactions, i.e., K_{ion} and $K_{hyd.}$. These are respectively in the case of the HCO_3^-, $K_{ion}(HCO_3^-)$ and $K_{hyd.}(HCO_3^-) = K_{ion}(HOH)/K_{ion}(H_2CO_3)$. Since $K_{ion}(HCO_3^-) =$

4.3×10^{-7} and $K_{hyd.}(HCO_3^-) = 1 \times 10^{-14}/4.7 \times 10^{-11} = 2.1 \times 10^{-4}$, we see that the hydrolysis prevails and the solution will be basic. Table XVIII compares some other amphoteric ions:

TABLE XVIII
RELATIVE ACIDITY OF SOME AMPHOTERIC IONS

Salt	$K_{ion}(I)$	$K_{ion}(II)$	$K_{hyd.} = \dfrac{K_{ion}(HOH)}{K_{ion}(I)}$
NaHS	$1 \times 10^{-7}(H_2S)$	$1 \times 10^{-15}(HS^{-1})$	1×10^{-7} (basic)
NaHSO$_3$	$1.2 \times 10^{-2}(H_2SO_3)$	$1 \times 10^{-7}(HSO_3^{-1})$	8×10^{-13} (acidic)
Na$_2$HPO$_4$	$6.2 \times 10^{-8}(H_2PO_4^{-1})$	$1 \times 10^{-12}(HPO_4^{-2})$	1.6×10^{-7} (basic)
NaH$_2$PO$_4$	$7.5 \times 10^{-3}(H_3PO_4)$	$6.2 \times 10^{-8}(H_2PO_4^{-1})$	1.3×10^{-12} (acidic)
NaHB$_4$O$_7$	$1 \times 10^{-4}(H_2B_4O_7)$	$1 \times 10^{-9}(HB_4O_7^{-1})$	1×10^{-11} (acidic)

The quantitative calculation can be treated just as we treated the hydrolysis of the salt of a weak acid and a weak base. We combine the two equations and make the usual assumption that both reactions are roughly equal. Thus, for 0.4 M solution NaHCO$_3$:

$$2HCO_3^{-1} \rightleftarrows H_2CO_3 + CO_3^{-2}$$

This reaction is called a disproportionation.
Let X = concentration of HCO_3^{-1} used up. Then

$$C_{HCO_3^{-1}} = 0.4 - X; \quad C_{H_2CO_3} = C_{CO_3^{-2}} = \frac{X}{2}$$

$$K_{eq.} = \frac{C_{H_2CO_3} \times C_{CO_3^{-2}}}{(C_{HCO_3^{-1}})^2} = \frac{K_{ion}(HCO_3^{-1})}{K_{ion}(H_2CO_3)}$$

Substituting numbers:

$$\frac{\left(\dfrac{X}{2}\right)\left(\dfrac{X}{2}\right)}{(0.4 - X)^2} = \frac{4.7 \times 10^{-11}}{4.3 \times 10^{-7}} = 1.1 \times 10^{-4} = \frac{X^2}{4(0.4 - X)^2}$$

Taking square roots of both sides:

$$\frac{X}{2(0.4 - X)} = 1.05 \times 10^{-2}$$

and solving for X we have

$$X = (0.8)(1.05 \times 10^{-2}) - (2.1 \times 10^{-2})X$$

or

$$1.02X = 8.4 \times 10^{-3}$$

Thus

$$X = 8.2 \times 10^{-3} \text{ mole/l.} \rightarrow 8 \times 10^{-3} M$$

and

$$C_{HCO_3^{-1}} = 0.4 - X = 0.392 \text{ mole/l.}$$

and

$$C_{H_2CO_3} = C_{CO_3^{-2}} = \frac{X}{2} = 4 \times 10^{-3} \text{ mole/l.}$$

$$\% \text{ hydrolysis} = \frac{X}{0.4} \times 100 = \frac{8 \times 10^{-3}}{0.4} \times 100 = 2\%$$

We may calculate $C_{H_3O^{+1}}$ from either $K_{ion}(I)$ or $K_{ion}(II)$ since both will give the same answer.

$$\frac{C_{H_3O^{+1}} \times C_{HCO_3^{-1}}}{C_{H_2CO_3}} = K_{ion}(H_2CO_3)$$

Solving for $C_{H_3O^{+1}}$:

$$C_{H_3O^{+1}} = \frac{K_{ion}(H_2CO_3) \times C_{H_2CO_3}}{C_{HCO_3^{-1}}}$$

but

$$\frac{C_{H_2CO_3}}{C_{HCO_3^{-1}}} = \frac{C_{CO_3^{-1}}}{C_{HCO_3^{-1}}} = \sqrt{\frac{K_{ion}(II)}{K_{ion}(I)}}$$

Thus on substitution:

$$C_{H_3O^{+1}} = \sqrt{K_{ion}(I) \times K_{ion}(II)}$$

$$= \sqrt{4.3 \times 10^{-7} \times 4.7 \times 10^{-11}} = 4.5 \times 10^{-9} \text{ mole/l.}$$

Finally

$$C_{OH^{-1}} = \frac{K_{ion}(H_2O)}{C_{H_3O^{+1}}} = \frac{1 \times 10^{-14}}{4.5 \times 10^{-9}} = 2.2 \times 10^{-6} \text{ mole/l.}$$

The exact solution to this problem gives an almost identical result.

The equilibrium constant $K_{disp.}$ for disproportionation reactions of an amphoteric ion is always given by the ratio of the acid ionization constants for the next and preceding step as illustrated above for

HCO_3^-. Since, however, it turns out that the K_{ion} of a polyvalent acid generally decreases by some 4 to 6 powers of 10 with every succeeding ionization we see that $K_{disp.}$ will generally lie between 10^{-4} to 10^{-6} so that disproportionation is never very extensive (e.g., not more than 1%).

3. Dissociation of Complex Ions

We have thus far emphasized the dissociation of weak electrolytes that contain H^{+1} ions. There are a large group of complex radicals made from metal ions and other groups. Thus the Zn^{+2} ion can form the following complexes: $Zn(OH)_4^{-2}$, $Zn(NH_3)_4^{+2}$, and $Zn(CN)_4^{-2}$. Similarly, the Hg^{+2} ion forms: $HgCl_4^{-2}$, HgI_4^{-2}, and $Hg(CN)_4^{-2}$; and the Ag^{+1} ion forms $Ag(NH_3)_2^{+1}$, $AgCl_3^{-2}$, $Ag(S_2O_3)_2^{-3}$, and $Ag(CN)_2^{-1}$. The types of complex ion which a metal ion can form can be determined only by experiment. However, for every complex ion we can write an equilibrium equation showing its tendency to dissociate into its original components:

$$Zn(OH)_4^{-2} \rightleftarrows Zn^{+2} + 4OH^{-1}$$

$$Zn(NH_3)_4^{+2} \rightleftarrows Zn^{+2} + 4NH_3$$

$$Zn(CN)_4^{-2} \rightleftarrows Zn^{+2} + 4CN^{-1}$$

The law of mass action may be applied to these *homogeneous* equilibria, and we can write an equilibrium constant for each. Thus

$$K_{eq.}[Zn(OH)_4^{-2}] = \frac{C_{Zn^{+2}} \times C^4_{OH^-}}{C_{Zn(OH)_4^{-2}}}$$

This $K_{eq.}$ or dissociation constant, as it is frequently called, may be measured experimentally. Once known it is useful (just like other equilibrium constants) to calculate the concentrations of ions and complex ions in equilibrium with each other.

Example:

$$K_{eq.}[Zn(NH_3)_4^{+2}] = 2.6 \times 10^{-10}$$

What are the ions present in a 0.2 M solution of $Zn(NH_3)_4Cl_2$? What are their concentrations?

Answer: $Zn(NH_3)_4Cl_2$ is a salt. In a 0.2 M solution we will have $C_{Cl^{-1}} = 0.4\ M$ and $C_{Zn(NH_3)_4^{+2}} = 0.2\ M$. However, some of the $Zn(NH_3)_4^{+2}$ complex ions dissociate as follows:

$$Zn(NH_3)_4^{+2} \rightleftarrows Zn^{+2} + 4NH_3$$

Let X = moles per liter of $Zn(NH_3)_4^{+2}$ that dissociate.   Then we will have in solution

$$C_{Zn(NH_3)_4^{+2}} = 0.2 - X; \quad C_{Zn^{+2}} = X; \quad C_{NH_3} = 4X$$

By definition:

$$K_{eq.} = \frac{C_{Zn^{+2}} \times C^4_{NH_3}}{C_{Zn(NH_3)_4^{+2}}}$$

Substituting we have

$$2.6 \times 10^{-10} = \frac{(X)(4X)^4}{(0.2 - X)} = \frac{256X^5}{0.2 - X}$$

Let us neglect X compared to 0.2.   Our equation simplifies to

$$\frac{256X^5}{0.2} = 2.6 \times 10^{-10}$$

or

$$X^5 = \frac{0.2 \times 2.6 \times 10^{-10}}{256}$$

or

$$X^5 = 2.0 \times 10^{-13}$$

or

$$X^5 = 200 \times 10^{-15}$$

Taking the $\frac{1}{5}$ root of both sides:

$$X = 2.9 \times 10^{-3} \ M.$$

Thus

$$C_{Zn^{+2}} = 2.9 \times 10^{-3} \ M$$

$$C_{NH_3} = 4X = 1.16 \times 10^{-2} \ M$$

$$C_{Zn(NH_3)_4^{+2}} = 0.2 - 0.0029 = 0.197 \ M \ \rightarrow 0.2 \ M$$

The percentage of dissociation $= \dfrac{X}{0.2} \times 100 = \dfrac{0.0029}{0.2} \times 100 = 1.5\%$

---

4. Common Ion Effect for Complex Ions

In laboratory practice we generally deal not with pure salts of complex ions but with solutions containing the complex ion in the presence of an excess of the agent used to form the complex.

Thus the $Ag(CN)_2^{-1}$ complex ion is always formed in the presence of an excess of CN^{-1} ions.   In such cases the calculations are much simplified if we know the excess concentration of the complexing agent.

Example: A solution contains 0.3 M $Ag(CN)_2^{-1}$ ions and 0.04 M CN^{-1}.   What is the concentration of Ag^{+1} ions if $K_{eq}[Ag(CN)_2^{-1}] = 1 \times 10^{-21}$? *Note:* There are, of course, positive ions present also, but they are not of importance for the problem.

Answer: By definition:

$$K_{eq}[\text{Ag(CN)}_2{}^{-1}] = \frac{C_{\text{Ag}^{+1}} \times C^2{}_{\text{CN}^{-1}}}{C_{\text{Ag(CN)}_2{}^{-1}}}$$

Solving for $C_{\text{Ag}^{+1}}$ and substituting numbers:

$$C_{\text{Ag}^{+1}} = K_{eq} \times \frac{C_{\text{Ag(CN)}_2{}^{-1}}}{C^2{}_{\text{CN}^{-1}}} = 1 \times 10^{-21} \times \frac{0.3}{(0.04)^2}$$

$$= \frac{1 \times 10^{-21} \times 0.3}{0.0016} = 1.9 \times 10^{-19} \text{ mole/l.}$$

Example: What concentration of NH_3 is needed in a solution containing 0.03 M Ag^{+1} to suppress the $C_{\text{Ag}^{+1}}$ to 2×10^{-10} M?

$$K_{eq}[\text{Ag(NH}_3)_2{}^{+1}] = 6.8 \times 10^{-8}.$$

Answer: By adding enough NH_3 all 0.03 M Ag^{+1} ions will be converted to 0.03 M $\text{Ag(NH}_3)_2{}^{+1}$ ions except for a negligible 2×10^{-10} M.

$$K_{eq} = \frac{C_{\text{Ag}^{+1}} \times C^2{}_{\text{NH}_3{}^{+1}}}{C_{\text{Ag(NH}_3)_2{}^{+1}}}$$

Solving for C_{NH_3} and substituting numbers:

$$C^2{}_{\text{NH}_3} = \frac{K_{eq} \times C_{\text{Ag(NH}_3)_2{}^{+1}}}{C_{\text{Ag}^{+1}}} = \frac{6.8 \times 10^{-8} \times 3 \times 10^{-2}}{2 \times 10^{-10}} = 10.2$$

Taking the square roots of both sides:

$$C_{\text{NH}_3} = 3.2 \text{ moles/l.}$$

Then the total amount of NH_3 needed is $3.2 + 2 \times 0.03 = 3.26 \simeq 3.3$ M, since some of the NH_3 is added to the Ag^{+1}.

---

5. Simultaneous Equilibria

In the usual situation that occurs in laboratory and commercial practice, we find not one equilibrium but often two or even three simultaneous equilibria going on at once. A set of typical situations is outlined in the following examples:

A. Separation of Two Ions by Selective Precipitation with the Ions of a Weak Electrolyte. In a typical scheme in qualitative analysis, a solution may contain two ions such as Zn^{+2} and Cd^{+2} which form insoluble sulfide salts. However, the solubility of the sulfide salts are different, the ZnS being 300 times more soluble than CdS. This difference in solubility is sufficient to permit us to ppt. the less soluble CdS without pptg. the more soluble ZnS if we can control the S^{-2} ion concentration carefully. The S^{-2} ion is the ion of the weak acid H_2S, and its concentration may be controlled by controlling the concentration of the hydrogen ion (H_3O^{+1}) in the solution.

Example: A solution contains 0.04 M Cd^{+2} ions and 0.3 M Zn^{+2} ions. It is proposed to ppt. CdS by making the solution 0.1 M with H_2S. What concentration of H_3O^{+1} is needed to prevent the ZnS from pptg.?

$$K_{S.P.}(ZnS) = 4.5 \times 10^{-24}$$

$$K_{S.P.}(CdS) = 1.4 \times 10^{-28}$$

$$K_{ion}(H_2S) = 1.1 \times 10^{-22} \text{ (overall)}$$

Answer: Since we don't want ZnS to ppt., let us consider its equilibrium first:

$$K_{S.P.}(ZnS) = C_{Zn^{+2}} \times C_{S^{-2}}$$

Since the solution is 0.4 M in $C_{Zn^{+2}}$, we want to keep $C_{S^{-2}}$ sufficiently low so that the product $C_{Zn^{+2}} \times C_{S^{-2}}$ is less than $K_{S.P.}(ZnS)$. The maximum permissible $C_{S^{-2}}$ is obtained from the equation:

$$C_{S^{-2}} = \frac{K_{S.P.}(ZnS)}{C_{Zn^{+2}}} = \frac{4.5 \times 10^{-24}}{0.3} = 1.5 \times 10^{-23} \ M$$

Thus we must not have a $C_{S^{-2}}$ greater than $1.5 \times 10^{-23} \ M$. Now we turn to the relation between the S^{-2} ion and H^{+1} ion. Since we are given C_{H_2S} we use the equation for the overall ionization:

$$H_2S + 2H_2O \rightleftarrows 2H_3O^{+1} + S^{-2}$$

$$K_{ion}(H_2S) = \frac{C^2_{H_3O^{+1}} \times C_{S^{-2}}}{C_{H_2S}}$$

Solving for $C_{H_3O^{+1}}$:

$$C^2_{H_3O^{+1}} = \frac{C_{H_2S}}{C_{S^{-2}}} \times K_{ion}(H_2S)$$

Substituting:

$$C^2_{H_3O^{+1}} = \frac{0.1}{1.5 \times 10^{-23}} \times 1.1 \times 10^{-22} = 0.73$$

Taking square roots:

$$C_{H_3O^{+1}} = 0.85 \text{ mole}/l.$$

Thus, to keep $C_{S^{-2}}$ below $1.5 \times 10^{-23} \ M$, $C_{H_3O^{+1}}$ must be at least 0.85 M. Will this $C_{S^{-2}}$ be sufficient to ppt. CdS? We now turn to the $K_{S.P.}(CdS)$.

$$K_{S.P.}(CdS) = C_{Cd^{+2}} \times C_{S^{-2}}$$

Solving for $C_{Cd^{+2}}$:

$$C_{Cd^{+2}} = \frac{K_{S.P.}(CdS)}{C_{S^{-2}}}$$

Substituting:

$$C_{Cd^{+2}} = \frac{1.4 \times 10^{-28}}{1.5 \times 10^{-23}} = 9.3 \times 10^{-6} \ M$$

We conclude that $C_{S^{-2}} = 1.5 \times 10^{-23} \ M$ (which can be obtained by adjusting the acidity of the solution until $C_{H_3O^{+1}} = 0.85\ M$) will ppt. CdS effectively (compare $9.3 \times 10^{-6}\ M$ to the original concentration of $0.04\ M$) and still not ppt. ZnS.

Example: A solution contains $0.2\ M$ Mg^{+2} ions and $0.2\ M$ Ca^{+2} ions. Is it possible to separate Ca^{+2} by forming a ppt. of $CaCO_3$ without pptg. $MgCO_3$?

$$K_{S.P.}(CaCO_3) = 4.8 \times 10^{-9}$$

$$K_{S.P.}(MgCO_3) = 1 \times 10^{-5}$$

Answer: The maximum $C_{CO_3^{-2}}$ that we can have in the solution that will not ppt. $MgCO_3$ is given by

$$K_{S.P.}(MgCO_3) = C_{Mg^{+2}} \times C_{CO_3^{-2}}$$

Solving for $C_{CO_3^{-2}}$:

$$C_{CO_3^{-2}} = \frac{K_{S.P.}(MgCO_3)}{C_{Mg^{+2}}}$$

Substituting numbers:

$$C_{CO_3^{-2}} = \frac{1 \times 10^{-5}}{0.2} = 5 \times 10^{-5}\ \text{mole/l.}$$

Will this now be enough to ppt. the Ca^{+2} ion?

$$K_{S.P.}(CaCO_3) = C_{Ca^{+2}} \times C_{CO_3^{-2}}$$

Solving for $C_{Ca^{+2}}$:

$$C_{Ca^{+2}} = \frac{K_{S.P.}(CaCO_3)}{C_{CO_3^{-2}}}$$

Substituting numbers:

$$C_{Ca^{+2}} = \frac{4.8 \times 10^{-9}}{5 \times 10^{-5}} = 9.6 \times 10^{-5}\ \text{mole/l.}$$

Thus this concentration of CO_3^{-2} will ppt. practically all the original Ca^{+2} ions and the separation is possible. How can we obtain this $C_{CO_3^{-2}}$ in the solution? If we just add Na_2CO_3 reagent, we might easily add too much. The answer is that if the pH of the solution is controlled (i.e., control the H_3O^{+1}) we can control the CO_3^{-2} ion concentration even if excess Na_2CO_3 is added. This is generally done by using a buffer solution for the H_3O^{+1} ion, and it can be shown that an alkaline solution containing NH_3 and NH_4^{+1} ions can keep the $C_{CO_3^{-2}}$ within desired limits.

B. Selective Precipitation Using Complex Ions.   The control of the concentration of the reagent used to do the precipitating may be done by means of a complex ion rather than by the hydrogen ion. This is frequently the case when anions are separated.

Example: A solution contains $0.4\ M$ Cl^{-1} and $0.4\ M$ I^{-1}. Is it possible to separate the I^{-1} by forming a ppt. of AgI without pptg. $AgCl$?

$$K_{S.P.}(AgCl) = 1.7 \times 10^{-10}$$

$$K_{S.P.}(AgI) = 8.5 \times 10^{-17}$$

Answer: The maximum $C_{Ag^{+1}}$ that will not ppt. AgCl is given by

$$K_{S.P.}(AgCl) = C_{Ag^{+1}} \times C_{Cl^{-1}}$$

Solving for $C_{Ag^{+1}}$ and substituting:

$$C_{Ag^{+1}} = \frac{K_{S.P.}(AgCl)}{C_{Cl^{-1}}} = \frac{1.7 \times 10^{-10}}{0.4} = 4.3 \times 10^{-10} \, M$$

Is this sufficient to ppt. the I^{-1} ion?

$$K_{S.P.}(AgI) = C_{Ag^{+1}} \times C_{I^{-1}}$$

Solving for $C_{I^{-1}}$ and substituting numbers:

$$C_{I^{-1}} = \frac{K_{S.P.}(AgI)}{C_{Ag^{+1}}} = \frac{8.5 \times 10^{-17}}{4.3 \times 10^{-10}} = 2.0 \times 10^{-7} \, M.$$

Thus a $C_{Ag^{+1}} = 4.3 \times 10^{-10} \, M$ is sufficient to ppt. all the I^{-1} ions (compare 2.0×10^{-7} with 0.4) without pptg. the Cl^{-1} ion.

How can we adjust the $C_{Ag^{+1}}$ to this value? The normal stock solutions are about 0.01 M! The answer is that we must use a buffered solution of Ag^{+1}. A 0.01 M solution of $Ag(NH_3)_2^{+1}$ containing excess NH_3 may be used as a reagent. What excess of NH_3 must it have?

$$K_{eq.}[Ag(NH_3)_2^{+1}] = \frac{C_{Ag^{+1}} \times C^2_{NH_3}}{C_{Ag(NH_3)_2^{+1}}}$$

Solving for C_{NH_3}:

$$C^2_{NH_3} = C_{Ag(NH_3)_2^{+1}} \times \frac{K_{eq.}}{C_{Ag^{+1}}}$$

and substituting numbers:

$$C^2_{NH_3} = 0.01 \times \frac{6.8 \times 10^{-7}}{4.3 \times 10^{-10}} = 15.8$$

or

$$C_{NH_3} = 4 \, M$$

Thus a solution of Ag^{+1} in 4 M NH_3 may be used to ppt. the I^{-1} ion without pptg. the Cl^{-1} ion. With this reagent there is no danger that adding an excess will cause the Cl^{-1} ion to ppt.

Example: A solution contains 0.04 M $Cd(CN)_4^{-2}$ and 0.06 M $Cu(CN)_3^{-2}$ complex ions together with an excess of CN^{-1} ions = 0.2 M. If the solution is now treated with Na_2S until $C_{S^{-2}} = 1 \times 10^{-3}$, will either ion ppt. as the sulfide?

$$K_{S.P.}(CdS) = 1.4 \times 10^{-28} \qquad K_{S.P.}(Cu_2S) = 2.5 \times 10^{-50}$$

$$K_{eq.}[Cd(CN)_4^{-2}] = 1.4 \times 10^{-17} \qquad K_{eq.}[Cu(CN)_3^{-2}] = 5 \times 10^{-28}$$

Answer: Let us first determine the concentrations of Cu^{+1} and Cd^{+2} ions in the buffered solution.

$$K_{\text{eq.}} [Cu(CN)_3{}^{-2}] = \frac{C_{Cu^{+1}} \times C^3{}_{CN^{-1}}}{C_{Cu(CN)_3{}^{-2}}}$$

Solving for $C_{Cu^{+1}}$:

$$C_{Cu^{+1}} = \frac{C_{Cu(CN)_3{}^{-2}}}{(C_{CN^{-1}})^3} \times K_{\text{eq.}}$$

Substituting numbers:

$$C_{Cu^{+1}} = \frac{0.06}{(0.2)^3} \times 5 \times 10^{-28}$$

$$= \frac{6 \times 10^{-2}}{8 \times 10^{-3}} \times 5 \times 10^{-28}$$

$$C_{Cu^{+1}} = 3.8 \times 10^{-27} \ M$$

$$K_{\text{eq.}} [Cd(CN)_4{}^{-2}] = \frac{C_{Cd^{+2}} \times C^4{}_{CN^{-1}}}{C_{Cd(CN)_4{}^{-2}}}$$

Solving for $C_{Cd^{+2}}$:

$$C_{Cd^{+2}} = \frac{C_{Cd(CN)_4{}^{-2}}}{(C_{CN^{-1}})^4} \times K_{\text{eq.}}$$

Substituting numbers:

$$C_{Cd^{+2}} = \frac{0.04}{(0.2)^4} \times 1.4 \times 10^{-17}$$

$$= \frac{4 \times 10^{-2} \times 1.4 \times 10^{-17}}{1.6 \times 10^{-3}}$$

$$C_{Cd^{+2}} = 3.5 \times 10^{-16} \ M$$

Let us now see if the $C_{S^{-2}} = 1 \times 10^{-3} \ M$ is sufficient to ppt. either ion.

$C^2{}_{Cu^{+1}} \times C_{S^{-2}}$

$\quad = (3.8 \times 10^{-27})^2 \times 1 \times 10^{-3}$

$\quad = 14.5 \times 10^{-57}$

But this is much less than $K_{\text{S.P.}}(Cu_2S) = 2.5 \times 10^{-50}$. Thus Cu_2S will not ppt.

$C_{Cd^{+2}} \times C_{S^{-2}}$

$\quad = 3.5 \times 10^{-16} \times 1 \times 10^{-3}$

$\quad = 3.5 \times 10^{-19}$

But this is much greater than $K_{\text{S.P.}}(CdS) = 1.4 \times 10^{-28}$. Thus the CdS will ppt.

We conclude that under these conditions CdS will ppt. whereas Cu_2S will not, and so the separation of Cu^{+1} and Cd^{+2} is possible.

Example: A solution containing excess $NH_4{}^{+1}$ ions and NH_3 will be buffered with respect to the OH^{-1} ion. What should the composition of such a reagent be if it is to be used to separate the 0.05 M Mg^{+2} from the 0.08 M Ni^{+2} ion?

$$K_{\text{S.P.}}[Mg(OH)_2] = 5.5 \times 10^{-12}$$

$$K_{\text{S.P.}}[Ni(OH)_2] = 1.6 \times 10^{-14}$$

$$K_{\text{ion}}(NH_4{}^+) = 5.5 \times 10^{-10}$$

Answer: Since $Mg(OH)_2$ is more soluble than $Ni(OH)_2$, we want to have the maximum $C_{OH^{-1}}$ present that will not ppt. $Mg(OH)_2$.

$$K_{\text{S.P.}}[Mg(OH)_2] = C_{Mg^{+1}} \times C^2{}_{OH^{-1}}$$

Solving for $C_{OH^{-1}}$ and substituting:

$$C^2{}_{OH^{-1}} = \frac{K_{\text{S.P.}}}{C_{Mg^{+2}}} = \frac{5.5 \times 10^{-12}}{0.05} = 1.1 \times 10^{-10}$$

Taking square roots:

$$C_{OH^{-1}} = 1.05 \times 10^{-5} \ M$$

Thus we want our solution to contain $C_{OH^{-1}}$ of 1.05×10^{-5} M or less. Is this enough to ppt. $Ni(OH)_2$?

$$K_{S.P.}[Ni(OH)_2] = C_{Ni^{+2}} \times C^2_{OH^{-1}}$$

Solving for $C_{Ni^{+2}}$ and substituting:

$$C_{Ni^{+2}} = \frac{K_{S.P.}}{C^2_{OH^{-1}}} = \frac{1.6 \times 10^{-14}}{(1.05 \times 10^{-5})^2} = \frac{1.6 \times 10^{-14}}{1.1 \times 10^{-10}} = 1.45 \times 10^{-4} \; M$$

$$\% \; Ni^{+2} \; \text{left in solution} = \frac{1.45 \times 10^{-4}}{0.08} \times 100 = 0.18\%$$

That is the $C_{OH^{-1}} = 1.05 \times 10^{-5}$ M will ppt. all but 0.18% of the Ni^{+2} and will not ppt. any of the Mg^{+2} ions. This is certainly close to the limits which we would permit but is satisfactory for qualitative analysis.

Finally, what composition of NH_4^{+1}, NH_3 buffer should we use to give $C_{OH^{-1}} = 1.05 \times 10^{-5}$ M?

$$K_{hyd.}(NH_3) = \frac{C_{NH_4^{+1}} \times C_{OH^{-1}}}{C_{NH_3}} = \frac{K_{ion}(HOH)}{K_{ion}(NH_4^+)}$$

Solving for the ratio $C_{NH_4^{+1}}/C_{NH_3}$ and substituting numbers:

$$\frac{C_{NH_4^{+1}}}{C_{NH_3}} = \frac{K_{hyd.}}{C_{OH^{-1}}} = \frac{1.8 \times 10^{-5}}{1.05 \times 10^{-5}} = 1.7$$

Thus any solution in which $C_{NH_4^{+1}}/C_{NH_3} = 1.7$ will do. A solution containing $C_{NH_4^{+1}} = 1.7$ M and $C_{NH_3} = 1.0$ M will work.

6. Summary—Principles of Selective Precipitation

In the last section we have illustrated methods for separating ions based on differences in solubilities of their salts. The conditions under which these separations are possible are:

1. *If the solubilities of the salts differ sufficiently, then one of the salts can be made to precipitate, leaving the other in solution if the precipitating ion is not present in too high concentrations. Too high concentrations of the precipitating ion may be avoided if this ion can be buffered. It may be buffered if it will form a weak acid or complex ion.*

2. *If the solubilities of the salts of the ions do not differ sufficiently, then it may be possible to reduce the concentration of one of the ions to the point where it is not present in high enough quantities to precipitate. This is possible if one of the ions forms a stronger complex than the other.*

7. Problems

1. Write equations for the stepwise and overall ionizations of each of the following. Also write a K_{ion} for each equation:

(a) H_3PO_4.

(b) H_2SO_3.

(c) $H_2SiO_4^{-2}$(ion).

(d) $Al(OH)_2^{+1}$(ion).

2. What are the concentrations of the ions present in a 0.05 M solution of $H_2B_4O_7$, given $K_{ion}(H_2B_4O_7) = 1 \times 10^{-4}$, $K_{ion}(HB_4O_7^{-1}) = 1 \times 10^{-9}$?

3. What are the concentrations of the ions present in a 0.8 M solution of H_2Se, given $K_{ion}(H_2Se) = 1.7 \times 10^{-4}$, $K_{ion}(HSe^{-1}) = 1 \times 10^{-10}$?

4. What is the pH and the percentage of hydrolysis of a 0.05 M solution of $NaHCO_3$? $K_{ion}(H_2CO_3) = 4.3 \times 10^{-7}$; $K_{ion}(HCO_3^{-1}) = 4.7 \times 10^{-11}$?

5. What is the pH and the percentage of hydrolysis of a 0.02 M solution of $NaHS$? $K_{ion}(H_2S) = 1.1 \times 10^{-7}$; $K_{ion}(HS^{-1}) = 1 \times 10^{-15}$?

6. What is the concentration of PO_4^{-3} ions in a 0.06 M solution of Na_2HPO_4? $K_{ion}(H_2PO_4^{-1}) = 6.2 \times 10^{-8}$; $K_{ion}(HPO_4^{-2}) = 1 \times 10^{-12}$. What is the pH of the solution?

7. What is the concentration of Hg^{+2} ions in a 0.3 M solution of $K_2Hg(CN)_4$? $K_{eq.}[Hg(CN)_4^{-2}] = 4 \times 10^{-42}$.

8. What is the concentration of Ag^{+1} ions in a 0.6 M solution of $NaAg(CN)_2$? $K_{eq.}[Ag(CN)_2^{-1}] = 1 \times 10^{-21}$. What is the $C_{CN^{-1}}$?

9. What concentration of NH_3 will reduce the $C_{Cd^{+2}}$ to 1×10^{-6} M in a solution of 0.3 M $Cd(NH_3)_4^{+2}$ ions? $K_{eq.}[Cd(NH_3)_4^{+2}] = 1 \times 10^{-7}$.

10. A solution of 0.4 M $HgCl_4^{=}$ ions is made 0.3 M in Cl^{-1} ions. What is the $C_{Hg^{+2}}$ in this solution? $K_{eq.}(HgCl_4^{=}) = 6 \times 10^{-17}$.

11. What $C_{S^{-2}}$ ion will effect the maximum separation of 0.3 M Mn^{+2} ions and 0.05 M Cd^{+2} ions? $K_{S.P.}(CdS) = 1.4 \times 10^{-28}$; $K_{S.P.}(MnS) = 5.6 \times 10^{-16}$. What $C_{H_3O}^{+1}$ in a 0.1 M H_2S solution will produce this $C_{S^{-2}}$ ion? What percentage of Cd^{+2} will be left in solution?

12. For each of the following pairs of ions, find a single ion that will ppt. one in the presence of the other when both are present in 0.1 M concentration. (Use texts for data on properties.)

(a) Br^{-1} and Cl^{-1}. (d) SO_4^{-2} and CO_3^{-2}.
(b) Ag^{+1} and Cu^{+2}. (e) CrO_4^{-2} and SO_4^{-2}.
(c) Sn^{+2} and Hg^{+2}. (f) Zn^{+2} and Ni^{+2}.

13. A solution contains 0.03 M Zn^{+2} ions and 0.02 M Mn^{+2} ions. Can these be separated by pptg. ZnS? $K_{S.P.}(ZnS) = 4.5 \times 10^{-24}$; $K_{S.P.}(MnS) = 5.6 \times 10^{-16}$. What $C_{S^{-2}}$ is needed for best separation? If a 0.1 M solution of H_2S is used, what $C_{H_3O^{+1}}$ is needed to give this $C_{S^{-2}}$? $K_{ion}(H_2S) = 1.1 \times 10^{-22}$.

14. What $C_{CO_3^{-2}}$ is needed to ppt. the 0.003 M Cd^{+2} ion in the presence of the 0.04 M Ca^{+2} ion. $K_{S.P.}(CdCO_3) = 2.5 \times 10^{-14}$; $K_{S.P.}(CaCO_3) = 4.8 \times 10^{-9}$. What percentage of Cd^{+2} is left in solution? What $C_{H_3O^{+1}}$ must be present in a 0.1 M solution of $NaHCO_3$ to give this $C_{CO_3^{-2}}$? $K_{ion}(HCO_3^{-1}) = 4.7 \times 10^{-11}$.

15. A solution contains 0.08 M Ba^{+2} ions and 0.08 M Sr^{+2} ions. What $C_{CrO_4^{-2}}$ ion will ppt. the maximum amount of $BaCrO_4$ without pptg. $SrCrO_4$? $K_{S.P.}(BaCrO_4) = 2 \times 10^{-10}$; $K_{S.P.}(SrCrO_4) = 3.6 \times 10^{-5}$. What must the pH of a 0.2 M K_2CrO_4 solution be to give the $C_{CrO_4^{-2}}$ ion? $K_{ion}(HCrO_4^{-1}) = 3.2 \times 10^{-7}$.

16. Can 0.1 M Br^{-1} ions and 0.1 M I^{-1} ions be separated using Ag^{+1} as a reagent? What must be $C_{Ag^{+1}}$ be to effect maximum separation? What excess concentration of $S_2O_3^{-2}$ ions must be present in a solution containing 0.1 M $Ag(S_2O_3)_2^{-3}$ ions to buffer the Ag^{+1} to give the desired concentration? $K_{S.P.}(AgBr) = 3.3 \times 10^{-13}$; $K_{S.P.}(AgI) = 8.5 \times 10^{-17}$; $K_{eq.}[Ag(S_2O_3)_2^{-3}] = 1 \times 10^{-13}$.

17. What concentrations of Ac^{-1} ions and HAc would you use to give a buffer solution of OH^{-1} ions that would ppt. the 0.1 M Fe^{+3} ion as $Fe(OH)_3$ without pptg. the 0.1 M Cr^{+3} ion. $K_{S.P.}[Cr(OH)_3] = 6.7 \times 10^{-31}$; $K_{S.P.}[Fe(OH)_3] = 4 \times 10^{-38}$; $K_{ion}(NH_4^+) = 5.5 \times 10^{-10}$.

CHAPTER XV

Oxidation and Reduction

1. Multivalence

In Chapter VII we discussed the valences of elements and groups present in binary compounds, that is, compounds behaving as though they contained only two distinct groups.

It was pointed out at the end of this chapter that many elements display the property of multivalence. That is, they have different valences (combining power) in different binary compounds. Thus nitrogen (N) may form NO (valence = 2), N_2O (valence = 1); N_2O_3 (valence = 3); NO_2 (valence = 4); N_2O_5 (valence = 5).

There are many reactions in which the apparent binary valence undergoes a change. To these reactions, our principle of equivalence will not be directly applicable. In order to develop methods for handling such reactions conveniently we shall enlarge our simple theory of binary valence so that it may deal with reactions in which there are changes in valence. These reactions are called oxidation-reduction reactions and may be related to processes in which electrons are transferred.

2. A New System of Valences—Oxidation Numbers

In the new system of valences we shall now develop, we shall start by creating a standard and a set of rules for assigning valence. To distinguish these new valences from the old binary valences, we shall call them oxidation numbers.

Rules: 1. *All free elements shall be assigned the oxidation number of zero (to indicate that they are not in combination).*

    2. *When H is present in a compound, it will be assigned an oxidation number of $+1$ (except in hydrides, when it is -1).*

    3. *The sum of all oxidation numbers of the elements present in a compound must be zero (to indicate that all valences are used up).*

    4. *The sum of all oxidation numbers in a complex ion or radical must equal the charge on the ion or radical (to indicate the residual combining power of the ion or radical).*

From these rules we can now proceed to find the oxidation numbers of all the other elements. We see that hydrogen is our standard.

Thus, in H_2O the oxidation number of O is -2 since 2H make $+2$ and H_2O is a compound in which all oxidation numbers must be zero. In H_2O_2 the oxidation number of O is -1.

Corollary: In compounds, O will always have the oxidation number -2 except in peroxide, where it is -1.

Proceeding in this manner we see that Al is $+3$ in Al_2O_3; Mn is $+2$ in MnO, but $+4$ in MnO_2; Fe is $+2$ in FeO, but $+3$ in Fe_2O_3; S is $+4$ in SO_2, but -2 in H_2S.

To take a few radicals: Cl is $+1$ in ClO^{-1}, $+3$ in the ClO_2^{-1} ion, $+5$ in the ClO_3^{-1} ion, $+7$ in ClO_4^{-1}, but -1 in the Cl^{-1} ion. Cr is $+3$ in the Cr^{+3} ion, but $+6$ in the CrO_4^{-2} ion. Mn is $+2$ in the Mn^{+2} ion, but $+7$ in the MnO_4^{-1} ion.

These rules permit us to assign oxidation numbers to all elements in any type of compound, ion or radical. These numbers have a $+$ or $-$ sign, thus differing from our binary valence numbers.

The oxidation number indicates the combining power (based on H $= +1$) of the elements in any state of combination. The units of oxidation valence, so defined, are equivalents per mole. Thus the oxidation number of Zn in Zn^{+2} ion is $+2$ equivalents Zn^{+2} per mole Zn^{+2}.

Do problem 1 at the end of the chapter.

3. Oxidation and Reduction

Definition: When the oxidation number of an element increases in a reaction, the element is said to be oxidized. When the oxidation number of an element is lowered in a reaction, the element is said to be reduced.

Thus, when O_2 gas reacts with Zn (metal) to form ZnO:

$$2Zn + O_2 \rightarrow 2ZnO$$

zinc has increased in oxidation number (from zero in free Zn to $+2$ in ZnO) and oxygen has decreased (from zero in free O_2 to -2 in ZnO). We say that Zn has been oxidized and oxygen has been reduced. Because of the rule that all oxidation numbers in a compound must add to zero, we will find that oxidation and reduction always occur together and in equivalent amounts. If one element is oxidized, another element must always be reduced to maintain the balance of oxidation numbers.   Thus in the above case for every Zn atom that increased from 0 to $+2$, an equal number of O atoms had to decrease from 0 to -2 to preserve the balance of zero in ZnO.

Oxidation-reduction reactions (redox for short) may always be recognized by computing the valences of all the elements in each substance.   If one of these has undergone a change, then the reaction is a redox reaction.   If there is no such change, then the reaction is not a redox reaction.

In the following illustrations the oxidation numbers of the elements are written in parentheses.   The equations are skeleton equations and not balanced.

$$\underset{(+2)\ (-2)}{Ca\ O} + \underset{(+1)(-2)}{H_2\ O} \rightarrow \underset{(+2)\ (-2)(+1)}{Ca\,(O\ H)_2} \qquad \text{(not redox)}$$

$$\underset{(+1)(+5)(-2)}{Ag\ N\ O_3} + \underset{(+1)(-1)}{Na\ Cl} \rightarrow \underset{(+1)(-1)}{Ag\ Cl} \downarrow + \underset{(+1)(+5)(-2)}{Na\ N\ O_3} \qquad \text{(not redox)}$$

$$\underset{(0)}{Al} + \underset{(0)}{O_2} \rightarrow \underset{(+3)(-2)}{Al_2\ O_3} \qquad \text{(redox)}$$

$$\underset{(+2)(-2)}{N\ O} + \underset{(0)}{O_2} \rightarrow \underset{(+4)(-2)}{N\ O_2} \qquad \text{(redox)}$$

$$\underset{(0)}{Cu} + \underset{(+1)(+5)(-2)}{H\ N\ O_3} \rightarrow \underset{(+2)(+5)\ (-2)}{Cu\,(N\ O_3)_2} + \underset{(+2)(-2)}{N\ O} + \underset{(+1)(-2)}{H_2\ O} \quad \text{(redox)}$$

$$\underset{(+6)\ (-2)}{Cr\ O_4^{-2}} + \underset{(+1)\ (-2)}{H_3\ O^{+1}} \rightleftarrows \underset{(+6)\ (-2)}{Cr_2\ O_7^{-2}} + \underset{(+1)(-2)}{H_2\ O} \qquad \text{(not redox)}$$

$$\underset{(0)}{Cl_2} + \underset{(+1)(-2)}{H_2\ O} \rightleftarrows \underset{(+1)(-1)}{H\ Cl} + \underset{(+1)(+1)(-2)}{H\ Cl\ O} \qquad \text{(redox)}$$

$$\underset{(0)}{Zn^0} + \underset{(+1)}{Ag^{+1}} \rightarrow \underset{(+2)}{Zn^{+2}} + \underset{(0)}{Ag^0} \downarrow \qquad \text{(redox)}$$

4. Redox Reactions as Electron Transfers

The changes in oxidation numbers can be related to transfers of electrons and consequent changes in charge.   Thus when metallic Zn reacts with Ag^{+1} ions to liberate free Ag metal and produce Zn^{+2} ions, we can look at this process from the point of view of electron transfer.

The Zn metal (zero oxidation number) went to the Zn^{+2} ion ($+2$ oxidation number).   This occurred by each Zn atom's losing 2 negatively charged electrons:

$$Zn^0 \rightarrow Zn^{+2} + 2e^{-1}$$

Similarly:

$$1e^- + Ag^{+1} \rightarrow Ag^0$$

We will soon see that all such changes in oxidation number can be associated with transfers of electrons.   We thus expand our definition.

Definitions: Oxidation (the increase in oxidation number) is a loss of electrons. *Reduction* (the decrease in oxidation number) is a gain of electrons.

This association of oxidation and reduction with transfer of electrons leads to a completely consistent scheme for discussing these reactions. It has the added advantage of permitting us to write individual equations to represent the oxidation and reduction and thus directly indicating the electron transfers. Finally, it enables us to include in a single scheme not only redox reactions but also the individual electrode reactions which occur in electrolysis.

For these reasons, the concept of electron transfer has become a most valuable one for discussing redox reactions, and we shall now proceed to a discussion of the conventions employed in writing these individual electron-transfer equations or ion-electron equations as they are called.

5. Ion-Electron Reactions

We shall now develop a method of separating a skeleton redox equation into a pair of ion-electron equations representing the oxidation and the reduction. The procedure is as follows:

1. The skeleton equation must be given! This is generally obtained from direct experiment.
2. Write oxidation numbers above each element.
3. Pick out the ion, molecule, or radical containing the element undergoing oxidation and the ion, molecule, or radical containing the element in its oxidized form. This pair will form the basis for the oxidation equation.
4. Select the similar pair of substances containing the element undergoing reduction and its reduced form. These will form the basis for the reduction equation.
5. The individual oxidation and reduction equations are now balanced; first chemically and finally electrically by means of electrons.

Example:

$$\overset{(0)}{Al} + \overset{(+2)(-1)}{Cu\ Cl_2} \rightarrow \overset{(+3)(-1)}{Al\ Cl_3} + \overset{(0)}{Cu} \qquad \text{(skeleton equation)}$$

Al undergoes oxidation, going from zero to $+3$. Cu undergoes reduction, going from $+2$ to zero. We write:

Oxidation: $\qquad\qquad\qquad\qquad Al \rightarrow Al^{+3} \qquad\qquad$ (skeleton equation)

Note: We write the Al^{+3} ion rather than the $AlCl_3$ molecule, since it is the Al^{+3} ion that exists in water solution.

This equation is balanced chemically. It may be balanced electrically by adding 3 electrons (e^{-1}):

Oxidation: $Al \to Al^{+3} + 3e^{-1}$ (balanced ion-electron equation)

The equation means: 1 mole of Al (or 1 atom Al) yields 1 mole of Al^{+3} ion (or 1 ion of Al^{+3}) plus 3 moles of electrons (or 3 electrons). It is balanced electrically since the total charge is the same on both sides. In a similar fashion we can write:

Reduction: $2e^{-1} + Cu^{+2} \to Cu$ (balanced)

Example:

$$\overset{(+1)(-2)}{H_2\ S} + \overset{(0)}{Cl_2} \to \overset{(+1)(-1)}{H\ Cl} + \overset{(0)}{S} \quad \text{(skeleton equation)}$$
(reaction in water solution)

Proceeding as before, S is oxidized from -2 to zero. Cl is reduced from 0 to -1.

Reduction: $Cl_2 \to Cl^{-1}$ (skeleton equation)

Note: We write Cl_2 since it is Cl_2 gas which contains the Cl. Also, we write Cl^{-1} rather than HCl, since in water solution, HCl is ionized into Cl^{-1} ions.

The balanced equation is:

Reduction: $2e^{-1} + Cl_2 \to 2Cl^{-1}$

The oxidation equation is:

Oxidation: $H_2S \to S$ (skeleton equation)

Note: We write H_2S rather than S^{-2} ion, since H_2S is a weak electrolyte.

To balance this equation chemically we must add $2H^{+1}$ ions to the right side. (We will write H^{+1} instead of H_3O^{+1} for the sake of simplicity.) A look at the skeleton equation tells us that hydrogen ion is indeed a product since the acid HCl is produced. Thus

Oxidation: $H_2S \to S + 2H^{+1}$

To balance electrically we add 2 electrons.

Oxidation: $H_2S \to S + 2H^{+1} + 2e^{-1}$ (balanced)

---

6. Single Ion-Electron Reactions

From the preceding we see that it is always possible to write a single ion-electron equation if we know the state of the substance before and after the transfer of electrons.

Example: Write a balanced ion-electron equation for the oxidation of $FeCl_2$ to $FeCl_3$.

Answer:

$$Fe^{+2} \to Fe^{+3} + 1e^{-1} \qquad \text{(oxidation)}$$

Example: Write a balanced equation for the reduction of $Hg(NO_3)_2$ to $Hg_2(NO_3)_2$.

Answer:

$$2e^{-1} + 2Hg^{+2} \rightarrow Hg_2^{+2} \qquad \text{(reduction)}$$

Example: Write a balanced equation for the oxidation of H_2SO_4 to $H_2S_2O_8$.

Answer:

$$2SO_4^{-2} \rightarrow S_2O_8^{-2} + 2e^{-1} \qquad \text{(oxidation)}$$

Example: Write a balanced equation for the oxidation of Cr^{+3} ions to CrO_4^{-2} ions.

Answer:

Oxidation: $\qquad\qquad Cr^{+3} \rightarrow CrO_4^{-2} \qquad\qquad$ (skeleton)

Now clearly this cannot be balanced chemically unless we add something to supply the O atoms for the CrO_4^{-2} ion. In such cases we use the ever-present H_2O to supply this oxygen. But we will have H left over. We leave this as H^{+1} ions. The equation thus becomes:

Oxidation: $\qquad Cr^{+3} + 4H_2O \rightarrow CrO_4^{-2} + 8H^{+1} \quad$ (balanced chemically)

To balance electrically we need 3 electrons:

Oxidation: $\qquad Cr^{+3} + 4H_2O \rightarrow CrO_4^{-2} + 8H^{+1} + 3e^{-1} \qquad$ (balanced)

In the above examples, we can see illustrated a few conventions which are used in writing ion-electron equations. These are applicable to water solutions with which we shall be dealing exclusively.

1. Always write the element in the form of the species which actually exists in water solution (ion or molecule).
2. Since H^{+1}, OH^{-1}, and H_2O always exist in water solution, we can use them to aid in balancing equations chemically.

Example: Write the equation for the oxidation of H_2O to O_2 gas.

Answer:

Oxidation: $\qquad\qquad H_2O \rightarrow O_2\uparrow \qquad\qquad$ (skeleton equation)

$\qquad\qquad\qquad 2H_2O \rightarrow O_2\uparrow \qquad\qquad$ (step 1)

$\qquad\qquad\qquad 2H_2O \rightarrow O_2\uparrow + 4H^{+1} \qquad\qquad$ (step 2)

Oxidation: $\qquad\qquad 2H_2O \rightarrow O_2 + 4H^{+1} + 4e^{-1} \qquad\qquad$ (balanced)

Example: Write the equation for the reduction of $KMnO_4$ to $MnCl_2$.

Answer:

Oxidized species: $\overset{(+7)}{MnO_4^{-1}}$ ions $\qquad\qquad$ *Reduced species:* $\overset{(+2)}{Mn^{+2}}$ ions

Reduction: $MnO_4^{-1} \rightarrow Mn^{+2}$ (skeleton equation)

$MnO_4^{-1} \rightarrow Mn^{+2} + 4H_2O$ (step 1)

$8H^{+1} + MnO_4^{-1} \rightarrow Mn^{+2} + 4H_2O$ (step 2)

$5e^{-1} + 8H^{+1} + MnO_4^{-1} \rightarrow Mn^{+2} + 4H_2O$ (balanced)

Do problem 2 at the end of the chapter.

7. Balancing Redox Equations

The principle of equivalence may now be stated for redox equations:

The number of electrons produced by an oxidation must be equal to the number of electrons used by the reduction.

This principle now enables us to balance redox equations.   The procedure is as follows:

1. Label the oxidation states of all elements in the skeleton equation.
2. From these, pick out the oxidized and reduced forms of each element undergoing changes and write individual, balanced ion-electron equations for the oxidation and reduction.
3. Add the oxidation and reduction steps in such fashion that the electrons cancel on both sides.
4. Cancel quantities that appear on both sides.

Example: Balance the following skeleton equation:

$$HCl + KMnO_4 \rightarrow Cl_2\uparrow + KCl + MnCl_2 + H_2O$$

Answer: Oxidation: Cl^{-1} (-1) goes to Cl_2 (gas) (0).
Reduction: Mn $(+7)$ in the MnO_4^{-1} ion goes to the Mn^{+2} ion $(+2)$.

Oxidation		Reduction	
(step 1)	$Cl^{-1} \rightarrow Cl_2$	$MnO_4^{-1} \rightarrow Mn^{+2}$	(step 1)
(step 2)	$2Cl^{-1} \rightarrow Cl_2$	$MnO_4^{-1} \rightarrow Mn^{+2} + 4H_2O$	(step 2)
(balanced)	$2Cl^{-1} \rightarrow Cl_2 + 2e^{-1}$	$8H^{+1} + MnO_4^{-1} \rightarrow Mn^{+2} + 4H_2O$	(step 3)
		$5e^{-1} + 8H^{+1} + MnO_4^{-1} \rightarrow Mn^{+2} + 4H_2O$	(balanced)

We note now that each oxidation releases $2e^{-1}$ and each reduction requires $5e^{-1}$.   Thus 5 oxidation steps will provide just enough electrons (10) for 2 reduction steps.   We thus combine them in the ratio of 5 to 2:

Oxidation: $2Cl^{-1} \rightarrow Cl_2 + 2e^{-1}$ (multiplying by 5)

Reduction: $5e^{-1} + 8H^{+1} + MnO_4^{-1} \rightarrow Mn^{+2} + 4H_2O$ (multiplying by 2) and add:

$$10e^{-1} + 10Cl^{-1} + 16H^{+1} + 2MnO_4^{-1} \rightarrow 5Cl_2 + 2Mn^{+2} + 8H_2O + 10e^{-1}$$

and, canceling electrons, we obtain the balanced ionic equation:

$$10Cl^{-1} + 16H^{+1} + 2MnO_4^{-1} \rightarrow 5Cl_2 + 2Mn^{+2} + 8H_2O$$

Generally the balanced ionic equation is all we want or will use since the ionic equation tells us the ions and molecules that actually take part in the reaction. Thus the above equation says that $2MnO_4^{-1}$ ions are needed for 16 H^{+1} ions and 10 Cl^{-1} ions. The K^{+1} ions provided by $KMnO_4$ do not actually take part in the reaction, and we could equally well have used $NaMnO_4$ or $LiMnO_4$ to provide the MnO_4^{-1} ions. Similarly, although in the original equation HCl is the source of H^{+1} ions and Cl^{-1} ions, the ionic equation tells us that, of 16 molecules of HCl, only 10 Cl^{-1} ions and 16 H^{+1} ions will be used. The extra 6 Cl^{-1} do not take part in the reaction, and, indeed, we see from the skeleton equation that some of the original Cl^{-1} ions end up in KCl and $MnCl_2$. These latter Cl^{-1} ions have not been oxidized!

Example: Balance the following skeleton equation:

$$SO_2 + Na_2Cr_2O_7 + H_2SO_4 \rightarrow Na_2SO_4 + Cr_2(SO_4)_3 + H_2O$$

Answer: Oxidation: $S(+4)$ in SO_2 goes to $S(+6)$ in the SO_4^{-2} ion.
Reduction: $Cr(+6)$ in the $Cr_2O_7^{-2}$ ion goes to $Cr(+3)$ in the Cr^{+3} ion.

	Oxidation		Reduction	
(step 1)	$SO_2 \rightarrow SO_4^{-2}$	$Cr_2O_7^{-2} \rightarrow Cr^{+3}$	(step 1)
(step 2)	$2H_2O + SO_2 \rightarrow SO_4^{-2}$	$Cr_2O_7^{-2} \rightarrow 2Cr^{+3} + 7H_2O$	(step 2)
(step 3)	$2H_2O + SO_2 \rightarrow SO_4^{-2} + 4H^{+1}$	$14H^{+1} + Cr_2O_7^{-2} \rightarrow 2Cr^{+3} + 7H_2O$	(step 3)
(balanced)	$2H_2O + SO_2 \rightarrow SO_4^{-2} + 4H^{+1} + 2e^{-1}$	$6e^{-1} + 14H^{+1} + Cr_2O_7^{-2} \rightarrow 2Cr^{+3} + 7H_2O$	(balanced)

Oxidation: $2H_2O + SO_2 \rightarrow SO_4^{-2} + 4H^{+1} + 2e^{-1}$ $(\times 3)$

Reduction: $6e^{-1} + 14H^{+1} + Cr_2O_7^{-2} \rightarrow 2Cr^{+3} + 7H_2O$ $(\times 1)$

and add

$$6e^{-1} + 6H_2O + 3SO_2 + 14H^{+1} + Cr_2O_7^{-2} \rightarrow$$
$$3SO_4^{-2} + 12H^{+1} + 2Cr^{+3} + 7H_2O + 6e^{-1}$$

Canceling $6e^{-1}$ and subtracting $6H_2O$ and $12H^{+1}$ from both sides:

$$3SO_2 + 2H^{+1} + Cr_2O_7^{-2} \rightarrow 3SO_4^{-2} + 2Cr^{+3} + H_2O$$

Note: If we add $2Na^{+1}$ and $1SO_4^{-2}$ to both sides we can obtain the balanced molecular equation:

$$3SO_2 + H_2SO_4 + Na_2Cr_2O_7 \rightarrow Na_2SO_4 + Cr_2(SO_4)_3 + H_2O$$

Do problem 3 at the end of the chapter.

8. Principle of Equivalence for Redox Reactions

If we have the balanced ionic equation we can do problems with them just as we have done before.   However, the principle of equivalence permits us to do such calculations without the balanced equation if only we know the oxidized and reduced states of each substance. The principle is:

Equal numbers of equivalents of oxidizing and reducing substance will always react with each other to produce equal numbers of equivalents of oxidized and reduced products.

To use this law we must reexamine what we mean by an equivalent of oxidizing or reducing agent.

Definition: The *equivalent weight* of a substance taking part in a redox reaction is that weight of the substance which will accept (oxidized form) or lose (reduced form) 1 mole of electrons in the reaction.

Algebraically:

$$\text{Equivalent weight} = \frac{\text{Molecular weight}}{\text{Moles of electrons transferred per mole of substance}}$$

Definition:

Redox valence = Number of moles of electrons transferred per mole of substance

Combing these definitions, we can write a relation connecting the units of equivalents with units of moles for a redox material:

$$\text{Redox equivalents} = \text{Moles} \times \text{Redox valence}$$

Thus when 1 mole of Zn metal goes to 1 mole of the Zn^{+2} ion, it loses 2 moles of electrons.   Thus 1 equivalent of Zn metal will be half a mole.   Similarly in the reverse reaction (the reduction) $\frac{1}{2}$ mole of the Zn^{+2} ion = 1 equivalent Zn^{+2} ion.   The following table gives a number of such relations:

Oxidized Form	Reduced Form	Number of Equivalents in 1 Mole	
		Oxidized Form	Reduced Form
Cu^{+2} ion	Cu (metal)	2	2
Fe^{+3} ion	Fe^{+2} ion	1	1
Fe^{+3} ion	Fe (metal)	3	3
Cl_2 (gas)	Cl^{-1} ion	2	1
CrO_4^{-2} ion	Cr^{+3} ion	3	3
O_2 (gas)	H_2O	4	2
$Cr_2O_7^{-2}$ ion	Cr^{+3} ion	6	3
NO_3^{-1} ion	NO (gas)	3	3

We observe that the redox valence (the number of equivalents per mole) depends on the particular reaction. Thus, when the Fe^{+3} ion is reduced to the Fe^{+2} ion, it has a valence of 1, but when the Fe^{+3} ion is reduced to the Fe metal, it has a valence of 3. This is a great cause of ambiguity, so much so that many chemists no longer use the units of equivalents in speaking of redox reactions. If, however, the nature of the reaction is understood, the ambiguity disappears. We must always state the reaction when we speak of redox equivalents.

The above definition of equivalent provides us with a conversion factor for going from units of equivalents to units of moles, the conversion factor being the redox valence—that is, the number of moles of electrons transferred per mole of material.

Do problem 4 at the end of the chapter.

9. Redox Calculations

We are now ready to use our principle of equivalence to do chemical calculations. To refresh his memory, the reader should review Chapter VIII on solutions.

Example: How many grams of HCl will be oxidized to free Cl_2 by 60 g. of K_2CrO_4, the latter producing Cr^{+3} ions.

Answer: The valence change of the Cl is 1 (from -1 to 0). The valence change of the Cr is 3 (from $+6$ to $+3$). Applying our general method for such problems:

$$60 \text{ g. } K_2CrO_4 = 60 \text{ g. } K_2CrO_4 \left(\frac{1 \text{ mole } K_2CrO_4}{194 \text{ g. } K_2CrO_4}\right) \times \left(\frac{3 \text{ eq. } K_2CrO_4}{1 \text{ mole } K_2CrO_4}\right)$$

$$\times \left(\frac{1 \text{ eq. } HCl}{1 \text{ eq. } K_2CrO_4}\right)\left(\frac{1 \text{ mole } HCl}{1 \text{ eq. } HCl}\right)\left(\frac{36.5 \text{ g. } HCl}{1 \text{ mole } HCl}\right)$$

$$= \frac{60 \times 3 \times 36.5}{194} \text{ g. } HCl = 33.9 \text{ g. } HCl \rightarrow 34 \text{ g. } HCl$$

Example: How many liters STP of O_2 will be reduced to H_2O by 30 g. of SO_2 in acid solution? (SO_2 goes to the SO_4^{-2} ion.)

Answer: Valence change of O is 2 (from 0 to -2). Valence change of S is 2 (from $+4$ to $+6$)

$$30 \text{ g. } SO_2 = 30 \text{ g. } SO_2 \left(\frac{1 \text{ mole } SO_2}{64 \text{ g. } SO_2}\right)\left(\frac{2 \text{ eq. } SO_2}{1 \text{ mole } SO_2}\right)\left(\frac{1 \text{ eq. } O_2}{1 \text{ eq. } SO_2}\right)$$

$$\left(\frac{1 \text{ mole } O_2}{4 \text{ eq. } O_2}\right)\left(\frac{22.4 \text{ l. STP } O_2}{1 \text{ mole } O_2}\right)$$

$$= \frac{30 \times 2 \times 22.4}{64 \times 4} \text{ l. STP } O_2 = 5.25 \text{ l. STP } O_2 \rightarrow 5.3 \text{ l. STP } O_2$$

Note: There are 4 equivalents of O in 1 mole of O_2 since there are 2 moles of O in 1 mole of O_2 and each O has a valence of 2.

Doing problems dealing with water solutions, we employ the same methods as we used previously.

Example: How many milliliters of $1.36 M$ $CuSO_4$ solution will react with 1.70 g. Zn metal, the products being Cu metal and Zn^{+2} ions?

Answer:

$$1.70 \text{ g. } Zn = 1.70 \text{ g. } Zn \left(\frac{1 \text{ mole } Zn}{65.4 \text{ g. } Zn}\right)\left(\frac{2 \text{ eq. } Zn}{1 \text{ mole } Zn}\right)\left(\frac{1 \text{ eq. } Cu^{+2}}{1 \text{ eq. } Zn}\right)$$

$$\left(\frac{1 \text{ mole } Cu^{+2}}{2 \text{ eq. } Cu^{+2}}\right) \times \left(\frac{1 \text{ l. soln.}}{1.36 \text{ moles } Cu^{+2}}\right)\left(\frac{1000 \text{ ml.}}{1 \text{ l.}}\right)$$

$$= \frac{1.70 \times 2 \times 1000}{2 \times 65.4 \times 1.36} \text{ ml. soln. } = 19.1 \text{ ml. soln.}$$

Example: How many milliliters of $0.68 M$ $KMnO_4$ will react with 42 ml. of $0.16 M$ $NaHSO_3$, the products being Mn^{+2} ions and SO_4^{-2} ions.

Answer: Let us call the $KMnO_4$ solution A, and $NaHSO_3$ solution B.

$$42 \text{ ml. soln. } B = 42 \text{ ml. soln. } B\left(\frac{0.16 \text{ mmole } SO_3^{-2}}{1 \text{ ml. soln. } B}\right)\left(\frac{2 \text{ meq. } SO_3^{-2}}{1 \text{ mmole } SO_3^{-2}}\right)$$

$$\left(\frac{1 \text{ meq. } MnO_4^{-1}}{1 \text{ meq. } SO_3^{-2}}\right) \times \left(\frac{1 \text{ mmole } MnO_4^{-1}}{5 \text{ meq. } MnO_4^{-1}}\right)\left(\frac{1 \text{ ml. soln. } A}{0.68 \text{ mmole } MnO_4^{-1}}\right)$$

$$= \frac{42 \times 0.16 \times 2}{5 \times 0.68} \text{ ml. soln. } A = 3.95 \text{ ml. } KMnO_4 \rightarrow 4.0 \text{ ml. } KMnO_4$$

Note the use of millimoles and milliequivalents here for convenience.

Example: 15.8 ml. of 0.42 M $FeCl_3$ will oxidize 38.0 ml. of a solution of CaI_2 to I_2 (element). If the $FeCl_3$ is reduced to Fe^{+2}, what is the concentration of the CaI_2 solution?

Answer: Call the $FeCl_3$ solution A, and the CaI_2 solution B.

$$15.8 \text{ ml. soln. } A = 15.8 \text{ ml. soln. } A \left(\frac{0.42 \text{ mmoles Fe}^{+3}}{1 \text{ ml. soln. } A} \right) \left(\frac{1 \text{ meq. Fe}^{+3}}{1 \text{ mmole Fe}^{+3}} \right)$$

$$\times \left(\frac{1 \text{ meq. } I^{-1}}{1 \text{ meq. Fe}^{+3}} \right) \left(\frac{1 \text{ mmole } I^{-1}}{1 \text{ meq. } I^{-1}} \right)$$

$$= 15.8 \times 0.42 \text{ mmoles } I^{-1} = 6.64 \text{ mmoles } I^{-1}$$

$$\text{Molarity} = \frac{\text{mmoles } CaI_2}{\text{ml. soln.}} = \frac{6.64 \text{ mmoles } I^{-1} \left(\dfrac{1 \text{ mmole } CaI_2}{2 \text{ mmoles } I^{-1}} \right)}{38.0 \text{ ml. soln.}}$$

$$= \frac{6.64}{2 \times 38.0} \frac{\text{mmoles } CaI_2}{\text{ml. soln.}} = 0.087 \ M \ CaI_2$$

---

10. Problems

1. Calculate the oxidation numbers of each element in the following:

(a) Cr_2O_3.   (e) K_2O_2 (peroxide).   (i) $HClO_3$.
(b) HNO_2.   (f) MnO_2.   (j) K_2MnO_4.
(c) SbH_3.   (g) H_3PO_4.   (k) $Na_2B_4O_7$.
(d) Na_2SO_4.   (h) H_3PO_3.   (l) $H_2S_2O_7$.

2. Write balanced ion-electron equations for each of the following:
(a) Oxidation of H_2S to K_2SO_3.
(b) Oxidation of Bi metal to Na_3BiO_3.
(c) Reduction of MnO_2 to Mn^{+2}.
(d) Reduction of HPO_4^{-2} to HPO_3^{-2}.
(e) Oxidation of NH_3 to NO.
(f) Oxidation of Cl^{-1} to ClO_2 gas.
(g) Reduction of $Cr_2O_7^{-2}$ to Cr metal.
(h) Oxidation of NO gas to HNO_3.

3. Write balanced ionic equations for each of the following reactions (add H_2O if needed):
(a) $NO_2 + HClO \rightarrow HNO_3 + HCl$
(b) Na_2O_2 (s) $+ CrCl_3 \rightarrow Na_2CrO_4 + NaCl$
(c) $Cu + HNO_3 \rightarrow Cu(NO_3)_2 + NO + H_2O$
(d) $Mg + HNO_3 \rightarrow Mg(NO_3)_2 + N_2 + H_2O$
(e) $Cu + H_2SO_4 \rightarrow CuSO_4 + SO_2 + H_2O$
(f) $Zn + HNO_3 \rightarrow Zn(NO_3)_2 + NH_4NO_3 + H_2O$
(g) $NaIO_3 + H_2S \rightarrow I_2 + Na_2SO_3$
(h) $CuS + HNO_3 \rightarrow Cu(NO_3)_2 + NO_2 + H_2O + S$
(i) $HCl + HNO_3 \rightarrow NOCl + Cl_2 + H_2O$
(j) $CuCl_2 + KCN \rightarrow K_2Cu(CN)_3 + (CN)_2\uparrow + KCl$

4. Make the following conversions:

(a) 8 equivalents of $KMnO_4$ to moles of $KMnO_4$ (product is Mn^{+2}).

(b) 0.3 mole of HCl to equivalent of HCl (product is Cl_2).

(c) 0.46 equivalent of Zn metal to grams of Zn (product is Zn^{+2}).

(d) 14.2 g. of $KMnO_4$ to equivalents of $KMnO_4$ (product is MnO_2).

(e) 12 mg. of $KMnO_4$ to milliequivalents of $KMnO_4$ (product is Mn^{+2}).

(f) 68 mmoles $K_2Cr_2O_7$ to equivalents of $K_2Cr_2O_7$ (product is Cr^{+3}).

(g) 8.4 meq. HNO_3 to milligrams of HNO_3 (product is NO).

5. How many grams of I_2 can be oxidized to KIO_3 by 16 g. of HClO?   (Product is HCl.)

6. How many cubic centimeters STP of O_2 gas are needed to oxidize 1.8 g. of KI in solution to I_2?   (O_2 is reduced to H_2O.)

7. How many liters STP of Cl_2 gas can be made from NaCl by oxidation with 24 g. of $K_2Cr_2O_7$?   (Product is Cr^{+3}.)

8. How many milligrams of CuS will be oxidized to free S by 210 mg. of HNO_3? (Product is NO gas.)

9. 90 ml. of a 0.43 M solution of H_2O_2 will oxidize 64 ml. of a solution of NaI to I_2.   If the H_2O_2 is reduced to H_2O, what is the concentration of the NaI?

10. 16.4 ml. of a 0.33 M solution of $K_2Cr_2O_7$ will oxidize 24.0 ml. of a solution of $FeCl_2$ to $FeCl_3$.   If Cr^{+3} is the product what is the molarity of the $FeCl_2$ solution?

11. How many milliliters of a 0.24 M solution of Na_2SO_3 will be oxidized to Na_2SO_4 by 180 ml. of 0.32 M $KMnO_4$ solution?   (The product is Mn^{+2}.)

12. How many milliliters of a 0.085 M solution of $Na_2S_2O_3$ will be oxidized to $Na_2S_4O_6$ by a 180 ml. of a 0.16 M solution of $KClO_3$?   (The product is the Cl^{-1} ion.)

CHAPTER XVI

Predicting Redox Reactions

1. Reversibility of Redox Reactions

Any single ion-electron equation is reversible. That is, under the proper circumstances it can be made to proceed in either direction. Thus the equation for the oxidation of Zn metal to Zn^{+2} ion:

(oxidation) $\qquad\qquad Zn^0 \rightleftharpoons Zn^{+2} + 2e^{-1} \qquad\qquad$ (reduction)

The reverse reaction, Zn^{+2} ion going to Zn metal, is a reduction.

If we put Zn metal in $CuSO_4$ solution the Zn will be oxidized to the Zn^{+2} ion by the Cu^{+2} ion which in turn goes to Cu metal.

$$Zn^0 + Cu^{+2} \rightarrow Zn^{+2} + Cu^0$$

If, however, we put a piece of Mg metal in a solution of $ZnSO_4$, then the Zn^{+2} will be reduced to Zn metal while the Mg metal is itself oxidized to Mg^{+2} ion.

$$Mg^0 + Zn^{+2} \rightarrow Zn^0 + Mg^{+2}$$

The familiar electromotive series of metals is nothing more or less than an expression of the relative tendency of the metals to displace each other from ionic solution. The active metals at the top of the list (Li, Na, K, Mg, etc.) will displace the less active metals below them. That is, the tendency of Li metal to lose an electron ($Li^0 \rightarrow Li^{+1} + 1e^{-1}$) and become Li^{+1} ion is greater than the tendency of any other metal to lose an electron. Or, putting it in reverse, Li^{+1} ion has the least attraction for electrons of any of the metal ions.

This tendency of metals to lose electrons (become oxidized) can be measured quantitatively as well as qualitatively.

2. Measurement of Oxidation Potentials—A Chemical Battery

If a piece of Zn rod is placed in a solution containing Zn^{+2} ions (e.g., $ZnCl_2$) and a piece of Cu rod is placed in a solution containing

202

Cu^{+2} ions (e.g., $CuCl_2$) and the two solutions are connected by a tube containing any electrolyte (e.g., KCl), then there will be a voltage between the Zn and Cu rods.   If a wire is connected to the two rods, a current will flow through it.   The current will continue to flow until all the Zn metal or all the Cu^{+2} ions are exhausted.   This is an example of a chemical battery.   Electrons flow through the wire from the anode to cathode.

In the solution Cl^{-1} ions (or other anions) move from the $CuCl_2$ beaker where Cu^{+2} ions are disappearing, through the salt bridge, to the $ZnCl_2$ beaker to neutralize the Zn^{+2} ions which are being formed.

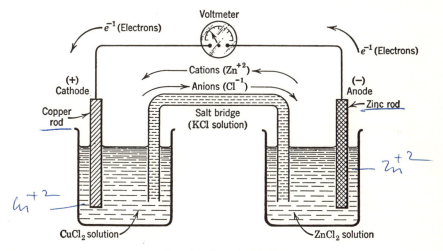

FIG. 3.   Chemical battery.

At the cathode:	$Cu^{+2} + 2e^{-1} \rightarrow Cu$ (metal)	(reduction)
At the anode:	Zn (metal) $\rightarrow Zn^{+2} + 2e^{-1}$	(oxidation)

The only function of the salt bridge is to allow the anions to move and thus complete the circuit.   (*Note:* Cations (+) move in the reverse direction.)

The voltage which is measured in this set-up is a direct measure of the tendency of the Zn metal to displace Cu^{+2} from solution.   If a more active metal than Zn is used, then the voltage is higher.   If a less active metal than Cu is used, the voltage will also be higher.

The greater the difference in activity between two metals, the greater will be their voltage in a chemical battery.

The information gained by measuring such voltages can be expressed in terms of a chart if we can choose a standard against which to measure such voltages.

3. Standard Oxidation Potentials—The Hydrogen Electrode

The standard which has been chosen is the hydrogen electrode. A beaker of a strong acid containing $C_{H_3O^{+1}} = 1.00\ M$ is taken. A piece of rough platinum foil is placed in this solution, and hydrogen at 1 atm. pressure is bubbled through the solution. This constitutes the hydrogen electrode. If we replaced the beaker of $CuCl_2$ and Cu metal in Figure 3 by this hydrogen electrode and measured the voltage, we would find that it would be precisely 0.762 v. (if the Zn^{+2} ion concentration were 1 M). The zinc is more active than hydrogen and tends to displace it from solution. In this case we can write the electrode reaction:

At the anode: $\quad Zn^0 \to Zn^{+2} + 2e^{-1}$ $\qquad\qquad$ (oxidation)

At the cathode: $\quad 2H^{+1} + 2e^{-1} \to H_2^0\uparrow$ $\qquad\qquad$ (reduction)

Overall reaction: $\quad Zn^0 + 2H^{+1} \to H_2^0\uparrow + Zn^{+2}$

The total voltage which is measured (0.762 v.) can be looked upon as being made up of two parts. One is the voltage due to reaction at the anode (oxidation of Zn metal). The other is the voltage due to the reaction at the cathode (reduction of the H^{+1} ion). We can never measure the absolute potential of either reaction separately. However, if we arbitrarily take as our standard that the hydrogen potential is zero, then the zinc potential must be +0.762 v.

$$E(\text{total}) = E(\text{reduction of } H^{+1}) + E(\text{oxidation of Zn})$$

Thus: $\underset{\text{(observed)}}{0.762\ \text{v.}} = \underset{\text{(choice)}}{0} \qquad + E(\text{oxidation of Zn})$

Potentials measured in this manner against the hydrogen electrode, at 25°C. and with all concentrations = 1 M, are called standard oxidation potentials ($E°_{ox.}$). They are positive if the reaction displaces H^{+1} from solution as H_2 gas. They are negative if the H_2 gas is more active and displaces the other substance.

Table XIX shows some potentials measured in this manner.

This table will now give the potential for any combination of electrode couples. Thus, if we measure the $Zn|Zn^{+2}$ electrode against the $Sn|Sn^{+2}$ electrode, the total voltage is the algebraic difference between their standard oxidation potentials:

$$E(\text{total}) = E°_{ox.}(\text{higher}) - E°_{ox.}(\text{lower})$$

$$E(Zn|Zn^{+2}||Sn^{+2}|Sn) = (+0.762) - (+0.136) = 0.626\ \text{v.}$$

TABLE XIX*

SOME STANDARD OXIDATION POTENTIALS

Oxidation Reaction	$E^\circ_{ox.}$ (Standard Oxidation Potential)
Na $\to$ Na^{+1} + $1e^{-1}$	+2.712 v.
Mg $\to$ Mg^{+2} + $2e^{-1}$	+2.34 v.
Al $\to$ Al^{+3} + $3e^{-1}$	+1.67 v.
Zn $\to$ Zn^{+2} + $2e^{-1}$	+0.762 v.
Fe $\to$ Fe^{+2} + $2e^{-1}$	+0.440 v.
Sn $\to$ Sn^{+2} + $2e^{-1}$	+0.136 v.
H$_2$ $\to$ 2H^{+1} + $2e^{-1}$	0.000 v. (standard)
Cu $\to$ Cu^{+2} + $2e^{-1}$	−0.345 v.
2I^{-1} $\to$ I$_2$ + $2e^{-1}$	−0.535 v.
Fe^{+2} $\to$ Fe^{+3} + $1e^{-1}$	−0.771 v.
Ag $\to$ Ag^{+1} + $1e^{-1}$	−0.800 v.
2Cl^{-1} $\to$ Cl$_2$ + $2e^{-1}$	−1.358 v.

* For the purpose of such a table it doesn't matter which electrode we choose as standard or what value we give it. The differences between two electrodes, which is all we care about, will still be the same.

and since Zn is higher it will displace Sn, the reaction being

$$Zn + Sn^{+2} \to Sn + Zn^{+2} \quad (E = 0.626 \text{ v.})$$

Example: What voltage will be generated by the battery consisting of a Zn|Zn^{+2} electrode and a Cu|Cu^{+2} electrode?

Answer:

$$E(Zn|Zn^{+2}||Cu^{+2}|Cu) = (+0.762) - (-0.345)$$

$$= 1.107 \text{ v.}$$

Since Zn is higher, it will displace Cu, the reaction being

$$Zn + Cu^{+2} \to Cu + Zn^{+2}$$

Do problem 1 at the end of the chapter.

4. Summary

1. *Almost every ion-electron reaction is reversible.*

2. *We can measure the tendency of any ion-electron reaction to proceed as an oxidation or reduction by placing the substances (at a 1 M concentration) in a vessel and measuring the voltage of this electrode compared to the hydrogen electrode.*

3. *If the reaction displaces* H^{+1} *from solution as* H$_2$ *gas then the ion-electron equation goes as an oxidation and its standard potential is positive.* ($E^\circ_{ox.} > 0$)

4. *If the reaction permits* H$_2$ *gas to go into solution as* H^{+1} *ions then the reaction proceeds as a reduction and its standard oxidation potential is negative.* ($E^\circ_{ox.} < 0$)

5. *If any two, individual ion-electron reactions are compared, the one with the higher standard oxidation potential will proceed as an oxidation.* The one with the lower standard oxidation potential will be forced to proceed as a reduction.

6. *The total potential generated by any two ion-electron reactions will be given by the algebraic difference of their standard oxidation potentials.*

From the above we see that if we have a complete list of standard oxidation potentials we can predict the direction that the redox reaction corresponding to these reactions will take.

Example: If a piece of Ag metal is put into a solution of $SnCl_2$, will it displace the Sn?

Answer: The $E^\circ_{ox.}$ for $Ag|Ag^{+1}$ is -0.800 v. The $E^\circ_{ox.}$ for $Sn|Sn^{+2}$ is $+0.136$ v. Thus Sn has a greater tendency to go to Sn^{+2} than Ag has to go to Ag^{+1}. No reaction occurs.

---

5. Effect of Concentration—The Nernst Equation

In most laboratory experiments we seldom have solutions at concentrations of $1\ M$. They are almost always higher or lower. How does this affect our ability to predict reactions?

Let us take a typical ion-electron reaction:

$$Zn \rightleftarrows Zn^{+2} + 2e^{-1} \qquad E^\circ_{ox.} = +0.762 \text{ v.}$$

What happens if we put a piece of Zn into a solution whose concentration of Zn^{+2} ion is less than $1\ M$? By applying Le Châtelier's principle we see that the equilibrium is shifted to the right, that is, by removing Zn^{+2} we increase the tendency of the forward reaction. The oxidation tendency is increased. If we were to measure the potential of this electrode with $C_{Zn^{+2}}$ less than $1\ M$, we would find it higher than $+0.762$ v.

Conversely, if we used a $C_{Zn^{+2}}$ greater than $1\ M$, we would shift the equilibrium in the opposite direction and *decrease* the oxidation tendency. The observed potential of such an electrode would be smaller.

There is an equation which tells us precisely to what extent the EMF is changed by changing the concentration. It is known as the Nernst equation:

$$E_{ox.}(\text{observed}) = E^\circ_{ox.} - \frac{2.3RT}{nF} \log \frac{C_{ox.}}{C_{red}}$$

in which R is the universal gas constant, T is the absolute temperature, F is the number of coulombs in 1 faraday (96,500) and n is the number of electrons transferred in the oxidation. $C_{ox.}$ is the concentration of the oxidized forms, and $C_{red.}$ is the concentration of the reduced forms. At room temperature, substituting numbers for the constants, this equation becomes

$$E_{ox.}(\text{obs.}) = E^{\circ}_{ox.} - \frac{0.06}{n} \log \frac{C_{ox.}}{C_{red.}} \qquad \text{(at 25°C.)}$$

Thus an electrode consisting of a rod of Zn in a solution in which $C_{Zn^{+2}} = 0.01\ M$ would be ($n = 2$ for this electrode):

$$E_{ox.}(\text{obs.}) = 0.762 - \frac{0.06}{2} \log 0.01 \qquad (\log 0.01 = -2)$$

$$= 0.762 - 0.03 \times (-2) = 0.762 + 0.06$$

$$= +0.822 \text{ v.}$$

Note: We omit $C_{red.}$ when the reduced form is a solid, as zinc is.

Example: What is the $E_{ox.}$ of an electrode consisting of Cu metal in a solution of $1 \times 10^{-6}\ M$ Cu^{+2} ions?
Answer:

$$E_{ox.}(\text{obs.}) = E^{\circ}_{ox.} - \frac{0.06}{2} \log C_{Cu^{+2}}$$

$$= -0.345 - 0.03 \log (1 \times 10^{-6})$$

$$= -0.345 - 0.03(-6) = -0.345 + 0.18$$

$$= -0.165 \text{ v.}$$

That is, as we should expect, the oxidation tendency of the reaction $Cu \rightarrow Cu^{+2} + 2e^{-1}$ is increased (the voltage is less negative).

In general we can say that, if n is the number of electrons transferred, then: (1) *For every power of 10 by which the concentration of an oxidized species decreases, the $E^{\circ}_{ox.}$ is increased by $0.06/n$ volts.* (2) *For every power of 10 by which the concentration of a reduced species decreases, the $E^{\circ}_{ox.}$ is decreased by $0.06/n$ volts.*

6. More Complex Ion-Electron Equations

The treatment we have just described is applicable not only to simple metal|metal ion reactions, but also to any reversible ion-electron equations.

We find for the oxidation of Mn^{+2} to MnO_4^{-1} in acid solution:

$$Mn^{+2} + 4H_2O \rightleftarrows MnO_4^{-1} + 8H^{+1} + 5e^{-1} \qquad (E^{\circ}_{ox.} = -1.52 \text{ volts})$$

The Nernst equation for such a reaction becomes

$$E_{ox.}(obs.) = (-1.52) - \frac{0.06}{5} \log \frac{C_{MnO_4^{-1}} \times C^8_{H^{+1}}}{C_{Mn^{+2}}} \quad \text{(as usual, water is left out)}$$

Example: What is the potential of the $Mn^{+2}|MnO_4^{-1}$ oxidation in a solution in which $C_{Mn^{+2}} = 1 \times 10^{-8}\ M$, $C_{H^{+1}} = 1 \times 10^{-5}\ M$, and $C_{MnO_4^{-1}} = 1 \times 10^{-2}\ M$?
    Answer:

$$E_{ox.}(obs.) = -1.52 - \frac{0.06}{5} \log \frac{(1 \times 10^{-2}) \times (1 \times 10^{-5})^8}{1 \times 10^{-8}}$$

$$= -1.52 - \frac{0.06}{5} \log (1 \times 10^{-34})$$

$$= -1.52 - \frac{0.06}{5} \times (-34) = -1.52 + 0.41$$

$$= -1.11\ v.$$

The tendency of the forward reaction is increased (oxidation). The tendency of the back reaction is decreased (reduction).

Do problem 3 at the end of the chapter.

7. Strength of Oxidizing and Reducing Agents

When an ion-electron reaction has a high positive $E^o_{ox.}$ we interpret this as meaning that it has a great tendency to go forward. Thus

$$Na \rightleftarrows Na^{+1} + 1e^{-1} \quad E^o_{ox.}(Na|Na^{+1}) = +2.71\ v.$$

We can interpret this as meaning that Na metal is easily oxidized (has small affinity for electrons). When we couple this reaction with any other ion-electron equation having a lower $E^o_{ox.}$.

$$Zn \rightleftarrows Zn^{+2} + 2e^{-1} \quad E^o_{ox.}(Zn|Zn^{+2}) = +0.76\ v.;$$

the higher $E^o_{ox.}$ of the $(Na|Na^{+1})$ reaction causes the $(Zn|Zn^{+2})$ to reverse and proceed as a reduction. This may also be interpreted as saying that Na metal is a *powerful reducing agent* since it causes other ion-electron reactions to proceed backwards, that is, as reductions.
    In a similar fashion, the couple $(Cl^{-1}|Cl_2)$ has a low $E^o_{ox.} = -1.36$ v. This means that the reaction $2Cl^{-1} \rightleftarrows Cl_2 + 2e^{-1}$ has a small tendency to proceed in the forward direction and, conversely, a great tendency to proceed in the reverse direction, that is, as a reduction. In terms of electron affinity, we say that the Cl^{-1} ion has a great affinity for its electrons. If we combine this with any ion-electron reaction having a higher $E^o_{ox.}$ (e.g., $2Br^{-1} \rightleftarrows Br_2 + 2e^{-1}$, $E^o_{ox.} = -1.07$ v.), the latter will proceed forward as an oxidation. Thus we would say that the $(Cl^{-1}|Cl_2)$ couple is a *good oxidizing agent* since it causes other couples to proceed as oxidations.

We can summarize these statements:

Standard Oxidation Potential	Tendency of Forward Reaction	Tendency of Reverse Reaction	Affinity for Electrons	Strength as Oxidizing Agent	Strength as Reducing Agent
High (positive)	High	Low	Small	Weak	Strong
Low (negative)	Low	High	Great	Strong	Weak

Do problem 8 at the end of the chapter.

8. Equilibrium Constants for Redox Reactions

When two redox reactions have the same value of E^o_{ox}. they have precisely the same tendency to proceed as oxidations; consequently, if the electrodes are connected as in a chemical battery, no voltage will be observed. If the substances are mixed, no reaction occurs. The system is in a state of equilibrium. If we know the concentrations of all materials, we can calculate the equilibrium constant. Thus it is possible from voltage measurements to calculate the equilibrium constants in a redox reaction.

If we consider the pair of equations:

$$Zn \rightleftarrows Zn^{+2} + 2e^{-1} \qquad E^o_{ox}.(Zn|Zn^{+2}) = +0.76 \text{ v.}$$

$$Fe \rightleftarrows Fe^{+2} + 2e^{-1} \qquad E^o_{ox}.(Fe|Fe^{+2}) = +0.44 \text{ v.}$$

$$\overline{\qquad\qquad E^o(\text{cell}) \qquad\qquad = \quad 0.32 \text{ v.}}$$

we see that the Zn metal is a more powerful reducing agent and will drive the Fe^{+2} ion out of solution. If we put a piece of Zn metal in a solution containing Fe^{+2} ions, the reaction will be:

$$Zn + Fe^{+2} \rightarrow Zn^{+2} + Fe$$

Suppose now we lower the concentration of Fe^{+2} ions and increase the $C_{Zn^{+2}}$. By Le Châtelier's principle, the above reaction, which is really an equilibrium, shifts to the left. If these concentrations are just right, we can reach equilibrium. The Nernst equation tells us what the concentrations are at this point:

$$E_{\text{obs.}}(Zn|Zn^{+2}) = E^o_{ox}.(Zn|Zn^{+2}) - \frac{0.06}{2} \log C_{Zn^{+2}}$$

$$E_{\text{obs.}}(Fe|Fe^{+2}) = E^o_{ox}.(Fe|Fe^{+2}) - \frac{0.06}{2} \log C_{Fe^{+2}}$$

When equilibrium is reached,

$$E_{obs.}(\text{Zn}|\text{Zn}^{+2}) = E_{obs.}(\text{Fe}|\text{Fe}^{+2})$$

At this point then

$$E^\circ_{ox.}(\text{Zn}|\text{Zn}^{+2}) - 0.03 \log C_{\text{Zn}^{+2}} = E^\circ_{ox.}(\text{Fe}|\text{Fe}^{+2}) - 0.03 \log C_{\text{Fe}^{+2}}$$

or, rearranging,

$$E^\circ_{cell} = E^\circ_{ox.}(\text{Zn}|\text{Zn}^{+2}) - E^\circ_{ox.}(\text{Fe}|\text{Fe}^{+2}) = 0.03 \log C_{\text{Zn}^{+2}}$$
$$- 0.03 \log C_{\text{Fe}^{+2}}$$

or

$$E^\circ_{cell} = 0.03 \log \frac{C_{\text{Zn}^{+2}}}{C_{\text{Fe}^{+2}}}$$

But $E^\circ_{cell} = 0.32$ v., and substituting:

$$\log \frac{C_{\text{Zn}^{+2}}}{C_{\text{Fe}^{+2}}} = \frac{0.32}{0.03} = 10.7$$

Taking antilogs:

$$\frac{C_{\text{Zn}^{+2}}}{C_{\text{Fe}^{+2}}} = 5 \times 10^{+10}$$

But, by definition $C_{\text{Zn}^{+2}}/C_{\text{Fe}^{+2}} = K_{eq.}$ for the overall reaction. Thus

$$K_{eq.} = 5 \times 10^{10}$$

Since $K_{eq.}$ is so large we can calculate that the reaction proceeds practically to completion.

The preceding illustration shows us how $E^\circ_{ox.}$ can be used to calculate equilibrium constants for redox reactions. Quite generally

$$E^\circ_{cell} = E^\circ_{ox.}(\text{higher}) - E^\circ_{ox.}(\text{lower}) = \frac{0.06}{n} \log K_{eq.}$$

If we know the individual $E^\circ_{ox.}$, we can always calculate $K_{eq.}$.

Example: Calculate the $K_{eq.}$ for $\text{Sn} + \text{Pb}^{+2} \rightleftarrows \text{Sn}^{+2} + \text{Pb}$.

$$E^\circ_{ox.}(\text{Sn}|\text{Sn}^{+2}) = +0.136 \text{ v.} \quad E^\circ_{ox.}(\text{Pb}|\text{Pb}^{+2}) = +0.126 \text{ v.}$$

Answer:

$$E^\circ_{cell} = E^\circ_{ox.}(\text{Sn}|\text{Sn}^{+2}) - E^\circ_{ox.}(\text{Pb}|\text{Pb}^{+2}) = 0.010 \text{ v.}$$

Thus

$$E^\circ_{cell} = \frac{0.06}{n} \log K_{eq.}$$

Solving for $\log K_{eq.}$ and substituting:

$$\log K_{eq.} = \frac{2}{0.06} \times (0.010) = 0.33$$

Therefore:

$$K_{eq.} = 2.1 \quad \left(\text{By definition, } K_{eq.} = \frac{C_{Sn^{+2}}}{C_{Pb^{+2}}}\right)$$

Example: Calculate the $K_{eq.}$ for $2Cr + 3Fe^{+2} \rightleftarrows 2Cr^{+3} + 3Fe$.

$$E^{\circ}_{ox.}(Cr|Cr^{+3}) = +0.71 \text{ v.}; \quad E^{\circ}_{ox.}(Fe|Fe^{+2}) = +0.44 \text{ v.}$$

Answer: We must be careful here since the n's are different for the two couples.

$$E^{\circ}_{cell} = E^{\circ}_{ox.}(Cr|Cr^{+3}) - E^{\circ}_{ox.}(Fe|Fe^{+2}) = \frac{0.06}{3} \log C_{Cr^{+3}} - \frac{0.06}{2} \log C_{Fe^{+2}}$$

Then

$$E^{\circ}_{cell} = 0.71 - 0.44 = \frac{0.06}{3 \times 2} \log C^2_{Cr^{+3}} - \frac{0.06}{2 \times 3} \log C^3_{Fe^{+2}}$$

Note: We want the same coefficients before the logs. We can multiply the bottoms by 2 and 3, respectively, and compensate for this by raising the concentrations to the 2 and 3 powers. Thus: $\log x = \frac{1}{2} \log x^2 = \frac{1}{3} \log x^3$.

Then

$$0.27 = 0.01(\log C^2_{Cr^{+3}} - \log C^3_{Fe^{+2}}) = 0.01 \log \frac{C^2_{Cr^{+3}}}{C^3_{Fe^{+2}}}$$

or

$$\log \frac{C^2_{Cr^{+3}}}{C^3_{Fe^{+2}}} = 27 \quad \text{and} \quad \frac{C^2_{Cr^{+3}}}{C^3_{Fe^{+2}}} = 10^{27}$$

Thus:

$$K_{eq.} = \frac{C^2_{Cr^{+3}}}{C^3_{Fe^{+2}}} = 10^{27}$$

The redox $K_{eq.}$ calculated in this fashion can now be used to tell how far towards completion a reaction will go.

Example: A small piece of Cr metal is put into a solution containing $C_{Fe^{+2}} = 1 \times 10^{-3} M$ and an excess of Cr^{+3} ions; $C_{Cr^{+3}} = 0.2 M$. What will the $C_{Fe^{+2}}$ be when equilibrium is reached? ($E^{\circ}_{ox.}$ are given.)

Answer: The reaction is

$$2Cr + 3Fe^{+2} \rightarrow 2Cr^{+3} + 3Fe$$

From the previous example we see that

$$K_{eq.} = \frac{C^2_{Cr^{+3}}}{C^3_{Fe^{+2}}} = 1 \times 10^{27}$$

As the reaction proceeds towards equilibrium, $C_{Cr^{+3}}$ increases. However, since it is 0.2 M to start with, the additional Cr^{+3} ions from the displacement of Fe^{+2} will not increase this very much (i.e., $\frac{2}{3} \times 1 \times 10^{-3} M$ is negligible compared to 0.2). Thus set $C_{Cr^{+3}} = 0.2$. Then

$$C^3_{Fe^{+2}} = \frac{C^2_{Cr^{+2}}}{K_{eq.}} = \frac{(0.2)^2}{1 \times 10^{27}} = 4 \times 10^{-29}$$

Taking cube roots:
$$C_{Fe^{+2}} = 3.4 \times 10^{-10} \text{ mole/l.}$$

The foregoing can be represented in simple fashion by the following equation:
$$K_{eq.} = 10^{n\Delta E/0.06}$$

where n is the number of electrons transferred for each step in the overall balanced equation as written and ΔE is the difference in standard oxidation potentials of the two couples making up the redox reaction. If the reaction as written is favorable (more active couple displacing the less active couple), the sign of ΔE is taken as positive. If the reverse is true, the sign of ΔE is taken as negative. We should always check the magnitude of $K_{eq.}$ with our qualitative judgment to see that no sign error has been made.

Example: Calculate $K_{eq.}$ for $Sn^{\circ} + Pb^{++} \rightleftarrows Sn^{++} + Pb^{\circ}$. (See preceding examples.)

Answer: In the above equation $n = 2$ and $\Delta E = +0.010$ v. (more active Sn is displacing Pb^{++}).

Therefore $K_{eq.} = 10^{\dfrac{2 \times 0.010}{0.06}} = 10^{0.33}$

          $= 2.1$ (slide rule accuracy)

Example: Calculate $K_{eq.}$ for $2Cr^{\circ} + 3Zn^{++} \rightleftarrows 2Cr^{+3} + 3Zn^{\circ}$:

$E^{\circ}_{ox.}$ $(Cr/Cr^{+3}) = +0.71$ v. $E^{\circ}_{ox.}$ $(Zn/Zn^{++}) = +0.76$ v.

Answer: For the equation as written $n = 6$ since 6 electrons are transferred for every 2Cr oxidized (and also $3Zn^{++}$ reduced). $\Delta E = -0.05$ v. since as written the reaction is unfavorable, the less active Cr° replacing the more active Zn. Thus:
$$K_{eq.} = 10^{-6 \times 0.05/0.06} = 10^{-5}$$

As expected for the unfavorable reaction written $K_{eq.}$ is small.

Note: Although 10^{-5} is a small number and it would appear that not much Cr^{+3} would be formed, the occurrence of large coefficients ($2Cr^{+3}$, $3Zn^{+2}$) in the balanced equation offsets this and we will find that at equilibrium the ratio of $C_{Cr^{+3}}/C_{Zn^{+2}}$ is not very small.

Example: Using the results of the preceding example calculate the concentration of Cr^{+3} that will be in equilibrium with 0.2 M Zn^{++}.

Answer: $K_{eq.} = \dfrac{C^2_{Cr^{+3}}}{C^3_{Zn^{++}}}$

solving for $C_{Cr^{+3}}$

$$C^2_{Cr^{+3}} = (K_{eq.} \times C^3_{Zn^{++}})$$
$$= 10^{-5} \times 8 \times 10^{-3} = 8 \times 10^{-8}$$

Therefore $C_{Cr^{+3}} = 2.8 \times 10^{-4}\ M$

---

9. Problems

1. From tables of $E^\circ_{ox.}$ calculate the voltages of the following when used as chemical batteries.

    (a) $Zn + 2Ag^{+1} \rightarrow Zn^{+2} + 2Ag.$
    (b) $Mg + Cl_2 \rightarrow Mg^{+2} + 2Cl^{-1}.$
    (c) $Cl_2 + 2Br^{-1} \rightarrow Br_2 + 2Cl^{-1}.$
    (d) $Fe + 2H^{+1} \rightarrow Fe^{+2} + H_2.$
    (e) $3Pb + 8H^{+1} + 2NO_3^{-1} \rightarrow 3Pb^{+2} + 2NO + 4H_2O.$
    (f) $2MnO_4^{-1} + 10Cl^{-1} + 8H^{+1} \rightarrow 2Mn^{+2} + 5Cl_2 + 4H_2O.$
    (g) $Cu + SO_4^{-2} + 2H^{+1} \rightarrow Cu^{+2} + SO_3^{-2} + H_2O.$
    (h) $Cl_2 + H_2O \rightarrow H^{+1} + Cl^{-1} + HClO.$

2. Write $K_{eq.}$ for each of the reactions in problem 1, and compute $K_{eq.}$ from the standard $E^\circ_{ox.}$, using the Nernst equation.

3. Compute the observed potential for the following electrodes:

    (a) $(Zn|Zn^{+2})$ when $C_{Zn^{+2}} = 1 \times 10^{-6}\ M.$ $E^\circ_{ox.} = +0.76$ v.
    (b) $(Fe^{+2}|Fe^{+3})$ when $C_{Fe^{+2}} = 2M;\ C_{Fe^{+3}} = 4 \times 10^{-3}\ M.$ $E^\circ_{ox.} = -0.77$ v.
    (c) $(Cu|Cu^{+2})$ when $C_{Cu^{+2}} = 1 \times 10^{-14}.$ $E^\circ_{ox.} = -0.345$ v.
    (d) $(Cl^{-1}|Cl_2)$ when $C_{Cl^{-1}} = 6 \times 10^{-8}\ M.$ $E^\circ_{ox.} = -1.358$ v.

4. What is the oxidation potential of the electrode consisting of a piece of Zn metal placed into a saturated solution of ZnS? $E^\circ_{ox.} = +0.76$ v.; $K_{S.P.}(ZnS) = 4.5 \times 10^{-24}$.

5. What is the oxidation potential of the electrode consisting of a piece of Ag metal placed in a saturated solution of AgI? $E^\circ_{ox.} = -0.800$ v.; $K_{S.P.}(AgI) = 8.5 \times 10^{-17}$.

6. What is the oxidation potential of the electrode consisting of a piece of Ag metal in a saturated solution of AgI, containing an excess I^{-1} concentration of $1 \times 10^{-2}\ M.$ (Use data from problem 5.) (Hint: Calculate $C_{Ag^{+1}}$ in solution.)

7. A solution containing Cu^{+2} ions is saturated with H_2S at a $C_{H_3O^{+1}}$ of $1 \times 10^{-2}\ M.$ The C_{H_2S} is 0.10 $M.$ What oxidation potential will this solution have when a piece of Cu metal is placed in it? $E^\circ_{ox.}(Cu|Cu^{+2}) = -0.35$ v.; $K_{ion}(H_2S)$ overall $= 1.1 \times 10^{-22}$; $K_{S.P.}(CuS) = 4 \times 10^{-36}$.

8. From Table XIX in section 3, which substance is the strongest oxidizing agent? Which substance is the strongest reducing agent? Which set of reactions will give the largest voltage in a chemical battery? Which set of reactions will give the smallest voltage in a chemical battery?

9. (a) What will happen when Fe metal is placed in HCl solution?
    (b) What will happen when Cu metal is placed in HCl solution?
    (c) Will free I_2 oxidize Fe^{+2} ions when placed in a solution?
    (d) What will happen when Cu metal and free iodine are brought together in a solution?

10. When a bar of Ag is placed in a solution containing $1\ M$ Br^-, the oxidation potential of Ag is increased by 0.75 v. over the standard potential of Ag/Ag^+.

(a) What is the concentration of Ag^+ in the Br^- solution?

(b) Calculate $K_{S.P.}(AgBr)$.

11. When a bar of Pb is placed in 0.1 M H_2SO_4, $E_{ox.}$ (Pb/Pb^{++}) = $+0.326$ v. If $E^\circ_{ox.}(Pb/Pb^{++})$ = $+0.126$ v., what is $C_{Pb^{++}}$ in the H_2SO_4 solution? Calculate $K_{S.P.}(PbSO_4)$.

12. When very large concentrations of NaCl are added to a standard Cu/Cu^{++} half-cell, the oxidation potential is observed to rise though no ppt. forms. How might you explain this?

13. The standard oxidation potential for the oxidation $Fe^{++} \rightarrow Fe^{+3} + 1e^-$ is -0.77 v. On adding a little NaOH to a half-cell containing Fe^{++} and Fe^{+++} the potential becomes more positive. What does this tell you about the relative affinities of Fe^{++} and Fe^{+3} for OH^-? Explain.

CHAPTER XVII

Thermochemistry

1. Predicting the Course of Chemical Reactions

In the last chapter on redox reactions, we observed that our ability to measure quantitatively half cell-potentials gave us an accurate method for predicting the direction and extent of completeness of all redox reactions. Is there any way of extending such methods to chemical reactions in general, to reactions that are not redox reactions? The answer to this question is "yes," and the general method for quantitatively predicting the equilibrium constants for chemical reactions is called *thermochemistry*.

We have seen earlier (Chapter V) that all chemical reactions are accompanied by heat changes. Heat is either liberated or absorbed during a chemical change. But heat is a form of energy. In order for heat to be liberated during a chemical change, forces had to move through distances to do work. These forces are the forces of attraction between the atoms as they exist in the products of the reaction. A chemical reaction may be described as the process of breaking bonds between atoms of reactant molecules and rearranging them to form the new bonds in the product molecules. In an exothermic reaction, the total forces acting in the bonds of the product molecules must be greater than the forces acting in the bonds of the reactant molecules. The work that these forces do, in excess of the work done to overcome the forces in the reactant molecules, is the net energy of the reaction.

As an example, consider the exothermic reaction:

$$H_2(g) + Cl_2(g) \rightleftarrows 2HCl(g) + 44 \text{ kcal.}$$

The heat evolved from this reaction ($\Delta H = -44$ kcal.) may be looked on as the difference in the energy required to break the H-H bond in H_2 and the Cl-Cl bond in Cl_2, less that released in forming the two new H-Cl bonds in the product. It is, in fact, possible to measure the energies of these bonds and we do find that they are respectively H-H = 104 kcal./mole, Cl-Cl = 58 kcal./mole and H-Cl = 103 kcal./mole. The net change is $2 \times 103 - 104 - 58 = 44$ kcal., precisely the observed heat of reaction.

Although such a simple analysis is possible for simple diatomic molecules such as HCl, H_2, and Cl_2, it is not easy or useful for more complex molecules, or reactions, such as the combustion of C_2H_6:

$$2C_2H_6 + 7O_2 \rightarrow 4CO_2 + 6H_2O$$

However, the conclusion would appear to be the same—namely that all chemical reactions, if observed at constant temperatures, should proceed in the direction that liberates heat, that is, toward the direction of establishing the strongest forces of attraction between atoms. As we shall see shortly, such a conclusion is basically correct, but subject to modification.

2. Opposing Forces in Chemical Equilibria

It is our common experience that the earth exerts a gravitational force of attraction on all objects. The amount of this force is proportional to the mass of the object and the result of the force is to pull the object toward the earth. The natural tendency of all massive objects on the earth is to fall toward the earth and, eventually, to come to rest at the lowest position available to them.

However, there appear to be some exceptions to this. The atmosphere is made up of massive molecules that are certainly attracted to the earth. Nevertheless, these molecules remain spread out for some distance, about 40 miles or so, above the surface of the earth. Why don't they settle out?

The kinetic theory of matter provides a ready explanation. At ambient temperatures (about 20°C. on the average), air molecules (namely, N_2 and O_2) are moving with velocities of about 40,000 cm./sec. (~ 700 miles per hour). Despite their attraction to the earth, when they strike it, they must rebound. They can only fail to rebound if they lose their kinetic energy. But in order to lose their kinetic energy, they would have to be *cooled* to absolute zero. Thus, as long as the surface of the earth maintains a temperature above 0°K., the atmospheric molecules will not settle out but will remain distributed in space.

The gravitational force of attraction does manifest itself in that the density of atmospheric molecules is greatest at the surface of the earth and decreases by about 3% for every thousand feet.

The random, thermal, kinetic energy of molecular motion may be looked on as a disruptive, or repulsive, energy that opposes itself to attractive forms of energy. At absolute zero, where this kinetic energy has disappeared, all chemical reactions would indeed tend to proceed in the direction that liberated the most heat.

3. Entropy and Gibbs Free Energy

The random, kinetic energy of molecules that opposes attractive forces, brings about diffusional motion of molecules in gases and liquids. It causes gases to be uniformly distributed in containers in which they are placed and to exert pressures on the walls of these containers.

To confine a gas within a container, we must do work against the pressure exerted by the gas. The smaller the container for a fixed amount of gas, the greater is its pressure and the greater is the amount of work needed to confine it.

There is a physical property that can be used to measure quantitatively the disruptive or "repulsive" energy that molecules have as a result of their random thermal motion. It is called "entropy" and has the symbol S. It can be defined loosely as the average kinetic energy of a molecule (or mole) per degree Kelvin. The product of the absolute temperature, T, in degrees Kelvin, times entropy, TS, represents the contribution to what is called the total free energy of a molecule due to its random thermal energy. This Free Energy (symbol G) is defined as the difference:

$$G = H - TS$$

where H is the heat content or enthalpy (per mole) of a given molecular species.

The Free Energy or, more properly, Gibbs Free Energy, is named after the American physical chemist Willard Gibbs who developed the science of thermochemistry. It is of special interest since in any chemical mixture at constant temperature and pressure no chemical or physical changes can occur unless the changes produce a lowering of the Gibbs Free Energy for the entire system.

Furthermore, the maximum energy available from a chemical or physical change is not the change in enthalpy or heat content, ΔH, but the change in Gibbs Free Energy, ΔG. The reason for this is that the products of any chemical or physical change still have random, thermal energy, hence entropy, and this energy, TS, remains in the final system unused.

Entropy is a molecular property and can be measured for each substance. For gases, theoretical methods have been developed for calculating entropy from the details of molecular structure. Its usual units are calories/mole °K, just like molar heat capacity. Table XX lists the molar entropies of a number of common compounds. Unlike heat of formation or enthalpy, the absolute entropy of a compound can be measured since at 0°K the entropy of all perfect crystals is zero!

TABLE XX.

The Molar Entropies of

Some Common Compounds and Elements at 25°C. and 1 Atmosphere Pressure

Substance	S° (cal/mole-°K)	Substance	S° (cal/mole-°K)
H (g)	27	NO (g)	50
Cl (g)	39	NO$_2$ (g)	57
He (g)	30	H$_2$S (g)	49
Ar (g)	37	NH$_3$ (g)	46
H$_2$ (g)	31	CH$_4$ (g)	45
O$_2$ (g)	49	C$_2$H$_6$ (g)	55
N$_2$ (g)	46	C$_2$H$_2$ (g)	48
S (s)	8	C$_6$H$_6$ (l)	41
Cl$_2$ (g)	53	NaCl (s)	17
C (s)	1	NaCl (aq)	28
CO (g)	47	KClO$_3$ (s)	34
CO$_2$ (g)	51	H$_2$SO$_4$ (l)	38
SO$_2$ (g)	59	NaOH (s)	14
SO$_3$ (g)	61	NaOH (aq)	12
H$_2$O (g)	45	HCl (aq)	13
H$_2$O (l)	17	H$_2$SO$_4$ (aq)	5
HCl (g)	45		
O$_3$ (g)	57		

(g) = gas state; (l) = liquid state; (s) = solid state; (aq) = dissolved in water

4. Equilibrium Constants and Free Energy Changes

In any chemical or physical change, the change in Gibbs Free Energy, ΔG, is given by:

$$\Delta G = \Delta H - T\Delta S \tag{1}$$

where ΔH is the heat of reaction and ΔS is the entropy change in the reaction. If we know the ΔH for the reaction, and the ΔS for the reaction, we can calculate the Gibbs Free Energy change. It can be shown that the equilibrium constant for a chemical reaction is related to the Gibbs Free Energy change by the equation:

$$2.303 \, RT \log K_{eq.} = -\Delta G \tag{2}$$

where R is the universal gas constant = 2 cal./mole−°K., or in exponential form:

$$K_{eq.} = 10^{-\Delta G/2.303 \, RT} \tag{3}$$

Equations 1, 2, and 3 are exact, and if we know the equilibrium constant for a chemical or physical change at a given T and P, we can calculate ΔG for that change at the same T and P. Conversely, if we know or can calculate ΔG at some T and P, we can use equation 3 to calculate the value of $K_{eq.}$ at the same T and P.

In the preceding chapter on redox reactions, we gave a formula for calculating $K_{eq.}$ for a redox reaction from combinations of half-cell oxidation potentials. It can be shown that for redox reactions, the Gibbs Free Energy change is related to the cell oxidation potential E^0_{cell} by:

$$\Delta G = -nFE^0_{cell} \tag{4}$$

or when E^0_{cell} is in volts and ΔG is in kcal./mole:

$$\Delta G = -23\, nE^0_{cell} \tag{5}$$

Here n is the moles of electrons transferred for the written balanced redox equation, and F is the Faraday, 96,500 coulombs/mole electrons.

Example: Calculate ΔG and $K_{eq.}$ for the reaction:

$$\text{Zn}(s) + \text{Cu}^{++} \rightleftharpoons \text{Cu}(s) + \text{Zn}^{++}$$

Answer: From Table XIX of standard helf-cell potentials,

$$E^0_{cell} = E^0_{ox.}(\text{Zn/Zn}^{+2}) - E^0_{ox.}(\text{Cu/Cu}^{+2}) = 0.762 - (-0.345)$$
$$= 1.107 \text{ volts}$$

From Eq. 4:
$$\Delta G = -23 \times 2 \times 1.107 = -51 \text{ kcal.}$$

From Eq. 3:
$$K_{eq.} = 10^{-\Delta G/2.3RT}$$
$$= 10^{+51,000/4.6 \times 298} = 10^{37.2}$$

Note that ΔG is negative for the reaction as written, hence, the reaction will favor product formation. $K_{eq.}$ is a very large number, which reflects the very large value of E^0_{cell} and the large negative free energy charge. The equilibrium point lies far over towards products.

In the absence of precise values of ΔG at a given T and P, we can frequently make estimates from values at another T and P. Thus, in Tables VIII and XX, we have listed values for ΔH_f and S, respectively, at 25°C. and 1 atm. pressure. From these, we can calculate ΔG for 25°C. and 1 atm. pressure. However, we make only a small error, if we use these values of ΔH and ΔS to calculate ΔG at other temperatures not too different from 25°C.

Example: An important industrial process is the water-gas reaction to convert coal into gaseous fuels:

$$\text{C}(s) + \text{H}_2\text{O}(g) \rightleftharpoons \text{H}_2(g) + \text{CO}(g)$$

From values of ΔH_f in Table VIII (pg. 58) and S in Table XX (pg. 218), calculate ΔG at 25°C., and K_{eq} for this reaction at 1000°K.

Answer: From Table VIII: $\Delta H = \Delta H_f(\text{CO}) + \Delta H_f(\text{H}_2) - \Delta H_f(\text{H}_2\text{O}) - \Delta H_f(\text{C})$
$$= -26.4 + 0 - (-57.8) - 0$$
$$= 31.4 \text{ kcal.}$$

Note positive sign, hence endothermic reaction.

From Table XX: $\Delta S = S(CO) + S(H_2) - S(H_2O(g)) - S(C(s))$
$= 47 + 31 - 45 - 1$
$= 32$ cal./mole-°K.

At 25°C. (298°K.): $T\Delta S = \dfrac{298 \times 32}{1000} = 9.6$ kcal./mole

Then, $\Delta G = \Delta H - T\Delta S = 31.4 - 9.6 = 21.8$ kcal./mole

At 1000°K.: $T\Delta S = \dfrac{1000 \times 32}{1000} = 32$ kcal./mole

Then, $\Delta G_{1000} \approx 31.4 - 32 = -0.6$ kcal./mole (by significant figures, this is essentially not different from zero).

$$K_{(1000°K)} = 10^{-\Delta G/2.3RT}$$
$$2.3\,RT = \frac{2.3 \times 2.00 \times 1000}{1000} = 4.6 \text{ kcal./mole}$$

$$K_{(1000°K.)} = 10^{-(-0.6/4.6)} = 10^{0.6/4.6} = 10^{0.13}$$
$$= 1.3 \text{ atm.}$$

(Note: $K_{eq.} = \dfrac{C_{H_2} \times C_{CO}}{C_{H_2O}}$ with concentrations measured in units of atmospheres pressure at 1000°K.)

5. Problems

1. Calculate entropy changes for the following reactions (Table XX):
   (a) $H_2(g) + Cl_2(g) \rightleftarrows 2HCl(g)$
   (b) $H_2(g) + \frac{1}{2}O_2(g) \rightleftarrows H_2O(g)$
   (c) $2NO_2(g) \rightleftarrows 2NO(g) + O_2(g)$
   (d) $SO_2(g) + \frac{1}{2}O_2(g) \rightleftarrows SO_3(g)$
   (e) $HCl(g) + NaOH(s) \rightleftarrows NaCl(s) + H_2O(g)$
2. Using values of ΔH_f from Table VIII, calculate heats of reaction for the reactions shown in problem 1, and calculate ΔG for these reactions at 298°K.(standard conditions).
3. $K_{eq.}$ for the reaction, $H_2(g) + I_2(g) \rightleftarrows 2HI(g)$ is 64 at 400°C. Calculate ΔG at 400°C. for this reaction. If $\Delta S_{(400°C.)} = 4$ cal./mole-°K., calculate the heat of the reaction at 400°C.
4. Using the values of $E^0_{ox.}$ from Table XIX (pg. 000), calculate ΔG for the reaction:
$$H_2(g) + Cl_2(g) \rightleftarrows 2HCl(aq)$$
Using Table XX, calculate ΔS and then ΔH for the reaction.
5. For the reaction:
$$H_2(g) + C_2H_6(g) \rightleftarrows 2CH_4(g)$$
Calculate ΔS from values in Table XX. Calculate ΔH, using Table VIII and ΔG. Calculate $K_{eq.}$ at 1000°K.

CHAPTER XVIII

Rates of Chemical Reactions

1. Specific Reaction Rate

By "rate" of chemical reaction we mean the quantity of matter which is being used up or produced per unit of time. The units are generally moles of substances per second.

If the total volume of the reacting system is doubled, then, all other conditions being constant, the amount of matter produced per second is doubled also (i.e., the rate is doubled). To get around this dependence on volume we define an intensive property, the specific reaction rate.

Definition: The specific reaction rate is the quantity of matter being transformed per second, per unit volume of the reaction system.

The usual unit of specific reaction rate is moles per liter-second.

Example:
$$H_2 + I_2 \rightarrow 2HI.$$

In a 25-l. vessel at 280°C., it is found that the rate of production of HI is 3.0×10^{-2} mole HI/sec. What is the specific reaction rate for this reaction?
Answer: By definition:

$$S.R.R. = \frac{\text{Rate of reaction}}{\text{Volume}}$$

$$= \frac{3.0 \times 10^{-2} \text{ mole HI/sec.}}{25 \text{ l.}}$$

Thus
$$S.R.R. = 1.2 \times 10^{-3} \text{ mole HI/l.-sec.}$$

---

2. Dependence of Rate on Concentration—Law of Mass Action

There is no infallible method for predicting the manner in which the rate of any given reaction will depend on the concentrations of either the reacting materials or the products.   To this extent, the dependence of reaction rate on concentration is a fact to be determined experimentally for each given reaction over very specific ranges of temperature and concentration. As an example, the rate of decomposition of $C_2H_6 \rightarrow C_2H_4 + H_2$ varies as the first power of C_2H_6 concentration at $800°K$. and 2 atmosphere pressure, but varies as the square of the C_2H_6 concentration at $1100°K$. and 0.01 atmosphere pressure.

Nevertheless, it is found experimentally that for many reactions, the rate of reaction is proportional to the concentrations of the reacting species raised to powers given by the coefficients in the balanced equation. Although many reactions do not obey this law (especially in gas phase), it is of considerable value for the others. This rule is known as the *law of mass action.*

We can express this law as follows for the reaction of H_2 and I_2:

$$H_2 + I_2 \rightleftarrows 2HI$$

$$\text{Specific rate (production of HI)} = k_f \times C_{H_2} \times C_{I_2}$$

and for the reverse reaction:

$$\text{Specific rate (decomposition of HI)} = k_r \times C^2_{HI}$$

The constants, k_f and k_r, are called the *specific reaction rate constants.* When all concentrations are 1 mole/l., then the specific reaction rate is equal to the specific reaction rate constant.

These specific reaction rate constants may be measured experimentally.   Once known they can be used to compute the reaction rate at any concentration, if the particular reaction follows the law of mass action.   However, their values change with temperature, and so they can be used only for the temperature at which they have been measured.

Their units will depend on the particular equation. If the concentrations are all in moles per liter, then k_f above has the units of $\dfrac{\text{moles HI} \times \text{liters}}{\text{moles } H_2 \times \text{moles } I_2 \times \text{seconds}}$ and k_r has the units of $\dfrac{\text{liters}}{\text{moles HI-seconds}}$.   Since we shall always use concentrations of moles per liter, we shall not bother to insert units for the k's.

Example: The specific reaction rate constant for the formation of HI at 400°C. is 0.43. What is the specific rate of formation of HI in a vessel at 400°C. if $C_{H_2} = 0.04$ mole/l. and $C_{I_2} = 0.006$ mole/l.?

Answer:

$$\text{Specific rate} = k \times C_{H_2} \times C_{I_2} \qquad \text{(by law of mass action)}$$

$$= 0.43 \times (0.04) \times (0.006)$$

$$= 1.0 \times 10^{-4} \text{ mole HI/l.-sec.}$$

Example:

$$2NO + O_2 \rightarrow 2NO_2$$

The specific reaction rate constant for the formation of NO_2 at 0°C. is 3.9×10^{-4}. What is the specific rate of formation of NO_2 in a vessel in which $C_{NO} = 0.070$ mole/l. and $C_{O_2} = 0.32$ mole/l.?

Answer:

$$\text{Specific rate} = k \times C^2_{NO} \times C_{O_2} \qquad \text{(by law of mass action)}$$

$$= 3.9 \times 10^{-4} \times (0.07)^2(0.32)$$

$$= 3.9 \times 10^{-4} \times (4.9 \times 10^{-3}) \times (0.32)$$

$$= 6.1 \times 10^{-7} \text{ mole } NO_2/\text{l.-sec.}$$

Note: For every reaction, chemists must first determine experimentally whether the rate law follows the law of mass action!

3. Mechanism of Chemical Reactions

The fact that many reaction rates seem to follow the law of mass action, particularly reactions in solution, suggests a very simple interpretation of the way in which chemical reactions occur. For example, the fact that the reaction $H_2 + I_2 \rightarrow 2HI$ has a rate of HI production proportional to the product of the concentration of H_2 and I_2, namely rate (HI) = $C_{H_2} \times C_{I_2}$, suggests that the reaction actually occurs by H_2 and I_2 molecules colliding with each other. Such a conclusion would be erroneous, however.

Detailed study of this particular reaction shows that the reaction occurs by way of a rapid, reversible equilibrium formation of I atoms from I_2:

$$I_2 \rightleftarrows 2I \qquad \text{fast}$$

followed by a rapid termolecular collision of two I atoms with an H_2 molecule to form 2HI:

$$2I + H_2 \rightleftarrows [I \cdots H \cdots H \cdots I] \rightarrow 2HI$$

The very similar reaction $H_2 + Br_2 \rightarrow 2HBr$ has a rate law (when $C_{Br_2} > C_{HBr}$):

$$\text{Rate (HBr)} = kC_{H_2} \times C^{1/2}_{Br_2}$$

The decomposition of N_2O_5 into $N_2O_4 + \frac{1}{2} O_2$ has been shown to have a number of steps, despite the fact that the rate depends only on $C_{N_2O_5}$; Rate $(N_2O_4) = kC_{N_2O_5}$. These are:

$$N_2O_5 \rightleftarrows NO_3 + NO_2 \qquad \text{(slow)}$$
$$NO_3 + NO_2 \rightarrow NO_2 + isoNO_3(ONOO) \text{ (slow)}$$
$$isoNO_3 \rightleftarrows NO + O_2 \qquad \text{(fast)}$$
$$NO + NO_3 \rightarrow 2NO_2 \qquad \text{(fast)}$$
$$2NO_2 \rightleftarrows N_2O_4 \qquad \text{(fast)}$$

The reaction in acid solution, $2Br^- + H_2O_2 + 2H^+ \rightleftarrows 2H_2O + Br_2$ has a rate which is third order (i.e., depends on the third power of concentration):

$$\text{Rate (Br}_2) = kC_{H_2O_2} \times C_{H^+} \times C_{Br^-}$$

The mechanism is found to be:

$$H^+ + H\!-\!O\!-\!O\!-\!H \rightleftarrows \left(\begin{array}{c} H \\ H \end{array}\!\!>\!\!O\!-\!O\!\!<\!\!\begin{array}{c} H \\ H \end{array} \right)^{+} \qquad \text{fast}$$

$$\left(\begin{array}{c} H \\ H \end{array}\!\!>\!\!O\!-\!O\!\!<\!\!\begin{array}{c} H \\ H \end{array} \right)^{+} + Br^- \longrightarrow \begin{array}{c} H \\ H \end{array}\!\!>\!\!O + O\!\!<\!\!\begin{array}{c} Br \\ H \end{array} \qquad \text{slow}$$

$$H^+ + HOBr \rightleftarrows \left(\begin{array}{c} H \\ H \end{array}\!\!>\!\!O\!-\!Br \right)^{+} \qquad \text{fast}$$

$$Br^- + \left(\begin{array}{c} H \\ H \end{array}\!\!>\!\!O\!-\!Br \right)^{+} \rightleftarrows Br_2 + H_2O \qquad \text{fast}$$

When reactions occur in consecutive stages, it is the rate of the slowest irreversible step that determines the overall rate. Thus, in the N_2O_5 decomposition, it is the rate of the first two slow steps that determine the overall rate. Every iso-NO_3 species is rapidly converted into products. Similarly, in the HBr + H_2O_2 reaction, it is the second slow step that determines the overall rate. Every HOBr is rapidly turned into Br_2. modern science Chemical Kinetics consists of devising unique and sensitive methods to uncover the detailed mechanisms of chemical reactions.

4. Relation of Equilibrium Constants and Rate Constants

Every chemical reaction can, in principle, reach a state of dynamic equilibrium in which the forward and reverse rates are equal so that no net change in composition occurs. For the HI reaction, this can be written as:

$$\text{Forward Rate} = \text{Reverse Rate}$$
$$\|\qquad\qquad\qquad\qquad\|$$
$$k_f C_{(H_2)e} \times C_{(I_2)e} = k_r \times C^2_{(HI)e}$$

If we divide both sides of this equation by $k_r C_{H_2} C_{I_2}$, we find:

$$\frac{k_f}{k_r} = \frac{C^2_{(HI)e}}{C_{(H_2)e} \times C_{(I_2)e}}$$

But, this last expression is just the expression we would have written for the equilibrium constant, $K_{eq.}$. Hence, we conclude:

$$K_{eq.} = \frac{k_f}{k_r}$$

That is, the ratio of the rate constants for forward and reverse reactions must equal the equilibrium constant. This is true whether the individual rates, forward or back, obey the simple law of mass action. What is found experimentally is that if, for example, the forward rate does not follow the simple law of mass action, then the reverse rate will not either. However, the way in which they both depend on various concentrations will be such as to yield the equilibrium constant expression when the rates are set equal to each other.

For example, in the $H_2 + Br_2 \rightarrow 2HBr$ reaction, the forward rate is given by:

$$\text{Rate (HBr)} = k_f \times C_{H_2} \times C^{1/2}_{Br_2}$$

The reverse rate must then be given by:

$$\text{Reverse Rate} = k_r \times \frac{C^2_{HBr}}{C^{1/2}_{Br_2}}$$

y are set equal:

$$k_f C_{(H_2)e} \times C^{1/2}_{(Br_2)e} = k_r \frac{C^2_{(HBr)e}}{C^{1/2}_{(Br_2)e}}$$

we can rearrange the equation to obtain:

$$\frac{k_f}{k_r} = \frac{C^2_{(HBr)e}}{C_{(H_2)e} \times C^{1/2}_{(Br_2)e} \times C^{1/2}_{(Br_2)e}} = \frac{C^2_{(HBr)e}}{C_{(H_2)e} \times C_{(Br_2)e}}$$

5. Dependence of Rate on Temperature—Arrhenius Equation

It is found experimentally that a small increase in temperature will increase the rates of simple reactions by a large amount. (For a few complex reactions such as the reaction of $2NO + O_2 \rightarrow 2NO_2$, the effect of increasing temperature is to decrease the reaction rate.)

TABLE XXI
Factors by Which Reaction Rates Are Changed by a 10°C. Change in Temperature
(For reactions with different activation energies at different temperatures)

Temperature		Activation Energy				
(t, °C.)	T, °K.	18 kcal./mole	36 kcal./mole	54 kcal./mole	72 kcal./mole	90 kcal./mole
(27)	300	2	3	4	5	6
(127)	400	1.6	2.1	2.7	3.3	3.9
(227)	500	1.36	1.7	2.1	2.4	2.8
(327)	600	1.25	1.5	1.75	2.0	2.25
(427)	700	1.2	1.36	1.55	1.7	1.92

The magnitude of this effect of increasing temperature will depend on what we call the "activation energy of the reaction." The activation energy of a reaction may be interpreted as the minimum energy which the molecules must have in order to break their bonds and react. Activation energies are measured experimentally for each reaction. Those reactions that have a large activation energy will be more sensitive to temperature changes than reactions having a low activation energy.

Roughly it is found that many reactions will double their specific rates when the temperature is increased by 10°C. However, this will depend somewhat on the temperature and on the activation energy of the reaction. Table XXI shows how reactions having different activation energies will be affected when the temperature is raised 10°C.

From the chart we see that a reaction having an activation energy of 54 kcal./mole will have its speed increased 4-fold at 300°K. by a 10°C. rise in temperature but only 1.55-fold at 700°K. by the same temperature change.

The relation between the specific reaction rate constants at different temperatures was discovered by Arrhenius and may be expressed algebraically:

Arrhenius Equation:

$$k = A \times 10^{-E_{\text{act.}}/2.3RT} \tag{1}$$

If we measure the rate constant at two different temperatures, T_1 and T_2, it will have values k_1 and k_2 given by:

$$k_1 = A \times 10^{-E_{\text{act.}}/2.3RT_1}, \qquad k_2 = A \times 10^{-E_{\text{act.}}/2.3RT_2}$$

Dividing these two equations, the A cancels and we have:

$$\frac{k_1}{k_2} = 10^{-E_{\text{act.}}/2.3RT_1 \; - \; E_{\text{act.}}/2.3RT_2} \tag{2}$$

Taking logarithms of both sides, we find:

$$\log\left(\frac{k_1}{k_2}\right) = \frac{(T_1 - T_2) \times E_{\text{act.}}}{2.3RT_1T_2} \tag{3}$$

$E_{\text{act.}}$ is the activation energy in calories per mole and R is the universal gas constant $= 2$ cal./mole-°K.

The Arrhenius equation may be used to obtain rate constants at one temperature, if the rate constant at any other temperature is known.

Example: The specific rate constant for the reaction of H_2 with I_2 to form HI is 0.43 at 400°C. If the activation energy is 41 kcal./mole, what is the specific rate constant at 500°C.?

Answer: From the Arrhenius equation:

$$\log \frac{k_1}{k_2} = \frac{(T_1 - T_2)E_{act.}}{2.3 \times R \times T_1 \times T_2}$$

Substituting: $T_1 = 773°K.$, (500°C.); $T_2 = 673°K.$ (400°C.); $R = 2$ cal./mole-°K.; $E_{act.} = 4.0 \times 10^4$ cal./mole; and $k_2 = 0.43$.

$$\log \left(\frac{k_1}{0.43} \right) = \frac{(773°K - 673°K) \times 4.1 \times 10^4 \text{ cal./mole}}{2.3 \times 2 \text{ cal./mole-°K.} \times 673°K. \times 773°K.}$$

$$= \frac{100 \times 4.0 \times 10^4}{2.3 \times 2 \times 673 \times 773} = 1.71$$

Taking antilogs:

$$\frac{k_1}{0.43} = 51$$

Thus

$$k_1 = 51 \times 0.43 = 22$$

Note: If we used our crude rule of doubling the rate for every 10°C. rise in temperature, we would have an increase of $2^{10} \cong 1000$ since there are 10 increases of 10°C. going from 400°C. to 500°C. This method would give $k_1 = 430$.

If we used the nearest factor from Table XIX, namely 1.36, we would have $(1.36)^{10}$ for the increase, which $= 21.6$, which is only off by a factor of 2.

The experimentally observed increase is 54-fold, i.e., $k_1 = 23.2$. Thus the Arrhenius equation was in error by only 6%.

The activation energy, $E_{act.}$, can be interpreted as the average excess energy that the reactants require in order to be transformed into product molecules. It is an energy barrier to the reactant.

6. Dependence of Equilibrium Constants on Temperature

It is found that the Arrhenius type equation applies to many other constants besides reaction rate constants. Thus all equilibrium constants obey an Arrhenius type equation. In these cases the heat of reaction is used to replace the activation energy. The relation is named after the famous chemist van't Hoff.

van't Hoff Equation for Equilibrium Constants (see Chapter XVII, Sect. 4):

$$K_{eq.} = 10^{\Delta S/2.3R \ - \ \Delta H/2.3RT}$$

Here ΔS = the molar entropy change for the reaction and ΔH = the molar heat of reaction.   Over short ranges of temperature, ΔS and ΔH may be considered constant.

If we measure $K_{eq.}$ at two different temperatures, T_1 and T_2, and final values K_1 and K_2:

$$K_1 = 10^{\Delta S/2.3R \;-\; \Delta H/2.3RT_1}, \quad K_2 = 10^{\Delta S/2.3R \;-\; \Delta H/2.3RT_2}$$

Dividing these two equations, we find the ΔS term cancels and:

$$\frac{K_1}{K_2} = 10^{-\Delta H/2.3R \,(1/T_1 \;-\; 1/T_2)}$$

Taking logarithms of both sides, we find:

$$\log \frac{K_1}{K_2} = \frac{(T_1 - T_2)\,(\Delta H)}{2.3RT_1T_2}$$

As an example we may apply it to the $K_{ion}(H_2O)$.   At 25°C., $K_{ion}(H_2O) = 1.0 \times 10^{-14}$.   The heat of the reaction:

$$2H_2O \rightleftarrows OH^{-1} + H_3O^{+1}$$

is exactly equal to minus the heat of neutralization of a strong acid and a strong base, namely 13 kcal./mole.   Let us calculate $K_{ion}(H_2O)$ at 50°C.

$$\log \frac{K_1}{K_2} = \frac{(323°K. - 298°K.) \times 13{,}000 \text{ cal./mole}}{2.3 \times 2 \text{ cal./mole-°K.} \times 298°K. \times 323°K.}$$

$$= \frac{25 \times 1.3 \times 10^4}{2.3 \times 2 \times 298 \times 323} = 0.735$$

Taking antilogs:

$$\frac{K_1}{K_2} = 5.4$$

Therefore

$$K_1 = 5.4 \times 1 \times 10^{-14} = 5.4 \times 10^{-14}$$

The observed value for K at 50°C. is 5.47×10^{-14}, an excellent check.

The Arrhenius equation for rate constants and the van't Hoff equation for equilibrium constants can be used to deduce a relation between activation energies for forward and reverse reactions:

$$\frac{k_f}{k_r} = K_{eq.}$$

$$\downarrow$$

$$\frac{A_f 10^{-E_{act.\,(f)}/2.3RT}}{A_r 10^{-E_{act.\,(r)}/2.3RT}} = 10^{\frac{\Delta S}{2.3R} - \Delta H/2.3RT}$$

$$E_{act.\,(f)} - E_{act.\,(r)} = \Delta H$$

$$\text{and } \frac{A_f}{A_r} = 10^{\Delta S/2.3R}$$

These relations can be given a simple interpretation when we are dealing with an elementary reaction between atoms or molecules. In such a reaction, the atoms or molecules must change their geometry and pass through a critical configuration called the "Transition State." If they reach this configuration, they will react; if they don't, they cannot react.

The value of $E_{act.\,(f)}$ is the heat of reaction to form this transition state from the reactants. $E_{act.\,(r)}$ is the heat of reaction to form this same transition state from products. The difference in these two activation energies is then precisely the difference in enthalpies of reactants and products or the heat of reaction.

In similar fashion, we can speak of an entropy of activation, $\Delta S_f^{\ddagger}$, which is the entropy change in going from reactant molecules to the transition state and $\Delta S_r^{\ddagger}$, which is the entropy change in going to the same transition state from products. Then, $\Delta S_f^{\ddagger} - \Delta S_r^{\ddagger} = \Delta S$. The Arrhenius A-factors can be related to these entropies of activation by a simple constant:

$$A = \frac{eRT}{N_{av.}h} 10^{\Delta S^{\ddagger}/2.3R}$$

where h is Planck's constant (6.6×10^{-27} erg-sec./molecule), $N_{av.}$ is Avogadro's number 6.02×10^{23} molecules/mole and $e = 2.72$.

7. First Order Reactions—Half Lives

For a large number of important reactions, the rate of the reaction is proportional to the first power of the concentration of *one* of the reactants. Because the concentration appears to the first power these are called first-order reactions. (*Note:* for reactions whose rates depend on the square or second power of concentrations, the rates are called second order, etc.)

For first-order reactions, the specific rate constants have units of reciprocal time (sec.$^{-1}$) and are thus independent of concentration units. A very important category of such first-order reactions are the spontaneous decompositions of radioactive nuclei of elements such as ^{238}U or ^{226}Ra. One property of such reactions is that their fractional or percentage rates of decomposition are independent of the quantity or concentration of material present. Thus for any sample of Ra, whether it is a ton or a milligram, it will be found that 10% of it will always take the same time to decompose.

Because of this property it is convenient to discuss these reactions in terms of their *half-lives*.

The half-life of a first-order reaction is equal experimentally to the time that would be required for half of the original quantity of material to decompose. The abbreviation of half-life is $t_{1/2}$. It can be demonstrated that the specific rate constant for a first order reaction is equal to $0.69/t_{1/2}$:

$$k_{1st\ order} = \frac{0.69}{t_{1/2}}$$

Example: It is found that 1.0 g. of ^{238}U gives off 1.3×10^4 alpha particles per second. Calculate the half-life of ^{238}U.

Answer: By definition

$$\text{Rate of reaction} = k \times (U)$$

Therefore

$$k = \frac{\text{Rate of reaction}}{(U)}$$

Assuming that 1 U atom is destroyed for every alpha particle produced:

$$k = \frac{1.3 \times 10^4 \text{ atoms U/sec.}}{1.0 \text{ g. U}\left(\dfrac{1 \text{ mole U}}{238 \text{ g. U}}\right)\left(\dfrac{6 \times 10^{23} \text{ atoms U}}{1 \text{ mole U}}\right)}$$

$$= \frac{1.3 \times 238 \times 10^4}{6 \times 10^{23}} \text{ sec.}^{-1}$$

$$= 5.1 \times 10^{-18} \text{ sec.}^{-1}\left(\frac{3600 \text{ sec.}}{1 \text{ hr.}}\right)\left(\frac{24 \text{ hr.}}{1 \text{ day}}\right)\left(\frac{365 \text{ days}}{1 \text{ yr.}}\right)$$

$$= 1.6 \times 10^{-10} \text{ yr.}^{-1}$$

therefore $t_{1/2} = \dfrac{0.69}{k} = \dfrac{0.69}{1.6 \times 10^{-10} \text{ yr.}^{-1}} = 4.3 \times 10^9 \text{ yr.}$

8. Problems

1. H_2 and Br_2 will react to form HBr at a rate of 1.5×10^{-4} mole HBr/sec. in a 300-cc. flask. What is the specific rate of this reaction?

2. Under given conditions the specific rate of reaction of Sn^{+2} ions with Hg^{+2} ions to produce Sn^{+4} and Hg_2^{+2} ions is 1.7×10^{-3} mole Sn^{+4} ions/l.-sec. What is the rate under the same conditions, if the reactants fill a 12-l. vessel?

3. Write expressions showing the dependence of the following reactions on concentration (assume law of mass action).

(a) $2NO + Cl_2 \rightarrow 2NOCl$.

(b) $2N_2O_5 \rightarrow 4NO_2 + O_2$.

(c) $Cl_2 + 2HBr \rightarrow 2HCl + Br_2$.

(d) $Cl_2 + Br_2 \rightarrow 2ClBr$.

4. If the specific reaction rate constant of reaction 3(a) is 2.1×10^{-3}, calculate the specific rate when $C_{NO} = 1.8$ moles/l. and $C_{Cl_2} = 0.6$ mole/l.

5. It is found that the specific rate of reaction 3(c) is 3.5×10^{-4} when $C_{Cl_2} = 0.04\ M$ and $C_{HBr} = 2.6\ M$. Calculate the specific reaction rate constant for this reaction.

6. Compute the change in specific reaction rate constant for the following, using the Arrhenius equation:

(a) $k = 3.5 \times 10^{-4}$ at 200°C. What is it at 150°C. if $E_{act.} = 24$ kcal./mole?

(b) $k = 2.7 \times 10^{-2}$ at -80°C. What is it at 30°C. if $E_{act.} = 12$ kcal./mole?

(c) $k = 7.5 \times 10^{-8}$ at 400°C. What is it at 600°C. if $E_{act.} = 63$ kcal./mole?

Compare these calculated values with the value predicted by the rough rule of doubling the rate for every 10°C. rise.

7. For the reaction, $CO_2\ (g) + H_2\ (g) \rightleftarrows CO\ (g) + H_2O\ (g)$, it is found that $K_{eq.} = 0.534$ at 686°C. and 1.571 at 986°C. Using the Arrhenius equation, compute the heat of this reaction. Is the reaction exothermic or endothermic?

8. ^{226}Ra has a half-life of 1620 yr. Calculate the specific rate constant in units of sec.$^{-1}$. How many disintegrations will occur in 1 sec. in 1 g. of Ra? (*Note:* This latter defines 1 curie of radiation.)

9. ^{218}Rn has a half-life of 0.019 sec. How many disintegrations will be observed from 1 microgram of Rn in 0.001 sec.?

10. The specific rate constant for the disintegration of ^{199}Bi is 4.2×10^{-4} sec.$^{-1}$. Calculate its half-life. How many disintegrations will occur from 1 mg. of Bi in 1 min.?

11. At 30°C., the half-life for the decomposition of N_2O_5 is 600 sec. Calculate the specific rate constant in units of hr.$^{-1}$. What fraction of a given sample of N_2O_5 will decompose in 10 sec.?

12. ^{40}K occurs naturally to the extent of 0.01% in normal K. Its half-life is 1.8×10^9 yr. If the average human body contains about 5 g. of normal K, how many disintegrations will occur in an average person in one day?

13. For the reaction, $2NO + Br_2 \rightleftarrows 2NOBr$, the rate law is:
$$\text{Rate (NOBr)} = k_f C^2_{NO} \times C_{Br_2}$$
What is the rate law for the reverse reaction?

14. For the reaction, $2NO + 2H_2 \rightarrow N_2 + 2H_2O$, the rate law is:
$$\text{Rate (N}_2) = k_f C^2_{NO} \times C^{1/2}_{H_2}$$
What is the rate law for the reverse reaction?

15. For the reaction, $CH_3CH_2Br \rightarrow CH_2 = CH_2 + HBr$, the rate law is:
$$\text{Rate (HBr)} = k_f C_{CH_3CH_2Br}$$

What is the rate law for the reverse reaction? If that reaction is endothermic by 18 kcal. and $E_{act.(f)} = 54$ kcal., what is $E_{act.}$ for the reverse reaction?

16. The rate law for the reaction, $Br + H_2 \rightleftarrows HBr + H$, is Rate $(HBr) = k_f C_{Br} \times C_{H_2}$ and k_f has $E_{act.} = 19$ kcal. If the reaction is endothermic by 17 kcal., what is $E_{act.}$ for the reverse reaction? What is the reverse rate law?

17. For the reaction, $2NO + O_2 \rightleftarrows 2NO_2 + 27.4$ kcal., the activation energy for the forward reaction is negative, -1 kcal. What is $E_{act.}$ for the reverse reaction?

APPENDIX I

Some Mathematical Definitions and Operations

1. Algebraic Operations

Algebra is an abstract science which deals with defined entities, numbers, and defined operations which may be performed on them. The great power of algebra lies in its system of abbreviation, that is, its compact language. This compactness permits us to express very complex relations very briefly.

Most of the operations of algebra have associated with them inverse operations. Thus addition $(+)$ has the inverse operation, subtraction $(-)$. Performing an operation and its inverse in succession leaves the quantity operated on unchanged. Thus, if we add a number to a quantity and then subtract the same number, the original quantity is unchanged. This may be expressed compactly by means of an equation. If we let $X =$ the original quantity and $N =$ the number added, then

$$X + N - N = X$$

Multiplication $(\times)$ is a compact method for expressing a continued addition. Thus 2×5 means $5 + 5$; 3×4 means $4 + 4 + 4$. Division $(\div)$ is the inverse of multiplication. If we multiply a quantity (X) by a number (N) and then divide by N, the original quantity is unchanged:

$$X \times N \div N = X$$

Some operations commute with other operations. That is, if the order of operations is reversed, the result is the same.

Operations commute with their inverse operations and with themselves. Thus

$$5 \times 4 \times 3 = 5 \times 3 \times 4 = 3 \times 5 \times 4, \text{ etc.}$$

Also

$$(5 \times 4) \div 2 = (5 \div 2) \times 4$$

Addition and subtraction do not commute with multiplication or division. Thus $(12 + 2) \times 3$ is not equal to $(12 \times 3) + 2$.

2. Exponents

The operation of continued multiplication by a certain quantity is abbreviated by an exponent. Thus 4^3 means $4 \times 4 \times 4$; X^5 means $X \cdot X \cdot X \cdot X \cdot X$. From this we can deduce some simple rules of operations on such quantities:

Combining Exponents.

$$5^3 \times 5^2 \text{ means } (5 \times 5 \times 5) \times (5 \times 5) = 5^5$$

or, in general,

$$X^a \cdot X^b = X^{a+b}$$

When multiplying quantities having different exponents we add the exponents. Note, however, we may not simplify $5^3 \times 4^2$ further. The base quantities having different exponents must be the same.

$$(5^3)^4 \text{ means } (5^3) \times (5^3) \times (5^3) \times (5^3) = 5^{12}$$

Thus when we raise a quantity (expressed as an exponent) to a power, the two exponents are multiplied. In general $(X^a)^b = X^{a \cdot b}$. Manipulation involving exponents can always be checked if the definitions are kept in mind.

Rule: Any quantity raised to exponent zero is 1. Thus $X^0 = 1$.

Negative Exponents. The operation of continued division by a given quantity is abbreviated by a negative exponent. Thus if we want to divide a quantity by 10 three times, this operation may be represented by multiplication by

$$10^{-3} = \frac{1}{10} \times \frac{1}{10} \times \frac{1}{10} = \frac{1}{10^3}$$

The reciprocal of a number is that number with a negative exponent. Thus

$$\frac{1}{2} = 2^{-1}; \quad \frac{1}{X} = X^{-1}; \quad \frac{1}{10} = 10^{-1}; \quad \frac{1}{X^4} = X^{-4}$$

Note that the inverse of the operation of multiplication by X^a is division by X^a or multiplication by X^{-a}; the application of both these operations cancel each other: $(X^a) \cdot (X^{-a}) = 1$.

Fractional Exponents. The inverse operation to raising a quantity to a given power is called taking the root of the quantity. This may be expressed by a fractional exponent.

Thus $9^{\frac{1}{2}}$ (the $\frac{1}{2}$ root or square root of 9) means the number which when multiplied by itself twice will give 9.   The answer is, of course 3.

The operations of squaring and taking square roots are inverse and commute.   Thus:

$$(3^2)^{\frac{1}{2}} = (3^{\frac{1}{2}})^2 = 3^1$$

$$(5^3)^{\frac{1}{3}} = (5^{\frac{1}{3}})^3 = 5^1.$$

The same rules for combination apply to integral and fractional exponents.   Thus

$$(3^{\frac{1}{2}}) \times 3^2 = 3^{2+\frac{1}{2}} = 3^{5\frac{1}{2}}.$$

3. Numbers Expressed as Powers of 10

It is a great convenience to express both large and small numbers as powers of 10.   Thus

$$1,000,000,000 = 1 \times 10^9$$

$$176,000 = 176 \times 10^3 = 17.6 \times 10^4 = 1.76 \times 10^5$$

$$0.001 = 1 \times 10^{-3}$$

$$0.000047 = 47 \times 10^{-6} = 4.7 \times 10^{-5}$$

To convert such a number expressed as an exponent back to a number, we have a simple rule:

For Positive Exponents: Shift the decimal point forward by the number of places indicated by the exponent.
For Negative Exponents: Move the decimal place back as many places as the exponent.

The use of exponents allows us to make operations much less cumbersome.

Whenever we have a group of numbers to multiply and divide, if the numbers are very large or small, express all numbers as numbers between 1 and 10 multiplied by some power of 10.   Group all the numbers and powers of 10 separately and then perform the operations.

Examples:

$$\frac{720,000}{48,000} = \frac{7.2 \times 10^5}{4.8 \times 10^4} = \frac{7.2}{4.8} \times 10^1 = 15$$

$$\frac{12,000 \times 0.006}{0.036 \times 250} = \frac{1.2 \times 10^4 \times 6 \times 10^{-3}}{3.6 \times 10^{-2} \times 2.5 \times 10^2} = \frac{1.2 \times 6 \times 10^1}{3.6 \times 2.5}$$

$$= \frac{7.2}{9.0} \times 10^1 = 0.8 \times 10 = 8.$$

4. Logarithms to the Base 10

We define the logarithm (log) of a number to the base 10 as that exponent to which 10 must be raised to give the number. Thus: $10^x = N$, may be taken to mean that the log of N is X: $\log N = X$ (in the base 10).

For very simple numbers:

$$\log 10 = \log 10^1 = 1$$

$$\log 100 = \log 10^2 = 2$$

$$\log 0.01 = \log 10^{-2} = -2$$

$$\log 1 = \log 10^0 = 0 \qquad \text{(since } 10^0 = 1\text{)}$$

Theorem: The logarithm of a product of two or more numbers is equal to the sum of the logarithms of the numbers. Thus

$$\log (N \times M) = \log N + \log M$$

Example:

$$\log (10 \times 100) = \log 10 + \log 100 = 1 + 2 = 3$$

Theorem: The logarithm of a number raised to some exponent is equal to the exponent times the log of the number. Thus

$$\log (X^a) = a \cdot (\log X)$$

Examples:

$$\log (10^4) = 4 \cdot \log 10 = 4 \times 1 = 4$$

$$\log \frac{1}{1000} = \log (1000^{-1}) = -1 \cdot \log 1000 = -3$$

$$\log \left(\frac{N}{M}\right) = \log (N \times M^{-1}) = \log N + \log M^{-1}$$

$$= (\log N) - 1 \cdot (\log M) = \log N - \log M$$

$$\log \left(\frac{100}{1000}\right) = \log 100 + \log (1000^{-1}) = \log 100 - \log 1000$$

$$= 2 - 3 = -1$$

Tables have been prepared for finding the logarithms of numbers that are not simple powers of 10. If we wish to find the logarithm of such a number it is simplest first to express the number as a power of 10.

To find the logarithm of 230 we observe that 230 lies between 100 (log = 2) and 1000 (log = 3). Then log 230 will be between 2 and 3. The number before the decimal point of a logarithm is called the *characteristic*. The remainder of the logarithm after the decimal point is called the *mantissa*.

The *characteristic* of log 230 will be 2. To find the mantissa we look in our logarithm tables under 23 or 230 (we will find the same mantissa for each). We obtain 0.3617 (four-place log tables) or 0.36173 (five-place log tables). We add the characteristic to the mantissa and obtain log 230 = 2.3617.

Examples:

$$\log 36{,}400 = \log (3.64 \times 10^4) \qquad \text{(characteristic} = 4)$$

$$= 4.561$$

$$\log 0.0075 = \log (7.5 \times 10^{-3}) \qquad \text{(characteristic} = -3)$$

$$= -3 + 0.8751$$

$$= -2.125$$

---

5. Antilogarithms

The inverse operation to finding the logarithm, given the number, is to find the number (antilogarithm), given the logarithm. Thus we may be told that the logarithm of a number is 2 and asked to find the number. This is, of course, simple since the log of 100 is 2. Thus the number whose log is 2 is 100, or conversely, the antilog 2 = 100. Note that the operations log and antilog commute and cancel each other.

To find the antilog of a number we must express the number with a positive mantissa since the log tables give values only for positive mantissas.

Example: Find the antilog of 1.301.

Answer: First note that log 10 = 1 and log 100 = 2. Thus the number lies between 10 and 100. We write

$$1.301 = 1 + 0.301$$

We can now look up 0.301 in the log tables, and we find that 200 has the value 0.301 as its mantissa. The number is thus $2.00 \times 10^1 = 20.0$.

Example: Find the antilog of -2.730.

Answer: We first write $-2.730 = -3 + 0.270$, that is, we write the number with a positive mantissa. We can now look up this positive mantissa and find that

186 is the nearest number having the mantissa 0.270.   The antilog of -2.730 is then $1.86 \times 10^{-3} = 0.00186$.

We can always check:

$$\log (0.00186) = \log (1.86 \times 10^{-3})$$
$$= \log 1.86 + \log (10^{-3})$$
$$= \log 1.86 - 3 = 0.270 - 3$$
$$= -2.730$$

6. Using Logarithms To Do Problems

We may use logarithms to simplify calculations for us:

Example: Find X, given:

$$X = \frac{27.6 \times 873}{0.076 \times 96,200}$$

Answer: We can, of course, work this out by longhand.   However, it is simpler if we use logarithms and exponents.   Let us write it using exponents:

$$X = \frac{2.76 \times 10^1 \times 8.73 \times 10^2}{7.6 \times 10^{-2} \times 9.62 \times 10^4} = \frac{2.76 \times 8.73}{7.6 \times 9.62} \times 10^1$$

We now proceed to evaluate the fraction $\dfrac{2.76 \times 8.73}{7.6 \times 9.62}$ using logarithms:

$$\log \left(\frac{2.76 \times 8.73}{7.6 \times 9.62} \right) = \log 2.76 + \log 8.73 - \log 7.6 - \log 9.62$$

(by rule about products)

$$\begin{bmatrix} \log 2.76 = 0.4409 & \log 7.6 \ \ = 0.8808 \\ \log 8.73 = 0.9410 & \log 9.62 = 0.9832 \\ \overline{ 1.3819} & \overline{ 1.8640} \end{bmatrix}$$

$$= 1.3819 - 1.8640 = -0.4821$$
$$= -1 + 0.5179$$

Now by looking up the antilog of 0.5179 we find it to be 330.   Thus

$$\frac{2.76 \times 8.73}{7.6 \times 9.62} = 3.30 \times 10^{-1} = 0.330$$

and

$$X = \frac{2.76 \times 8.73}{7.6 \times 9.62} \times 10^1 = 0.330 \times 10^1 = 3.30$$

Example: What is the sixth root of 18.76?   We write:

$$X = 18.76^{1/6}$$

Taking logarithms of both sides:

$$\log X = \log (18.76)^{\frac{1}{6}} = \frac{1}{6} \times \log 18.76$$

by our rules about logarithms of exponents. Thus

$$\log X = \frac{1}{6} \times \log (1.876 \times 10^1)$$

$$= \frac{1}{6} \times (1.2732) = 0.2122$$

Taking antilogs of both sides:

$$X = 1.630 \quad (\text{i.e., } 1.630^6 = 18.76)$$

Note: If we want only 1% accuracy all of these types of problems may be performed with a slide rule. Log tables are needed for greater accuracy.

7. Getting the Decimal Point in a Problem

The method of writing numbers as powers of 10 provides us with a simple method of obtaining the decimal point in a calculation involving many operations.

To do this we write the numbers as powers of 10. We then make an approximate calculation in which we replace each coefficient by the nearest integer and perform the operations. This approximate answer will be close to the correct answer and tell us where the decimal belongs.

Example: Calculate the value of $X = \dfrac{394 \times 7.25 \times 0.0642}{0.0428 \times 9{,}680 \times 22.8}$.
We express this in powers of 10:

$$X = \frac{3.94 \times 10^2 \times 7.25 \times 6.42 \times 10^{-2}}{4.28 \times 10^{-2} \times 9.68 \times 10^3 \times 2.28 \times 10^1}$$

$$= \frac{3.94 \times 7.25 \times 6.42}{4.28 \times 9.68 \times 2.28} \times 10^{-2}$$

The approximation gives:

$$X = \frac{4 \times 7 \times 6}{4 \times 10 \times 2} \times 10^{-2} = \frac{42}{20} \times 10^{-2}$$

$$= 2.1 \times 10^{-2} = 0.021$$

The correct answer must be close to 0.021. If we now perform all operations on a slide rule without worrying about the decimal, we find $X = 1945$. The correct answer is then $X = 0.01945$.

8. Significant Figures

An experimental measurement is no better than the instrument used. If we are measuring a length with a ruler marked off in millimeters as the smallest division, then we express our result to the nearest millimeter.* The length 273 mm. measured in this fashion indicates that the actual length may lie anywhere between 272.5 mm. and 273.4 mm. It is closest to 273 mm. which is written.

The answer 273 mm. is said to be expressed to three significant figures. The number 58 contains two significant figures. In order to obtain a uniform system of expressing results so that the number of significant figures is apparent, many scientists will express their findings in powers of 10. The total number of digits given when this is done indicates the number of significant figures in the answer. The following list shows some results expressed this way.

Result	Number of Significant Figures	Expressed as a Power of 10
21,000	2	2.1×10^4
	3	2.10×10^4
	4	2.100×10^4
0.00035	2	3.5×10^{-4}
	3	3.50×10^{-4}
	4	3.500×10^{-4}
	5	3.5000×10^{-4}

When written in this fashion, as powers of 10, it will be immediately clear from the total number of digits how many significant figures are indicated.

9. Adding Significant Figures —Absolute Accuracy

The principal reason for reporting significant figures in scientific experiments is that we can then tell how far to go in interpreting and using the result.

If a chemist weighs the amount of Ag to be used in an experiment to the nearest gram and records 18 g., we interpret this as meaning

* Generally we would try to estimate the length to 0.1 mm.

that the true weight might be anywhere between 17.5 and 18.4 g.
If now another chemist weighs some more silver in a more sensitive
balance and records the weight as 1.25 g. and adds it to the first
sample, how shall we record the total weight? If we add we obtain
19.25 g. However, the first result was recorded only to the nearest
gram, and it is not meaningful to repeat the total weight to any more
significant figures. It should be recorded simply as 19 g.

The numbers 11 and 99 both contain two significant figures. They
imply that the physical measurements, which were made to obtain them,
had the same *absolute* accuracy, namely $\pm$ 0.5 units. When we add
or subtract numbers, the final result has an absolute inaccuracy at
least as large as that of the least accurate number in the sum or difference.

When we add or subtract numbers that have the same absolute
accuracy, then the result usually has actually a lesser absolute accuracy
than the individual accuracies. The reason for this is that it is a matter
of chance whether the uncertainties in each number have the same sign
or opposite sign. Hence, the errors may cancel, or they may augment
each other.

For example, suppose we measure a result 20 and a result 18. Since
each has two significant figures, each has the same absolute inaccuracy,
namely $\pm$ 0.5 units. Their sum is 38, but the inaccuracy is now ($\pm$ 0.5)
+ ($\pm$ 0.5) units. This may be as big as $\pm$ 1, if we have been unlucky,
or as small as zero, if we have been lucky. The rule that is used in
adding or subtracting experimental numbers is to *add* their uncertainties
vectorially. That is, we add the squares of the uncertainties and take
the square root of the sum.

Example: Add the measured masses, 164 g., 18 g., and 23 g., and estimate the
uncertainty of the result.

Answer: Since each result is recorded to the nearest unit, the absolute uncertainty
in each is the same, namely $\pm$ 0.5 g.

	Measured Mass	Absolute Accuracy
	164 g.	$\pm$ 0.5 g.
	18 g.	$\pm$ 0.5 g.
	23 g.	$\pm$ 0.5 g.
Sum	205 g.	$\pm [(0.5)^2 + (0.5)^2 + (0.5)^2]^{1/2} = \pm 0.87$ g.
		$\rightarrow \pm 0.9$ g.

In adding significant figures, the result cannot be expressed with any more precision than the crudest measurement in the sum.

Example: Add the following lengths: 102.5 cm., 32.76 cm., 0.008 cm., and 915 cm.

Answer:

$$
\begin{array}{l}
102.5 \\
32.76 \\
0.008 \\
915. \qquad \leftarrow \text{crudest measurement} \\
\hline
1050.268
\end{array}
$$

The answer is 1050 cm. since the crudest figure (915 cm.) is not reported to better than 1 cm.

---

10. Multiplying and Dividing Significant Figures —Relative Accuracy

Although the numbers 11 and 99 have the same number of significant figures, and also the same *absolute* accuracy, they differ in their *relative* accuracy. An uncertainty of $\pm$ 0.5 units out of 11 units is $\frac{1}{2}$ part in 11, or 1 part in 22 parts.   In contrast, an uncertainty of $\pm$ 0.5 parts in 99 is $\frac{1}{2}$ part in 99, or 1 part in 198 parts.   This latter has a much greater relative accuracy than the former, namely nine-fold greater.

When we *multiply* or *divide* measured numbers, the final result cannot have greater *relative* accuracy than that of the least accurate (relative) number.

Thus, if we divide 99 by 11, the result of 9.00 should be written as 9. This implies a relative accuracy of $\pm$ 0.5 or 1 part in 18, which is about the same as that of 11.

If we multiply 99 by 11, the result is 1089.   However, since the result should only be expressed to the same relative accuracy as that in the least accurate figure 11, we shall record it as 11×10^2:

$$
99 \times 11 = 1089 = 10.89 \times 10^2 \rightarrow 11 \times 10^2
$$

In multiplying or dividing significant figures, the final result cannot contain any more significant figures than the least number of significant figures in any of the quantities used.   Thus

$$\frac{982 \times 22 \times 14}{362 \times 756} = 0.55259 \quad \text{(by longhand division)}$$

However, the answer should be recorded as 0.55 since there is a number (actually two, 14 and 22) which is recorded to only two significant figures.

Classification of the Properties of Pure Substances

I. Elements

PHYSICAL PROPERTIES

A. Metals	B. Non-metals	C. Amphoteric Metals
(e.g., Na,K,Ca,Fe)	(e.g., C,N,O,Cl)	(e.g., Al,Zn,Pb,Sb, Ge,Hg,At,Te)
1. Good conductors of heat and electricity.	1. Usually poor or non-conductors of heat and electricity.	1. Have physical properties which may be intermediate between metals and non-metals, but usually they resemble metals.
2. Usually solids at room temperatures with high boiling points, although occasionally occur as liquids.	2. Usually gases, liquids, or easily volatilized solids at room temperature. Occasionally occur as high-boiling solids.	
3. Have luster and are soft, malleable, ductile and possess high tensile strength.	3. Are hard and brittle, with no luster and very low tensile strength.	

CHEMICAL PROPERTIES

1. Tend to lose electrons in reacting with non-metals to form positively	1. Tend to gain electrons in reacting with metals to form negatively charged	1. Tend to share electrons in reactions with non-metals to form covalent

A. Metals	B. Non-metals	C. Amphoteric Metals
(e.g., Na,K,Ca,Fe)	(e.g., C,N,O,Cl)	(e.g., Al,Zn,Pb,Sb, (Ge,Hg,At,Te)

charged ions (cations). The compound formed when a metal and non-metal react to form a solid ionic compound is called a salt.	ions (anions). 2. Tend to share electrons in reacting with non-metals to form covalent compounds which are not salts.	salts (pseudo-electrolytes). 2. Will not generally react with metals.
2. Will not generally react with other metals.	3. Will react with oxygen to form acidic oxides.	3. Will react with oxygen to form oxides which are generally basic but may also be acidic.
3. Will react with oxygen to form basic oxides.	4. Usually act as oxidizing agents.	4. Usually act as reducing agents.
4. Usually act as reducing agents.		

(*Note:* Most elements may be prepared from their salts by electrolysis of the molten salts.)

II. Compounds

A. Salts	B. Acids	C. Bases (Salt)	D. Miscellaneous
(e.g., NaCl, CaSO₄)	(e.g., HCl, H₂SO₄)	(e.g., NaOH, Ba(OH)₂)	(e.g., CH₄, CO₂, CCl₄, PCl₃)
1. Solid substances made of positively charged ions of metals or radicals	1. Gases, liquids, or low-melting solids containing hydrogen re-	1. Solid substances made of positively charged ions of metals (cations) and	1. All other types of compounds other than acids, bases and salts.

A. Salts	B. Acids	C. Bases (Salt)	D. Miscellaneous
(e.g., NaCl, $CaSO_4$)	(e.g., HCl, H_2SO_4)	(e.g., NaOH, $Ba(OH)_2$)	(e.g., CH_4, CO_2, CCl_4, PCl_3)
(cations) and the negatively charged ions of non-metals or radicals (anions). 2. They are usually soluble in water and polar solvents to give solutions of anions and cations. 3. Usually have high melting and high boiling points. 4. Will conduct the electric current when molten or when dissolved in water. Salts which are non-conductors in the	placeable by a metal. 2. They dissolve in water to give solutions containing the hydronium ion. 3. In pure state they do not conduct electricity. 4. Water solutions will turn blue litmus red and conduct the electric current.	the negative hydroxyl ion $(OH)^{-1}$. 2. They dissolve in water to give solutions containing the hydroxyl ion $(OH)^{-1}$ and positive ions. 3. Water solutions will conduct the electric current and turn red litmus blue.	They are covalent compounds generally gases, liquids or low-melting solids. Most organic compounds fall into this group. 2. In pure state they will not conduct the electric current. 3. This type of compound also includes the product formed when non-metals react. 4. Are generally not soluble in water.

A. Salts	B. Acids	C. Bases (Salt)	D. Miscellaneous
(e.g., NaCl, $CaSO_4$)	(e.g., HCl, H_2SO_4)	(e.g., NaOH, $Ba(OH)_2$)	(e.g., CH_4, CO_2, CCl_4, PCl_3)
molten state are called pseudo-electrolytes.			
5. May be prepared by direct union of metal and non-metal or by neutralization of an acid with base.	*Note:* Equivalent amounts of acids and bases will neutralize each other to give a salt and water.		

III. Properties of Ionic and Covalent Compounds

We can divide chemical compounds into two categories: (A) ionic compounds and (B) covalent compounds.

A. Ionic Compounds	B. Covalent Molecular Compounds
1. Solid crystals at room temperature with high M.P. (above 400°C).	1. Gases, liquids, and solids at room temperature. Those that are solids usually have low M.P.
2. Crystals are made up of individual positive and negative ions. There are no individual molecules. The crystal is a single giant molecule with large electrical forces between the individual ions.	2. Are made up of individual molecules with small forces holding molecules together. These small forces account for their existence as volatile liquids or gases and the low melting points of the solids.

A. Ionic Compounds

B. Covalent

3. As solids they do not conduct the electric current (with the exception of metals). When in the molten state they do.

4. They dissolve in water by a process of hydration and are completely dissociated. Each ion is surrounded in the solution by water molecules which are attracted to it. The water solutions behave as though the ions were independent of each other. They conduct the electric current.

3. Some covalent compounds like diamond and quartz have very high melting points because the crystals are not made up of single molecules but rather is a giant continuous network of covalent bonds (i.e., polymers).

4. Most covalent compounds are not soluble in water.

5. Some compounds will dissolve in water to give ionic solutions (e.g., HCl, $SnCl_4$, Al_2Cl_6, Fe_2Cl_6). These compounds are called pseudo-electrolytes. Pseudo-electrolytes are sometimes only partially dissociated in solution and the per cent of ionization will depend on the concentration.

Note: As is usual with most dichotomies, many compounds are intermediate in their properties between these two extreme classes.

APPENDIX III

Some Important General Reactions

1. Metals and Non-Metals

Metals generally react with non-metals to product a salt. In these salts the metal will usually exist as a positive ion in one of its higher valence states, the non-metal a negative ion.

$$\text{Metal} + \text{Non-metal} \rightarrow \text{Salt (Metal ion}^+ - \text{Non-metal ion}^-)$$

Examples:

$$2Na + Cl_2 \rightarrow 2NaCl \ (Na^{+1}, Cl^{-1})$$

$$Ca + I_2 \rightarrow CaI_2 \ (Ca^{+2}, 2I^-)$$

$$Sn + O_2 \rightarrow SnO_2 \ (Sn^{+4}, 2O^=)$$

$$Sn + 2Br_2 \rightarrow SnBr_4 \text{ (covalent pseudo-electrolyte)}$$

$$3Ba + 2P \rightarrow Ba_3P_2 \ (3Ba^{++}, 2P^\equiv)$$

$$3Mg + N_2 \rightarrow Mg_3N_2 \ (3Mg^{++}, 2N^\equiv)$$

$$4Al + 3C \rightarrow Al_4C_3 \ (4Al^{+++}, 3C^{-4})$$

2. Combustion

Combustion is the reaction of a substance with oxygen. The usual products are the oxides of the elements present in the original substance in their higher valence states. When, N, Cl, Br, and I are present in original compound, they are usually released as free elements, not the oxides.

$$\text{Substance} + \text{Oxygen} \rightarrow \text{Oxides of Elements}$$

Examples:

$$4Fe + 3O_2 \rightarrow 2Fe_2O_3$$

$$2Zn + O_2 \rightarrow 2ZnO$$

250

$S + O_2 \rightarrow SO_2$ (exception—SO_3 not formed generally)

$C + O_2 \rightarrow CO_2$

$4FeS + 7O_2 \rightarrow 2Fe_2O_3 + 4SO_2$

$2CaC_2 + 5O_2 \rightarrow 2CaO + 4CO_2$

$2Mg_3N_2 + 3O_2 \rightarrow 6MgO + 2N_2$ (exception-N_2 formed, not oxide)

$2C_2H_6 + 7O_2 \rightarrow 4CO_2 + 6H_2O$

$4CH_5N + 9O_2 \rightarrow 4CO_2 + 2N_2 + 10H_2O$ (exception-N_2 formed)

3. Reactions of Oxides

A Metallic Oxide is called a basic oxide, since if it dissolves in water, it will react to form a base. Metallic oxides are also called basic anhydrides.

Metallic oxide + Water → Base (metallic hydroxide)

$Na_2O + H_2O \rightarrow 2NaOH$ (ionic)

$CaO + H_2O \rightarrow Ca(OH)_2$ (ionic)

A Non-Metallic Oxide is called an acidic oxide, since if it dissolves in water it will react to form an acid. Non-metallic oxides are also called acid anhydrides.

Non-metallic oxide + Water → Acid (non-metallic hydroxide)

$CO_2 + H_2O \rightarrow H_2CO_3$ (carbonic acid)   also $CO(OH)_2$

$SO_2 + H_2O \rightarrow H_2SO_3$ (sulfurous acid)   also $SO(OH)_2$

$SO_3 + H_2O \rightarrow H_2SO_4$ (sulfuric acid)   also $SO_2(OH)_2$

$N_2O_5 + H_2O \rightarrow 2HNO_3$ (nitric acid)   also $NO_2(OH)$

$P_2O_5 + 3H_2O \rightarrow 2H_3PO_4$ (ortho-phosphoric acid) also $PO(OH)_3$

A Metallic Oxide (basic) may neutralize an acid to form a salt and water. Similarly, a *non-metallic oxide (acidic)* may neutralize a base to form a salt and water.

Metallic oxide + Acid → Salt + Water

$$CaO + 2HCl \rightarrow CaCl_2 + H_2O$$

$$Al_2O_3 + 6HNO_3 \rightarrow 2Al(NO_3)_3 + 3H_2O$$

Non-metallic oxide + Base $\rightarrow$ Salt + Water

$$SO_2 + 2NaOH \rightarrow Na_2SO_3 + H_2O$$

$$CO_2 + 2KOH \rightarrow K_2CO_3 + H_2O$$

A metallic oxide (basic) may neutralize a *non-metallic oxide* (acid) at high enough temperatures, to form a salt directly. Both are anhydrides, thus no water is found.

Metallic oxide + Non-metallic oxide $\rightarrow$ Complex salt

---

$$CaO + CO_2 \rightarrow CaCO_3$$

$$Na_2O + SO_3 \rightarrow Na_2SO_4$$

The oxides of the lower valence states of elements are usually basic, whereas the oxides of higher valence states are usually acidic. Thus the lower valence state may neutralize the higher valence state to form a salt in which the metal exists in a radical (CrO_4^{-2}, MnO_4^{-1}, etc.).

This explains such peculiar formulae as Fe_3O_4 in which Fe appears to have a valence of $2\frac{2}{3}$. It is actually a salt of the two oxides, $Fe_2O_3 + FeO$, or ferrous ferrite.

$$\begin{array}{lll} FeO & + Fe_2O_3 & \rightarrow Fe(FeO_2)_2 \quad (or\ Fe_3O_4) \\ (base) & (acid) & (salt) \end{array}$$

$$\begin{array}{lll} 2PbO & + PbO_2 & \rightarrow Pb_2(PbO_4) \quad (or\ Pb_3O_4) \\ (base) & (acid) & (salt) \end{array}$$

$$\begin{array}{lll} Na_2O & + CrO_3 & \rightarrow Na_2(CrO_4) \\ (base) & (acid) & (salt) \end{array}$$

$$\begin{array}{lll} CaO & + Mn_2O_7 & \rightarrow Ca(MnO_4)_2 \\ (base) & (acid) & (salt) \end{array}$$

4. Reactions of Acids and Bases

A. Neutralization. An acid will neutralize a base to form a salt and water.

Acid + Base $\rightarrow$ Salt + Water

$$H_2SO_4 + 2KOH \rightarrow K_2SO_4 + 2H_2O$$

$$2HCl + Ca(OH)_2 \rightarrow CaCl_2 + 2H_2O$$

B. Reactions with Metals. Metals will react with acids in water solution to displace hydrogen and form salts.

$$\boxed{\text{Acid} + \text{Metal} \rightarrow \text{Salt} + \text{Hydrogen}}$$

$$2HCl + Zn \rightarrow ZnCl_2 + H_2$$

$$H_2SO_4 + Ni \rightarrow NiSO_4 + H_2$$

Amphoteric metals will react with bases in water solution to liberate hydrogen and form salts in which the amphoteric metal is in an oxide radical, i.e., complex ion with oxygen.

$$\boxed{\text{Bases} + \text{Metal} \rightarrow \text{Salt} + \text{Hydrogen}}$$

$$2NaOH + Zn \rightarrow Na_2ZnO_2 + H_2\uparrow$$

$$6KOH + 2Al \rightarrow 2K_3AlO_3 + 3H_2\uparrow$$

$$4NaOH + Si \rightarrow Na_4SiO_4 + 2H_2\uparrow$$

$$2KOH + Sn \rightarrow K_2SnO_2 + H_2\uparrow$$

The complex ion of the metal thus formed usually has the metal in its lower valence state.

C. Reaction of Acids with Salts. When a strong acid reacts with a salt of a weak acid, the hydrogen ion of the strong acid combines with the anion of the weak acid to form the undissociated weak acid. This is usually described by saying that the strong acid drives the weak acid out of solution (displacement reaction).

$$\boxed{\text{Strong acid} + \frac{\text{Salt of}}{\text{weak acid}} \rightarrow \frac{\text{Salt of}}{\text{strong acid}} + \frac{\text{Weak}}{\text{acid}}}$$

$$HCl + NaAc \rightarrow Na^{+1} + Cl^{-1} + HAc$$

$$2HNO_3 + Na_2CO_3 \rightarrow 2NaNO_3 + H_2CO_3$$

If a non-volatile acid is allowed to react with the salt of a volatile acid and the mixture heated, the volatile acid will evaporate. This is used to manufacture volatile acids from non-volatile acids. The most frequently used non-volatile acids are H_2SO_4 and H_3PO_4. See Table XVI for classifications of common acids.

$$\boxed{\text{Non-volatile acid} + \frac{\text{Salt of}}{\text{volatile acid}} \xrightarrow{\Delta} \frac{\text{Salt of non-}}{\text{volatile acid}} + \frac{\text{Volatile}\uparrow}{\text{acid}}}$$

$$H_2SO_4 + 2NaCl \xrightarrow{\Delta} Na_2SO_4 + 2HCl\uparrow$$

$$H_3PO_4 + 3NaBr \xrightarrow{\Delta} Na_3PO_4 + 3HBr\uparrow$$

5. Reactions with Water

A. Solvation. When true salts of strong acids and bases dissolve in water, the ions of the salt are surrounded by a shell of water molecules which act to keep the ions free from each other. This process is termed solvation. All ions in solution are solvated by water molecules, usually $6H_2O$ molecules in first solvent shell.

$$Na^+Cl^{-1} + 12H_2O \rightarrow Na(H_2O)_6{}^+ + Cl^-(H_2O)_6$$

$$K^+OH^{-1} + 12H_2O \rightarrow K(H_2O)_6{}^+ + OH^-(H_2O)_6$$

B. Ionization. When pseudo-electrolytes dissolve in water, the covalent bonds of the pseudo-electrolyte are broken and ions are produced. These ions are always solvated. The reactions are generally reversible.

$$HCl + H_2O \rightleftarrows H(H_2O)^{+1} + Cl^{-1}(H_2O)_x \text{ (not balanced)}$$

$$H_2SO_4 + H_2O \rightleftarrows 2H(H_2O)^{+1} + SO_4{}^{-1}(H_2O)_x \text{ (not balanced)}$$

C. Hydrolysis. When pseudo-electrolytes which are salts of weak acids or weak bases, dissolve in water, the covalent bonds are broken and solvated ions are produced. These reactions are not always reversible. The ions formed make the solution either acidic or basic.

$$SnCl_2 (s) + 18H_2O \rightarrow Sn(H_2O)_6{}^{+2} + 2Cl(H_2O)_6{}^{-1} \text{ (not reversible)}$$

$$AlCl_3 (s) + 24H_2O \rightarrow Al(H_2O)_6{}^{+3} + 3Cl(H_2O)_6{}^{-1} \text{ (not reversible)}$$

If any of the products are insoluble or volatile, the reaction can proceed to completion:

$$Na_2S + HOH \rightarrow 2Na^{+1} + 2OH^{-1} + H_2S\uparrow$$

$$Al_4C_3 + 12HOH \rightarrow 4Al(OH)_3\downarrow + 3CH_4\uparrow$$

$$Mg_3N_2 + 6HOH \rightarrow 3Mg(OH)_2\downarrow + 2NH_3\uparrow$$

$$BiCl_3 + HOH \rightarrow BiOCl\downarrow + 2HCl\uparrow$$

$$CaC_2 + 2HOH \rightarrow Ca(OH)_2 + C_2H_2\uparrow \text{ (acetylene)}$$

Note: In such hydrolyses, the water splits into H^{+1} and OH^{-1}, the H^{+1} going to the non-metal and the OH^{-1} to the metal.
This is the complete reverse of the reaction of neutralization.

Salt + Water → Metal hydroxide (base) + Non-metal hydride (acid)

$$Ca_3P_2 + 6H_2O \rightarrow 3Ca(OH_2) + 2PH_3\uparrow$$

6. Prediction of Non-Redox Reactions in Solution

If we omit redox reactions, then there are some simple rules which we can use to predict the products of reactions in aqueous solutions. The following are helpful categories of reaction in applying these rules.

A. *A ppt. is formed* (see solubility rules).
B. *A weak acid is formed* (see tables of ionization constants of weak acids).
C. *A complex ion is formed* (see table of common complex ions).
D. *A gas or volatile product is formed.* (To decide this we must know something about the volatility of comon species.)

To aid us in deciding about the formation of ppts. we can make use of the following general rules concerning insolubility (i.e., less than 0.01 M in saturated solution at 20°C.):

I.   Salts of Na^+, K^+, NH_4^+ are soluble.
II. Salts of Ac^-, NO_3^- and ClO_3^- are soluble.
III. Salts of Cl^-, Br^- and I^- are soluble (except Pb^{++}, Hg_2^{++}, Ag^+).
IV. Salts of $SO_4^=$ are soluble (except Pb^{++}, Ca^{++}, Ba^{++}, Ra^{++}, Sr^{++}).
V.   Salts of $CO_3^=$ and $PO_4^=$ are insoluble (except category I).
VI. Salts of OH^-, $O^=$ and $S^=$ are insoluble (except category I).

For deciding about the formation of weak acids or displacement reactions of weak acids we can employ Table XVI which lists the ionization constants of weak acids. Most useful will be to characterize these acids according to the 5 groups given in Chapter XII, section 11.

The possibility of complex ion formation is much more involved and dependent on concentrations, and we shall not consider it here. However, a table of common complex ions characterized as we have done for the 5 groups of acids would be useful.

The possibility of volatile products in reactions at room temperature is limited by observing that only CO, NH_3, CO_2, SO_2, H_2S and NO are the common substances that are volatile under these conditions.

In applying these rules we must consider the possible combinations of all negative ions in the solution with all positive ions to see what combinations can be formed and then checking these off against the rules given.

Example: What reaction happens, if any, when excess 0.1 M $(NH_4)_2SO_4$ is mixed with 0.2 M KCl?

Answer: The ions here are NH_4^+, K^+, $SO_4^=$ and Cl^-. According to our solubility rules all NH_4^+ and K^+ salts are soluble; hence no ppts. can form. None of these substances are volatile. Discarding complex ion possibilities the last remaining category is that of weak acids. But $SO_4^=$ and Cl^- are conjugate bases of strong acids. Hence no reaction!

Example: What reaction occurs when 0.1 M $Ba(NO_3)_2$ and 0.3 M $ZnSO_4$ are mixed?

Answer: The ions are Ba^{++}, Zn^{++}, NO_3^-, and $SO_4^=$. No NO_3^- can ppt., but $BaSO_4$ is insoluble (rule IV). Hence the reaction is:

$$Ba^{++} + SO_4^= \rightarrow BaSO_4\downarrow$$

The following table of examples will give results obtained from these considerations:

Reactants	Ions and Molecules Present	Products
$AgNO_3$; $BaCl_2$	Ag^+, Ba^{++}, NO_3^-, Cl^-	$AgCl \downarrow$
NH_4Cl; $NaOH$	NH_4^+, Na^+, Cl^-, OH^-	$NH_3 + HOH$
KCN; HAc	K^+, CN^-, HAc	$HCN + Ac^-$
K_2S; $CuCl_2$	K^+, Cu^{++}, $S^=$, Cl^-	$CuS\downarrow$
$ZnAc_2$; H_2SO_4	Zn^{++}, H^+, Ac^-, $SO_4^=$	HAc

It is much more difficult to decide when heterogeneous reactions will occur. Thus $CaCO_3$ (s) will dissolve in HCl and HAc but not in HCN. The reason is that $K_{S.P.}(CaCO_3)$ is too small compared to the amount of H^+ released by HCN [see $(K_{ion.})$]. Similarly ZnS (s) will dissolve in 1 M HCl but not in 0.01 M HCl. Such reactions, like complex ion formation depend too specifically on reagent concentrations to lend themselves to easy rules.

APPENDIX IV

Table of Common Units

Length:

1 meter (m.) = 100 centimeters (cm.)

1 centimeter (cm.) = 10 millimeters (mm.)

1 centimeter (cm.) = 100,000,000 (1×10^8) angstroms (A)

1 inch (in.) = 2.54 centimeters = 25.4 millimeters

1 yard (yd.) = 3 feet (ft.)

1 mile = 5280 ft.

Area:

1 square centimeter (cm.^{2}) = 1×10^{-4} square meters (m.^{2})

1 square centimeter (cm.^{2}) = 100 square millimeters (mm.^{2})

1 square inch (in.^{2} or sq. in.) = 6.452 square centimeters (cm.^{2})

Volume:

1 liter (l.) = 1000 milliliters (ml.)

1 milliliter (ml.) = 1 cubic centimeter (cc.) (cm.^{3}) (to three significant figures)

1 cubic centimeter (cc.) = 1000 cubic millimeters (mm.^{3})

1 cubic inch (in.^{3}) = 16.387 cubic centimeters

1 quart (qt.) = 2 pints (pt.) = 0.25 gallon (gal.) = 32 ounces (fluid, oz.) = 0.946 l.

Mass:

1 kilogram (kg.) = 1000 grams

1 gram (g.) = 1000 milligrams

1 milligram (mg.) = 1000 micrograms (μg.)

1 pound (lb.) = 454 grams (three significant figures) = 16 ounces

1 ton = 907 kilograms = 9.07×10^{11} micrograms

Density:

Density = $\dfrac{\text{Mass}}{\text{Volume}}$ (usually expressed as g./cm.^{3})

1 g./cm.^{3} = 62.4 lb./ft.^{3}

257

Pressure:

Pressure $= \dfrac{\text{Force}}{\text{Area}}$ (usually expressed as atmospheres; mm. of mercury; lb./sq. in.)

1 atmosphere (atm.) = 14.7 lb./sq. in. = 760 mm. Hg.
$$= 760 \text{ torr}$$

Temperature: Temperature may be expressed in any of four scales, centigrade (°C.), Rankine (°R.), Fahrenheit (°F.), or absolute (°K.). The standards for these scales are the freezing point of water at 1 atm. pressure and the boiling point of water at 1 atm. pressure.

	Rankine	Fahrenheit	Centigrade	Absolute
Boiling point	672°R.	212°F.	100°C.	373°K.
Freezing point	492°R.	32°F.	0°C.	273°K.

$$°K. = °C. + 273$$
$$°F. = \tfrac{9}{5}°C. + 32$$
$$°R. = °F. + 459$$

Energy: *Heat energy* is generally expressed in calories:

1000 calories (cal.) = 1 kilocalorie (kcal.)

Mechanical energy is expressed in foot-pounds (ft.-lb.) or ergs.

1 foot-pound = 0.32 calorie

1 erg = 1 dyne-centimeter

1×10^7 ergs = 1 joule = 0.24 calorie

Electrical energy is usually expressed in joules.

1 electron-volt = 23.05 kilocalories/mole

1 joule = 1 watt-second = 0.24 calorie

1 kilowatt-hour = 1000 watt-hours

1 kilowatt-hour = 864 kilocalories = 864,000 calories

Force:

1 dyne = force necessary to give 1 gram of mass an acceleration of 1 cm./sec.^{2}

Force of gravity = 980 dynes/gram

Electricity: The most common unit used for expressing the *difference in potential energy* between two bodies or two points in space is the *volt*. A position of low voltage is at a low potential energy compared to a position of high voltage.

It always requires energy to move electrical charges from low-potential energy levels to high-potential energy levels.

Charges will tend naturally to move from a high potential to a low potential, and in doing this they produce kinetic energy. (An analogy can be drawn between electrical potential energy and gravitational potential energy.)

Masses always tend naturally to move closer to the center of the earth, and in doing this they release energy (e.g., waterfalls—as the water descends it moves faster).

The *motion of charges* is an *electric current*. This is most frequently the motion of electrons which possess negative charges, but it may be the motion of any body which possess a charge, e.g., ions in solution.

The common unit for measuring the number of electric charges is the coulomb.

$$1 \text{ equivalent of charge } = 96,500 \text{ coulombs } = 1 \text{ faraday}$$

The faraday represents 1 mole of electrons or 1 equivalent of electric charges.

In 1 faraday there are 6.02×10^{23} units of charge. If we consider the electron which has one unit of electric charge:

$$1 \text{ faraday } = 1 \text{ mole of electrons } = 6.02 \times 10^{23} \text{ electrons}$$

When electric charges are transported in a closed circuit at the rate of 1 coulomb of charge per second a current of 1 ampere is said to be flowing in the circuit.

$$1 \text{ ampere } = 1 \text{ coulomb/second } = 1.04 \times 10^{-5} \text{ moles electrons/sec.}$$

Coulombs = Amperes × Seconds

Concentration:

Density = Mass/Volume   (usually in g./cc.)

$$\text{Per cent by weight } = \frac{\text{Mass of solute}}{\text{Mass of total solution}} \times 100$$

$$\begin{cases} \text{Molarity} = \dfrac{\text{Moles of solute}}{\text{Liters of solution}} \\ \text{Moles of Solute} = \text{Molarity} \times \text{Liters of solution} \end{cases}$$

$$\begin{cases} \text{Normality} = \dfrac{\text{Equivalents of solute}}{\text{Liters of solution}} \\ \text{Equivalents of solute} = \text{Normality} \times \text{Liters of solution} \end{cases}$$

Molality = Moles of solute/1000 g. of solvent

APPENDIX V

Answers to Problems

Chapter I

1. (a) 0.2 milliinches.
   (b) 10 millimiles.
   (c) 12 kilodresses.
2. (a) 3500 bucks.
   (b) 0.000075 meters.
   (c) 42,000 watts.
   (d) 160,000 pounds.
   (d) 100 megayears.
   (e) 7000 megawatts = 7 gigawatts.
   (f) 3.5 microtons.
   (e) 0.000095 gram.
   (f) 0.0027 gallons.
   (g) 1600 sheep.
3. Intensive: (a), (c), (e); extensive: (b), (d), (f).
4. 0.055 lb.
5. 16×10^{-6} kg.
6. 1.32×10^{-7} ton.
7. 2.3×10^7 dollars = 23 megabucks.
8. 1.67×10^5 cm.^{2}.
9. 40.8 m.$^3 \rightarrow$ 41 m.^{3}.
10. 2.3×10^{-5} in. = 5.9×10^{-5} cm. = 0.59 micron.
11. 1.35 tons/yd.^{3}.
12. 25 miles/hr. = 1.1×10^3 cm./sec.
13. 980 cm./sec.^{2} = 3.5×10^4 m./min.^{2}.
14. 35 kg.-m./min.^{2}.
15. 28 in.^{3} = 450 cm.^{3}.
16. 7.75 g./cm.^{3}.
17. 22,000 mg.
18. Rate = 0.893 egg/hen-day; time = 56 days.
19. 1.70×10^{26} atoms O.
20. (a) 1.0×10^{15} molecules; (b) 6×10^{13} molecules.
21. 0.58 g./cm.^{3}.
23. 1.9×10^{10} tons.
24. 940 lb./in.^{2}.
25. 1250 lb./in.$^2 \rightarrow 1.3 \times 10^3$ lbs./in.^{2}.
26. 1.2×10^3 lbs./in.^{2}.
27. 1.6×10^{-13} erg.
28. 2.0×10^{17} cm./sec.^{2}. = 6.6×10^{15} ft./sec.^{2} (2×10^{14} G$_l$).

Chapter II

1. (a) 97.4 g.
   (b) 78.0 g.
   (c) 159.6 g.
   (d) 158.1 g.
2. (a) 0.35 mole.
   (b) 0.17 g.
   (c) 840 g.
   (d) 0.14 mmole.
   (e) 0.59 mole.
   (e) 249.7 g.
   (f) 78.1 g.
   (g) 562.0 g.
   (h) 96.1 g.
   (f) 6.2×10^3 mmoles = 6.2 moles.
   (g) 0.096 mole.
   (h) 0.020 ml. STP.
   (i) 1.8×10^4 tons H$_2$.
   (j) 2.3×10^{11} molecules Cl$_2$.

261

3. (*a*) 4 mmoles HNO_3.   (*b*) 16 g. NaCl.   (*c*) 4 moles Ca $(NO_3)_2$.
4. (*a*) 1.4×10^{23} molecules.   (*d*) 8×10^{24} molecules.
    (*b*) 1.0×10^{21} molecules.   (*e*) 3×10^{-24} ton.
    (*c*) 1.15×10^{-7} mole.
5. 0.59 mmole S.   **6.** 1.9×10^{-4} mole NO_2.
7. 1.55×10^{-22} cc.
8. (*a*) 1.2 mmoles Cl_2.   (*c*) 11 cc. STP H_2S gas.
    (*b*) 7800 l. STP NO gas.   (*d*) 0.22 kg. SF_6.
9. 79 g./mole.   **10.** 140 g./mole.
11. 502.72 g.   **12.** 83.5 g./mole.
13. 70.0 g./mole.
14. (*a*) 1.28×10^{-19} cm.^{3}.   (*b*) 1.60×10^{-22} cm.^{3}.   (*c*) 5.4 A.
16. 1.28×10^{-23} cm.^{3}; 14.7 cm.^{3}/mole.
17. 2.2 Kg.   **18.** 1.63×10^{-22} g.
19. 2.7×10^{19} molecules/cm.^{3}.   **20.** 28.9 g./mole.
21. Dry Air.   **22.** 28.3 g./mole.

Chapter III

1. True formulae are (*a*), (*f*), and (*h*).   **2.** 3.36 moles O.
3. 7.12 moles S.   **4.** 1.56 moles O.
5. 1.6×10^{-3} mole.   **6.** 190 mmoles B.
7. 189 mmoles B.   **8.** 0.40 mole B.
9. 280 mmoles S.   **10.** 4.1 g. O.
11. 104 mg. Na.   **12.** 3.5×10^5 g. Ca.
13. 7.1×10^{20} atoms O.
14. (*a*) Na = 22.4%; Br = 77.6%.
    (*b*) Ca = 43.5%; C = 26.1%; N = 30.4%.
    (*c*) K = 41.1%; S = 33.7%; O = 25.2%.
    (*d*) Na = 16.1%; C = 4.2%; O = 72.7%; H = 7.0%.
    (*e*) C = 40.0%; H = 6.7%; O = 53.3%.
    (*j*) C = 58.5%; H = 4.1%; N = 11.4%; O = 26.0%.
15. (*a*) $CaBr_2$.   (*d*) Sr_2SiO_4.
    (*b*) C_3O_2.   (*e*) Mn_3S_4.
    (*c*) $Al_2(CO_3)_3$.   (*f*) $CaCl_2 \cdot 6H_2O$.
16. C_6H_{12}; H_2O_2; $C_5H_{10}O_5$; Hg_2Cl_2; H_4F_4.
17. $CuSO_4 \cdot 5H_2O$.   **18.** Cr_2S_3.
19. $PtCl_4$.   **20.** $CuCl_2 \cdot 6NH_3$.
21. $C_{12}H_{22}O_{11}$.   **22.** Formula is $HNO_3 \cdot H_2O$.
23. CHF; $C_5H_5F_5$.   **24.** 80 g./mole; $B_3N_3H_6$.
25. (*a*) 66.7.   (*b*) 25.   (*c*) 50.   (*d*) 25.   (*e*) 39.5%; 46.5% H.   (*f*) 21.
26. 78.   **27.** 75.

Chapter IV

1. Refer to text.   **2.** 0.70 mole HCl.
3. 1.00 moles Cl_2.   **4.** 0.025 mole K_3PO_4.
5. 1800 cc. STP SO_2.
6. (*a*) 259 g. FeS.
    (*b*) 66 l. STP H_2S gas.
7. 6.9 l. STP O_2 gas.   **8.** 2.04 moles NaOH.

9. 32.2 cc. STP O_2 gas.   **10.** 3000 kg. H_2SO_4.

11. 1600 l.   **12.** 1.4×10^4 cc. STP CO_2.

13. 0.564 g. $KMnO_4$.

14. 536 g. Al_4C_3; 803 g. H_2O; 803 cc. H_2O.

15. 20 kg. C_6H_{14}.   **16.** 48 kg. hydrate.

17. (a) 1.2 moles Zn; 0.8 mole $ZnCl_2$; 0.8 mole H_2.

   (b) 2.4 g. HCl; 29 g. $ZnCl_2$; 0.43 g. H_2.

18. (a) 5.3 moles HNO_3; 4 moles $Cu(NO_3)_2$; 2.7 moles NO; 5.3 moles H_2O.

   (b) 19.5 g. Cu; 13 g. $Cu(NO_3)_2$; 1.4 g. NO; 1.7 g. H_2O.

19. 6600 ml. STP O_2.   **20.** 59.2 ml. STP H_2.

Chapter V

2. (a) 9.8×10^5 ergs.   (b) 443 cm./sec.

3. (a) 2.1×10^{19} ergs.   (c) 1.9 cal.

   (b) 0.014 kcal.   (d) 6 microjoules.

4. (a) 423°K.   (g) 200°K.

   (b) −460°F.   (h) 509.6°R.

   (c) −273°C.   (i) −449.6°F.

   (d) 727°C.   (j) 1391.6°R.

   (e) 441°F.   (k) 50°K.

   (f) −40°F.

5. 15°K.; 27°F.   **6.** 184 cal.

7. 4.5×10^4 cal.   **8.** 1.06×10^{10} ergs/lb. °F.

9. 1480 cal.   **10.** Low temperature.

11. 0.25 cal./g.-°C.   **12.** 40 a.m.u.; Ca.

13. 0.10 cal./g.-°C.   **14.** 1.3×10^{-14} erg.

15. (a) −27.4 kcal. (exothermic).   (d) −74.4 kcal. (exothermic).

   (b) −212.7 kcal. (exothermic).   (e) −67.4 kcal. (exothermic).

   (c) 31.4 kcal. (endothermic).   (f) −189.1 kcal. (exothermic).

16. (a) −96.5 kcal./mole.   (c) +34.5 kcal./mole.

   (b) −297.8 kcal./mole.   (d) −291 kcal./mole.

17. 93 l. STP $O_2(g)$.   **18.** 188 kcal.

19. 108,000 cal.; 9.72 kcal./mole.   **20.** (a) 1440 cal./mole.

          (b) 40,000 cal.

          (c) 36,000 cal.

21. Less by heats of vaporization of H_2 and O_2.

22. More by heat of condensation of H_2O. Note that this is 9% of total heat of combustion.

23. Very cold foods (less than 50°F.) if we neglect work done by body in cooling very hot foods (135°F.) to body temperature (98.6°F.).

24. 7.5×10^5 ft. lb. $= 1.0 \times 10^{13}$ ergs $= 240$ kcal.

25. 880 cal.

26. 19.8 kcal./mole.   **27.** $\Delta H = -104.0$ kcal./mole.

28. $\Delta H_{ion}(HAc) = 3.4$ kcal./mole.

29.

	C_p(cal./mole-°K.)	$\Delta H_f°$(kcal./mole)
(a)	23.3	−30.1
(b)	15.3	−23.9
(c)	11.9	− 6.4
(d)	15.7	−26.0

30. (a) $\Delta H = -171.0$ kcal./mole.
    (b) $\Delta H = -295.9$ kcal./mole.

Chapter VI

1. 0.53 atm.

2. 25 cm. Hg; 0.33 atm.

3. 81.6 cm. Hg; 1.07 atm.

4. (a) 3.33×10^5 dynes/cm.^{2}.
    (b) 0.46 atm.

    (c) 15.5 cm. Hg.

    (d) 6.0×10^{-5} atm.

5. 6.4×10^4 dynes/cm.^{2}.

6. 1790 lb./sq. in.

7. 2 l.

8. 225 cm. Hg.

9. 32 l.

10. 118°K. $= -155$°C.

11. 350 cc.

12. 246 cc. STP.

13. (a) Initial volume.
    (b) Initial temperature.
    (c) Final volume.

14. (a) 6.24×10^4 cc.-mm. Hg/mole-°K.
    (b) 1.21 l.-lb./sq. in./mole-°K.
    (c) 8.31×10^7 ergs/mole-°K.

15. 72 g./mole.

16. 3.6 l.

17. 35.8 g./mole.

18. 108 g./mole.

19. 1.15 g./l.

20. 2.14 mg./cc.

21. 2×10^{11} molecules/cc.

22. 2.86 g./l.; 2.63 g./l.

23. 80 g./mole.

24. 58.1 g./mole; 2.59 g./l.

25. 585 torr N_2; 158 torr O_2; 7.5 torr Ar.

26. (a) 0.87 atm.
    (b) $H_2 = 0.10$ atm.; $H_2O = 0.51$ atm.

27. HCl has two extra rotational degrees of freedom.

28. 21 kcal.

29. 5 cal./mole-°C.; linear molecule.

30. 4 cal./mole-°C.; 3.0 g./mole.

31. Internal vibrations must be storing energy.

32. 112°C., 1.41 atm.

33. 350 cal.

34. (a) Gas does work, hence has less energy, thus colder.
    (b) More heat at constant pressure.

35. 560 mm. Hg.

36. 40.5 torr Cl_2 and NOCl.

37. 0.58 atm.

38. 5.2 l. at 80°C.

39. 440 calories; 40 liters.

Chapter VII

1. 29.5 g.

2. 100.0 g.

3. 15.5 g.

4. (a) 6 eq. $FeCl_3$.
    (b) 0.783 eq. Zn.
    (c) 26 meq. Pb.
    (d) 0.096 eq. $CaSO_4$.
    (e) 1.8 eq. CCl_4.

    (f) 0.225 mole H_2S.
    (g) 57 g. $AlBr_3$.
    (h) 16 g. $CuSO_4$.
    (i) 0.0607 eq. O_2.
    (j) 15.7 cc. Cl_2STP.

5. (a) 4.9 g. $CuSO_4$.
    (b) 21 g. NaOH.
    (c) 12.6 g. $Ca_3(PO_4)_2$.

    (d) 24.5 g. H_2SO_4.
    (e) 9.0 g. Zn.

6. 47.3 g./eq.

7. 28.0 g./eq.

8. 262 g./eq.

9. 160 g./eq.

10. (*a*) 1.   (*e*) 6.   (*i*) 3.
      (*b*) 4.   (*f*) 2.   (*j*) 4.
      (*c*) 3.   (*g*) 7.   (*k*) 2⅔.
      (*d*) 5.   (*h*) 4.

Chapter VIII

1. Mass; volume.   **2.** Density.
3. (*a*) 6 N.   (*f*) 98 g. /l.
      (*b*) 0.05 M.   (*g*) 0.043 N.
      (*c*) 0.035 M.   (*h*) 0.9 Molal.
      (*d*) 0.125 M.   (*i*) 5.5%.
      (*e*) 0.5 mmole/ml.
4. 0.060 mole $Al_2(SO_4)_3$.   **5.** 3.6 g. = 48 mmoles = 96 meq.
6. 1.2 g. = 0.018 eq. = 3.0 mmoles.
7. 41 cc.   **8.** 5.7 ml.
9. Add water to 43 g. $Ca(NO_3)_2$ until the final volume is 150 ml.
10. Add water to 0.91 g. N_2CO_3 until final volume is 240 ml.
11. Add 15 ml. water to 10 ml. stock solution.
12. Add 13.6 ml. water to 1.4 ml. stock solution.
13. 2.0 M K_2SO_4; 0.70 M $NaNO_3$.
14. Add 70 cc. water to 10 cc. stock solution.
15. 3.2 g. Cu = 50 mmoles Cu.
16. 9.2 ml. water for every 1.0 ml. stock solution.
17. 12 mmoles Zn.   **18.** 0.36 g. Al.
19. 4.5 g. $BaCl_2$; 5.0 g. $BaSO_4$.   **20.** 720 cc. STP HCl gas; 32 mmoles KCl.
21. 11.3 ml.; 68 meq. $CaSO_4$.   **22.** 123 ml.; 9.4 g. Na_3PO_4.
23. 78 g./eq.   **24.** 0.51 M.
25. 13.8 M; 13.8 N; 0.904.   **26.** 12.0 M; 12.0 N; 10.2 Molal.
27. HNO_3 is 15.2 N; 4.2 ml.   **28.** 1.43 g./l.; 0.0446 M.
29. 1.03 atm.   **30.** 0.0082 M.
31. 0.080 M Al^{+3}; 0.120 M $SO_4^=$ **32.** 0.050 M H^+; 0.050 M Cl^-
33. 0.065 M Zn^{++}; 0.130 M Cl^-
34. 0.04 M Ca^{++}; 0.13 M Al^{+3}; 0.48 M Cl^-
35. 0.12 M Na^+; 0.06 M Cl^-; 0.06 M OH^-
36. 0.055 M Cu^{++}; 0.003 M H^+; 0.11 M NO_3^-

Chapter IX

1. (*a*) 0.047 mole fraction; 2.7 Molal.   (*c*) 0.030 mole fraction; 1.7 Molal.
      (*b*) 0.23 mole fraction; 5.0 Molal.   (*d*) 0.026 mole fraction; 1.5 Molal.
2. Vapor pressure = 38.6 mm. Hg (Mole fraction phenol = 0.121).
3. Vapor pressure = 146.9 torr (Mole fraction glucose = 0.016).
4. 2.5 Molal solution.   **5.** 135 g./mole.
6. 58 g./mole.   **7.** 230 g. diethylene glycol.
8. 113 g./mole.   **9.** 24 atm.
10. 2700 g./mole.   **11.** 1.67°C.
12. 0.52°C.   **13.** 6.8°C.-kg./mole.
14. $K_{F.Pt}$(NaCl) = 5.9°C./Molal; $K_{F.Pt}$(KCl) = 5.7°C./Molal.
15. 600 g. Sn.   **16.** 20°C.-kg.Sn./mole.
17. 34.5°C.-kg./mole.

18. (*a*) To lower the melting point of $MgCl_2$.

(*b*) K is more active than Mg; Ca and Zn are not.

19. (*a*) 193 g. polymer.   (*b*) 0.027°C.   (*c*) No.

20. No, depressions are too small.

Chapter X

3. $C_{CO} = C_{H_2O} = 0.19$ mole/l. $C_{CO_2} = C_{H_2} = 0.21$ mole/l.

4. $C_{NO_2} = 0.42$ mole/l.   **5.** $K_{eq.} = 29$.

6. $C_{H_2} = C_{I_2} = 7.0 \times 10^{-3}$ mole/l. $C_{HI} = 5.6 \times 10^{-2}$ mole/l.

7. $K_{eq.} = C_{H_2O (g)}$; no change.   **8.** 27.11 M; no.

9. (*a*) $K_{eq.} = C_{CCl_4 (g)}$.   (*b*) 10.3 M.   (*c*) Decreases.   (*d*) Nothing.

10. Consult text.

11. (*a*) $K_{eq.} (10c) = [K_{eq.}(10d)]^4 = 1.3 \times 10^{27}$.

(*b*) See text.

12. (*a*) $K_{eq.} (10e) = [K_{eq.}(10f)]^2 = 4900$.   (*b*) See text.

13. See text.   **14.** $2CO \rightleftharpoons CO_2 + C$; $K_{eq.} = K_b \cdot K_c$.

15. $2H_2O + C \rightleftharpoons CO_2 + 2H_2$; $K_{eq.} = K_b/K_c$.

16. See problem 14 above.

17. (*a*) See text. (*b*) $K = K(II) \times K(III)$.

18. (*a*) See text.   (*b*) $K_{eq.} = K_{ion}(HAc)/K_{ion}(HCN)$.

19. $K_{III} = K_{II}^{1/4}/K_I^{1/2}$.

Chapter XI

2. (*a*) 4.97×10^{-3} faraday.   (*e*) 1.9×10^4 coulombs.

(*b*) 8×10^{-9} coulomb.   (*f*) 4800 coulombs.

(*c*) 1.4×10^5 coulombs.   (*g*) 2.4 amp.-hr.

(*d*) 6.56×10^4 coulombs.

3. 5.2 g. Cr; 3.3 l.STP Cl_2 gas.   **4.** 0.41 mmole H_2; 0.21 mmole O_2.

5. 1780 sec. = 30 min.; 0.093 mole O_2.

6. (*a*) 34 amp.   **7.** (*a*) 14.6 g. Zn.

(*b*) 7.1 l.STP O_2/hr.   (*b*) 28,600 sec.

8. (*a*) 330 sec.   (*b*) 0.69 l.STP Cl_2.   **9.** 14.1 g. V.

10. (*a*) $3e^- + Al^{+++} \rightleftharpoons Al^\circ$ (cathode).

$O^= + C \rightleftharpoons CO + 2e^-$ (anode).

Note: actual species are AlO^+ and AlO_2^-.

(*b*) 18.6 l.STP CO.

(*c*) 0.55 mole Al.

11. (*a*) $3e^- + Au(CN)_4^- \rightleftharpoons Au + 4CN^-$ (anode is reverse of cathode).

(*b*) 0.109 hr.

12. (*a*) $Cl^- + 4HOH \rightleftharpoons ClO_4^- + 8H^+ + 8e^-$.

(*b*) 0.25 mole ClO_4^-.

Chapter XII

2. 1.4×10^{-5} M.   **3.** 2.1×10^{-3} M.

4. 2.9×10^{-9} M.

5. (*a*) 3.4×10^{-13}.   (*c*) 3.0×10^{-16}.

(*b*) 4.9×10^{-9}.   (*d*) 1.5×10^{-72}.

6. (*a*) 5.0×10^{-9} M.   **7.** (*a*) 4.3×10^{-4} M.

(*b*) 8.0×10^{-10} M.   (*b*) 9.3×10^{-5} M.

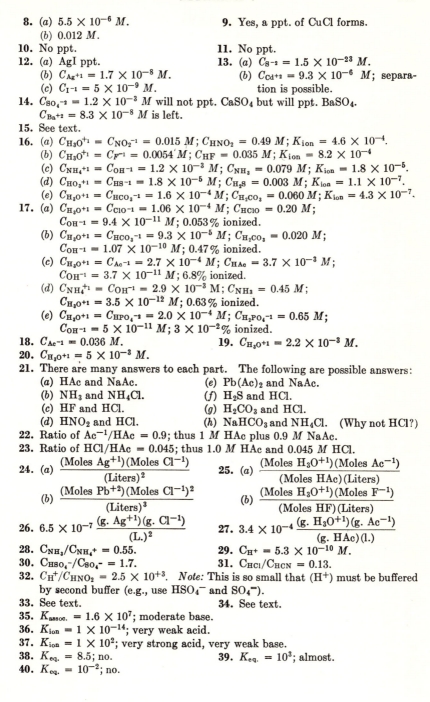

8. (a) 5.5×10^{-6} M.

    (b) 0.012 M.

9. Yes, a ppt. of CuCl forms.

10. No ppt.

11. No ppt.

12. (a) AgI ppt.

    (b) $C_{Ag^{+1}} = 1.7 \times 10^{-8}$ M.

    (c) $C_{I^{-1}} = 5 \times 10^{-9}$ M.

13. (a) $C_{S^{-2}} = 1.5 \times 10^{-23}$ M.

    (b) $C_{Cd^{+2}} = 9.3 \times 10^{-6}$ M; separation is possible.

14. $C_{SO_4^{-2}} = 1.2 \times 10^{-3}$ M will not ppt. CaSO$_4$ but will ppt. BaSO$_4$. $C_{Ba^{+2}} = 8.3 \times 10^{-8}$ M is left.

15. See text.

16. (a) $C_{H_3O^{+1}} = C_{NO_2^{-1}} = 0.015$ M; $C_{HNO_2} = 0.49$ M; $K_{ion} = 4.6 \times 10^{-4}$.

    (b) $C_{H_3O^{+1}} = C_{F^{-1}} = 0.0054$ M; $C_{HF} = 0.035$ M; $K_{ion} = 8.2 \times 10^{-4}$

    (c) $C_{NH_4^{+1}} = C_{OH^{-1}} = 1.2 \times 10^{-3}$ M; $C_{NH_3} = 0.079$ M; $K_{ion} = 1.8 \times 10^{-5}$.

    (d) $C_{HO_3^{+1}} = C_{HS^{-1}} = 1.8 \times 10^{-5}$ M; $C_{H_2S} = 0.003$ M; $K_{ion} = 1.1 \times 10^{-7}$.

    (e) $C_{H_3O^{+1}} = C_{HCO_3^{-1}} = 1.6 \times 10^{-4}$ M; $C_{H_2CO_3} = 0.060$ M; $K_{ion} = 4.3 \times 10^{-7}$.

17. (a) $C_{H_3O^{+1}} = C_{ClO^{-1}} = 1.06 \times 10^{-4}$ M; $C_{HClO} = 0.20$ M;

    $C_{OH^{-1}} = 9.4 \times 10^{-11}$ M; 0.053% ionized.

    (b) $C_{H_3O^{+1}} = C_{HCO_3^{-1}} = 9.3 \times 10^{-5}$ M; $C_{H_2CO_3} = 0.020$ M;

    $C_{OH^{-1}} = 1.07 \times 10^{-10}$ M; 0.47% ionized.

    (c) $C_{H_3O^{+1}} = C_{Ac^{-1}} = 2.7 \times 10^{-4}$ M; $C_{HAc} = 3.7 \times 10^{-3}$ M;

    $C_{OH^{-1}} = 3.7 \times 10^{-11}$ M; 6.8% ionized.

    (d) $C_{NH_4^{+1}} = C_{OH^{-1}} = 2.9 \times 10^{-3}$ M; $C_{NH_3} = 0.45$ M;

    $C_{H_3O^{+1}} = 3.5 \times 10^{-12}$ M; 0.63% ionized.

    (e) $C_{H_3O^{+1}} = C_{HPO_4^{-2}} = 2.0 \times 10^{-4}$ M; $C_{H_2PO_4^{-1}} = 0.65$ M;

    $C_{OH^{-1}} = 5 \times 10^{-11}$ M; $3 \times 10^{-2}\%$ ionized.

18. $C_{Ac^{-1}} = 0.036$ M.

19. $C_{H_3O^{+1}} = 2.2 \times 10^{-3}$ M.

20. $C_{H_3O^{+1}} = 5 \times 10^{-3}$ M.

21. There are many answers to each part. The following are possible answers:

    (a) HAc and NaAc.

    (b) NH$_3$ and NH$_4$Cl.

    (c) HF and HCl.

    (d) HNO$_2$ and HCl.

    (e) Pb(Ac)$_2$ and NaAc.

    (f) H$_2$S and HCl.

    (g) H$_2$CO$_3$ and HCl.

    (h) NaHCO$_3$ and NH$_4$Cl. (Why not HCl?)

22. Ratio of $Ac^{-1}/HAc = 0.9$; thus 1 M HAc plus 0.9 M NaAc.

23. Ratio of $HCl/HAc = 0.045$; thus 1.0 M HAc and 0.045 M HCl.

24. (a) $\dfrac{(\text{Moles } Ag^{+1})(\text{Moles } Cl^{-1})}{(\text{Liters})^2}$

    (b) $\dfrac{(\text{Moles } Pb^{+2})(\text{Moles } Cl^{-1})^2}{(\text{Liters})^3}$

25. (a) $\dfrac{(\text{Moles } H_3O^{+1})(\text{Moles } Ac^{-1})}{(\text{Moles HAc})(\text{Liters})}$

    (b) $\dfrac{(\text{Moles } H_3O^{+1})(\text{Moles } F^{-1})}{(\text{Moles HF})(\text{Liters})}$

26. $6.5 \times 10^{-7} \dfrac{(\text{g. } Ag^{+1})(\text{g. } Cl^{-1})}{(\text{L.})^2}$

27. $3.4 \times 10^{-4} \dfrac{(\text{g. } H_3O^{+1})(\text{g. } Ac^{-1})}{(\text{g. HAc})(\text{l.})}$

28. $C_{NH_3}/C_{NH_4^+} = 0.55$.

29. $C_{H^+} = 5.3 \times 10^{-10}$ M.

30. $C_{HSO_4}-/C_{SO_4}- = 1.7$.

31. $C_{HCl}/C_{HCN} = 0.13$.

32. $C_{H^+}/C_{HNO_2} = 2.5 \times 10^{+3}$. *Note:* This is so small that (H$^+$) must be buffered by second buffer (e.g., use HSO$_4^-$ and SO$_4^=$).

33. See text.

34. See text.

35. $K_{assoc.} = 1.6 \times 10^7$; moderate base.

36. $K_{ion} = 1 \times 10^{-14}$; very weak acid.

37. $K_{ion} = 1 \times 10^2$; very strong acid, very weak base.

38. $K_{eq.} = 8.5$; no.

39. $K_{eq.} = 10^3$; almost.

40. $K_{eq.} = 10^{-2}$; no.

Chapter XIII

1. (a) $\dfrac{(\text{Moles } H_3O^{+1})(\text{Moles } OH^{-1})}{(\text{Liters})^2}$.

   (b) $3.2 \times 10^{-12} \dfrac{(\text{Grams } H_3O^{+1})\ \text{Grams } OH^{-1})}{(\text{Liter})^2}$.

2. (a) $Cl^{-1} = H_3O^{+1} = 0.005\ M;\ 2 \times 10^{-12}\ M\ OH^{-1}$.

   (b) $Na^{+1} = Cl^{-1} = 0.34\ M;\ H_3O^{+1} = OH^{-1} = 1 \times 10^{-7}\ M$.

   (c) $1.20\ M\ K^{+1};\ 0.60\ M\ SO_4^{-2};\ H_3O^{+1} = OH^{-1} = 1 \times 10^{-7}\ M$.

   (d) $Na^{+1} = OH^{-1} = 0.75\ M;\ H_3O^{+1} = 1.3 \times 10^{-14}\ M$.

   (e) $K^{+1} = OH^{-1} = 3 \times 10^{-5}\ M;\ H_3O^{+1} = 3.3 \times 10^{-10}\ M$.

   (f) $Zn^{+2} = 2 \times 10^{-4};\ Cl^{-1} = 4 \times 10^{-4};\ H_3O^{+1} = OH^{-1} = 1 \times 10^{-7}$.

3. (a) $pH = 3.0;\ pOH = 11.0$.   (e) $pH = 0.92;\ pOH = 13.08$.

   (b) $pH = 2.70;\ pOH = 12.30$.   (f) $pH = 13.72;\ pOH = 0.28$.

   (c) $pH = 10.0;\ pOH = 4.0$.   (g) $pH = 12.83;\ pOH = 1.17$.

   (d) $pH = 3.47;\ pOH = 10.53$.   (h) $pH = 3.14;\ pOH = 10.86$.

4. (a) 5.60.   (d) 2×10^{-10}.

   (b) 11.52.   (e) 5×10^{-7}.

   (c) 15.10.   (f) 2.3×10^{-9}.

5. $pH = 6.92$.   6. See text.

7. $Na^{+1} = 0.4\ M;\ ClO^{-1} \cong 0.4\ M;\ HClO = OH^{-1} = 8.8 \times 10^{-5}\ M;$
   $H_3O^{+1} = 1.14 \times 10^{-10}\ M;$

8. $K^{+1} = 0.07\ M;\ CN^{-1} = 0.069\ M;\ HCN = OH^{-1} = 0.0014\ M;$
   $H_3O^{+1} = 7 \times 10^{-12}\ M$.

9. $Na^{+1} = 0.25\ M;\ HCO^{-3} \cong 0.25\ M;\ H_2CO_3 = OH^{-1} = 7.5 \times 10^{-5}\ M;$
   $H_3O^{+1} = 1.3 \times 10^{-10}\ M$.

10.

	$K_{hyd.}$	pH	% Hydrolysis
(a)	5.5×10^{-10}	8.72	0.01%
(b)	5.5×10^{-10}	5.68	0.026%
(c)	1.6×10^{-7}	10.37	0.072%
(d)	10.0	13.68	96%
(e)	2.5×10^{-5}	11.60	0.62%
(f)	5×10^{-11}	5.76	0.0029%
(g)	3.1×10^{-5}	7.0	0.56%
(h)	1.4	9.34	54%

11. (a) $pH \cong 4.6$.   (c) $pH \cong 10.6$.

    (b) $pH \cong 11.7$.   (d) $pH \cong 14.5$.

12. $C_{Na^+} = 0.10\ M;\ C_{HAc} = C_{OH^-} = 7.5 \times 10^{-4}\ M;\ C_{H^+} = 1.3 \times 10^{-11}\ M$.

13. (a) $C_{Na^+} \approx C_{F^-} = 0.17\ M;\ C_{HF} = 0.10\ M$.   (b) $pH = 3.4$

14. $C_{Cl^-} = 0.27\ M;\ C_{Na^+} = 0.12\ M;\ C_{H^+} = 0.15\ M;\ C_{OH^-} = 6.7 \times 10^{-14}\ M$.
    $pH = 0.82$.

15. $pH = 10.5$.

16. 3.1% hydrolysis of $Fe^{+++};\ pH = 2.2$.

17. 1.2% hydrolysis of $Al^{+++};\ pH = 3.3$.

18. $pH(\text{base}) = 11.5;\ pH(\text{final}) = 2.7;\ pH(\text{acid}) = 0.30$.

19. (a) 11 to 13, etc. See text.

    (e) Not useful because of extensive hydrolysis of $S^=$.

20. Hydrolysis significant in b, c, and f.

21. $K_{eq.} = 1.0 \times 10^{-12}$.

22. $3.7 \times 10^{-6}\ M$, OH^- from HOH not important.

23. (a) $BaCO_3 + HAc \rightleftarrows Ba^{++} + HCO_3^- + Ac^-$.

    (b) $K_{eq.} = 1.9 \times 10^{-3}$.

    (c) Yes, since $K_{eq.}$ is not very small.

24. (a) NO_2^-/HNO_2, etc. See text.   **25.** 96% hydrolyzed; $pH = 5.2$.

26. 0.56% hydrolyzed; $pH = 7.0$.   **27.** 100% hydrolyzed; $pH = 9.7$.

28. 0.09% hydrolyzed; $pH = 6.2$.

29. At half-way point $C_{ion} = C_{acid}$ ∴ $C_{H^+} = K_{ion}$.

30. $C_{OH^-} = 10^{-3.8} \ M$; $C_{CN^-} \cong 0.5 \ M$; $C_{HCN} = 10^{-3.1} \ M$ before addition. After addition $C_{HCN} \approx 10^{-3.2} \ M$, so that C_{OH^-} changes to $10^{-3.9} \ M$, and pH becomes 11.1.

31. Their concentrations are very small.

32. (a) NH_2^- (b) $3 \times 10^{-15} \ M$.

Chapter XIV

2. $C_{H_3O^{+1}} = C_{HB_4O_7^{-1}} = 2.2 \times 10^{-3} \ M$; $C_{H_2B_4O_7} = 0.048 \ M$;
$C_{OH^{-1}} = 4.5 \times 10^{-12} \ M$; $C_{B_4O_7^{-2}} = 1 \times 10^{-9} \ M$.

3. $C_{H_3O^{+1}} = C_{HSe^{-1}} = 0.012 \ M$; $C_{H_2Se} = 0.78 \ M$;
$C_{OH^{-1}} = 8.5 \times 10^{-13} \ M$; $C_{Se^{-2}} = 1 \times 10^{-10} \ M$.

4. 2.1%; $pH = 8.35$.   **5.** 0.019%; $pH = 10.98$.

6. $C_{PO_4^{-3}} = C_{H_2PO_4^{-1}} = 2.4 \times 10^{-4}$; $pH = 9.60$.

7. $C_{Hg^{+2}} = 1.4 \times 10^{-9} \ M$.

8. $C_{Ag^{+1}} = 5.3 \times 10^{-8} \ M$; $C_{CN^{-1}} = 1.06 \times 10^{-7} \ M$.

9. $C_{NH_3} = 0.42 \ M$.   **10.** $C_{Hg}^{+2} = 3 \times 10^{-15} \ M$.

11. $C_{S^{-2}} = 1.9 \times 10^{-15} \ M$; $C_{H_3O^{+1}} = 7 \times 10^{-5}$; Cd^{+2} left $= 1.5 \times 10^{-10}\%$.

13. Yes; $C_{S^{-2}} = 2.8 \times 10^{-14} \ M$; $C_{H_3O^{+1}} = 2.0 \times 10^{-5} \ M$.

14. $C_{CO_3^{-2}} = 1.2 \times 10^{-7} \ M$; Cd^{+2} left $= 0.007\%$; $C_{H_3O^{+1}} = 3.9 \times 10^{-5} \ M$.

15. $C_{CrO_4^{-2}} = 4.5 \times 10^{-4} \ M$; $pH = 3.85$.

16. Yes; $C_{Ag^{+1}} = 3.3 \times 10^{-12} \ M$; $C_{S\,O_3^{-2}} = 0.055 \ M$.

17. $C_{OH^{-1}} = 1.9 \times 10^{-10} \ M$; ratio of $Ac^{-1}/HAc = 0.34$.

Chapter XV

1. (a) $Cr \ (+3)$; $O \ (-2)$.

    (b) $H \ (+1)$; $N \ (+3)$; $O \ (-2)$.

    (c) $Sb \ (-3)$; $H \ (+1)$.

    (d) $Na \ (+1)$; $S \ (+6)$; $O \ (-2)$.

    (e) $K \ (+1)$; $O \ (-1)$.

    (f) $Mn \ (+4)$; $O \ (-2)$.

    (g) $H \ (+1)$; $P \ (+5)$; $O \ (-2)$.

    (h) $H \ (+1)$; $P \ (+3)$; $O \ (-2)$.

    (i) $H \ (+1)$; $Cl \ (+5)$; $O \ (-2)$.

    (j) $K \ (+1)$; $Mn \ (+6)$; $O \ (-2)$.

    (k) $Na \ (+1)$; $B \ (+3)$; $O \ (-2)$.

    (l) $H \ (+1)$; $S \ (+6)$; $O \ (-2)$.

2. See text.   **3.** See text.

4. (a) 1.6 moles $KMnO_4$.

    (b) 0.3 eq. HCl.

    (c) 15.0 g. Zn.

    (d) 0.270 eq. $KMnO_4$.

    (e) 0.38 meq. $KMnO_4$.

    (f) 0.408 eq. $K_2Cr_2O_7$.

    (g) 176 mg. HNO_3.

5. 15.5 g. I_2.   **6.** 60.8 cc. STP O_2 gas.

7. 5.49 l. STP of Cl_2 gas.   **8.** 478 mg. CuS.

9. 1.21 M.   **10.** 1.35 M $FeCl^2$.

11. 600 ml.   **12.** 2030 ml. $Na_2S_2O_3$ solution.

Chapter XVI

1. (a) 1.56 v.   (e) 1.09 v.
   (b) 3.70 v.   (f) 0.16 v.
   (c) 0.288 v.   (g) −0.15 v. (does not go).
   (d) 0.44 v.   (h) −0.27 v. (goes in reverse).

2. (a) $K_{eq.} = 10^{52}$.   (e) $K_{eq.} = 10^{109} = \dfrac{C^3_{Pb^{+2}} \times C^2_{NO}}{C^8_{H^{+1}} \times C^2_{NO_3^{-1}}}$.

   (b) $K_{eq.} = 10^{123}$.   (f) $K_{eq.} = 5 \times 10^{26}$.

   (c) $K_{eq.} = 4 \times 10^9 = \dfrac{C^2_{Cl^{-1}} \times C_{Br_2}}{C^2_{Br^{-1}} \times C_{Cl_2}}$.   (g) $K_{eq.} = 1 \times 10^{-5}$.

   (d) $K_{eq.} = 4.7 \times 10^{14}$.   (h) $K_{eq.} = 3.2 \times 10^{-5}$.

3. (a) $E = +0.94$ v.   (c) $E = +0.075$ v.
   (b) $E = -0.61$ v.   (d) $E = -1.791$ v.
4. $E = +1.11$ v.   5. $E = -0.317$ v.
6. $C_{Ag^{+1}} = 8.5 \times 10^{-15} M$. $E = +0.044$ v.
7. $C_{S^{-2}} = 1.1 \times 10^{-19} M$; $C_{Cu^{+2}} = 4 \times 10^{-17} M$; $E = +0.14$ v.
8. Strongest oxidizing agent is Cl_2 gas.   Strongest reducing agent is Na metal.
   Largest voltage is $Na^{+1}|Na\|Cl^{-1}|Cl_2 = 4.07$ v.
   Smallest voltage is $Fe^{+2}|Fe^{+3}\|Ag^{+1}|Ag = 0.029$ v.
9. See text.
10. (a) $C_{Ag^+} = 3 \times 10^{-13} M$; (b) $K_{SP} = 3 \times 10^{-13}$.
11. (a) $C_{Pb^{++}} = 2.2 = 10^{-7} M$; (b) $K_{SP} = 2.2 \times 10^{-8}$.
12. Cl^- complexes with Cu^{++}.   13. Fe^{+++} has greater affinity.

Chapter XVII

1. (a)   6 cal./mole-°K.   (d) −23 cal./mole-°K.
   (b) −11 cal./mole-°K.   (e)   3 cal./mole°-K.
   (c)   35 cal./mole-°K.

2. (a) $\Delta H = -44$ kcal./mole; $\Delta G = -46$ kcal./mole.
   (b) $\Delta H = -57.8$ kcal./mole; $\Delta G = -54.5$ kcal./mole.
   (c) $\Delta H = 27.4$ kcal./mole; $\Delta G = 16.9$ kcal./mole.
   (d) $\Delta H = -23.7$ kcal./mole; $\Delta G = -16.8$ kcal./mole.
   (e) $\Delta H = -42.5$ kcal./mole; $\Delta G = -43.4$ kcal./mole.
3. $\Delta G(400°C.) = -5.6$ kcal./mole; $\Delta H(400°C.) = -2.9$ kcal./mole.
4. $\Delta G = -62.5$ kcal./mole; $\Delta S = -71$ cal./mole-°K.; $\Delta H = -41.2$ kcal./mole.
5. $\Delta S = 4$ cal./mole-°K.; $\Delta H = -15.6$ kcal./mole; $\Delta G = -16.8$ kcal./mole;
   $\Delta G_{(1000°K.)} = -19.6$ kcal./mole; $K_{eq.(1000)} = 4 \times 10^4$.

Chapter XVIII

1. 5×10^{-4} mole HBr/l.-sec.   2. 2.04×10^{-2} mole Sn^{+4}/sec.
3. See text.   4. 4.08×10^{-3} mole NOCl/l.-sec.
5. 1.29×10^{-3}.
6. (a) $k_{150} = 1.72 \times 10^{-5}$. (b) $k_{30} = 2.19 \times 10^3$. (c) $k_{600} = 3.5 \times 10^{-3}$.
7. $H_R = +8.64$ kcal.; endothermic.
8. 1.35×10^{-11} sec.$^{-1}$; 3.6×10^{10} dis./sec.-g. Ra.
9. 1×10^{14} dis.   10. 1.6×10^3 sec.; 7.4×10^{16} dis.
11. 4.1 hr.$^{-1}$; 0.012.   12. 8×10^6 dis.
13. $k_r C^2_{NOBr}$ 14. $k_r C_{N_2} \times C^2_{H_2O}/C^{1/2}_{H_2}$
15. (a) $k_r C_{C_2H_4} \times C_{HBr}$ (b) 36 kcal.
16. (a) 2 kcal.   (b) $k_r C_{HBr} \times C_H$
17. 26.4 kcal.

APPENDIX VI

Logarithms

APPENDIX VI

LOGARITHMS

Natural Numbers.	0	1	2	3	4	5	6	7	8	9	Proportional Parts.								
											1	2	3	4	5	6	7	8	9
10	0000	0043	0086	0128	0170	0212	0253	0294	0334	0374	4	8	12	17	21	25	29	33	37
11	0414	0453	0492	0531	0569	0607	0645	0682	0719	0755	4	8	11	15	19	23	26	30	34
12	0792	0828	0864	0899	0934	0969	1004	1038	1072	1106	3	7	10	14	17	21	24	28	31
13	1139	1173	1206	1239	1271	1303	1335	1367	1399	1430	3	6	10	13	16	19	23	26	29
14	1461	1492	1523	1553	1584	1614	1644	1673	1703	1732	3	6	9	12	15	18	21	24	27
15	1761	1790	1818	1847	1875	1903	1931	1959	1987	2014	3	6	8	11	14	17	20	22	25
16	2041	2068	2095	2122	2148	2175	2201	2227	2253	2279	3	5	8	11	13	16	18	21	24
17	2304	2330	2355	2380	2405	2430	2455	2480	2504	2529	2	5	7	10	12	15	17	20	22
18	2553	2577	2601	2625	2648	2672	2695	2718	2742	2765	2	5	7	9	12	14	16	19	21
19	2788	2810	2833	2856	2878	2900	2923	2945	2967	2989	2	4	7	9	11	13	16	18	20
20	3010	3032	3054	3075	3096	3118	3139	3160	3181	3201	2	4	6	8	11	13	15	17	19
21	3222	3243	3263	3284	3304	3324	3345	3365	3385	3404	2	4	6	8	10	12	14	16	18
22	3424	3444	3464	3483	3502	3522	3541	3560	3579	3598	2	4	6	8	10	12	14	15	17
23	3617	3636	3655	3674	3692	3711	3729	3747	3766	3784	2	4	6	7	9	11	13	15	17
24	3802	3820	3838	3856	3874	3892	3909	3927	3945	3962	2	4	5	7	9	11	12	14	16
25	3979	3997	4014	4031	4048	4065	4082	4099	4116	4133	2	3	5	7	9	10	12	14	15
26	4150	4166	4183	4200	4216	4232	4249	4265	4281	4298	2	3	5	7	8	10	11	13	15
27	4314	4330	4346	4362	4378	4393	4409	4425	4440	4456	2	3	5	6	8	9	11	13	14
28	4472	4487	4502	4518	4533	4548	4564	4579	4594	4609	2	3	5	6	8	9	11	12	14
29	4624	4639	4654	4669	4683	4698	4713	4728	4742	4757	1	3	4	6	7	9	10	12	13
30	4771	4786	4800	4814	4829	4843	4857	4871	4886	4900	1	3	4	6	7	9	10	11	13
31	4914	4928	4942	4955	4969	4983	4997	5011	5024	5038	1	3	4	6	7	8	10	11	12
32	5051	5065	5079	5092	5105	5119	5132	5145	5159	5172	1	3	4	5	7	8	9	11	12
33	5185	5198	5211	5224	5237	5250	5263	5276	5289	5302	1	3	4	5	6	8	9	10	12
34	5315	5328	5340	5353	5366	5378	5391	5403	5416	5428	1	3	4	5	6	8	9	10	11
35	5441	5453	5465	5478	5490	5502	5514	5527	5539	5551	1	2	4	5	6	7	9	10	11
36	5563	5575	5587	5599	5611	5623	5635	5647	5658	5670	1	2	4	5	6	7	8	10	11
37	5682	5694	5705	5717	5729	5740	5752	5763	5775	5786	1	2	3	5	6	7	8	9	10
38	5798	5809	5821	5832	5843	5855	5866	5877	5888	5899	1	2	3	5	6	7	8	9	10
39	5911	5922	5933	5944	5955	5966	5977	5988	5999	6010	1	2	3	4	5	7	8	9	10
40	6021	6031	6042	6053	6064	6075	6085	6096	6107	6117	1	2	3	4	5	6	8	9	10
41	6128	6138	6149	6160	6170	6180	6191	6201	6212	6222	1	2	3	4	5	6	7	8	9
42	6232	6243	6253	6263	6274	6284	6294	6304	6314	6325	1	2	3	4	5	6	7	8	9
43	6335	6345	6355	6365	6375	6385	6395	6405	6415	6425	1	2	3	4	5	6	7	8	9
44	6435	6444	6454	6464	6474	6484	6493	6503	6513	6522	1	2	3	4	5	6	7	8	9
45	6532	6542	6551	6561	6571	6580	6590	6599	6609	6618	1	2	3	4	5	6	7	8	9
46	6628	6637	6646	6656	6665	6675	6684	6693	6702	6712	1	2	3	4	5	6	7	7	8
47	6721	6730	6739	6749	6758	6767	6776	6785	6794	6803	1	2	3	4	5	5	6	7	8
48	6812	6821	6830	6839	6848	6857	6866	6875	6884	6893	1	2	3	4	4	5	6	7	8
49	6902	6911	6920	6928	6937	6946	6955	6964	6972	6981	1	2	2	4	4	5	6	7	8
50	6990	6998	7007	7016	7024	7033	7042	7050	7059	7067	1	2	3	3	4	5	6	7	8
51	7076	7084	7093	7101	7110	7118	7126	7135	7143	7152	1	2	3	3	4	5	6	7	8
52	7160	7168	7177	7185	7193	7202	7210	7218	7226	7235	1	2	2	3	4	5	6	7	7
53	7243	7251	7259	7267	7275	7284	7292	7300	7308	7316	1	2	2	3	4	5	6	6	7
54	7324	7332	7340	7348	7356	7364	7372	7380	7388	7396	1	2	2	3	4	5	6	6	7

LOGARITHMS

Natural Numbers.	0	1	2	3	4	5	6	7	8	9	Proportional Parts.								
											1	2	3	4	5	6	7	8	9
55	7404	7412	7419	7427	7435	7443	7451	7459	7466	7474	1	2	2	3	4	5	5	6	7
56	7482	7490	7497	7505	7513	7520	7528	7536	7543	7551	1	2	2	3	4	5	5	6	7
57	7559	7566	7574	7582	7589	7597	7604	7612	7619	7627	1	2	2	3	4	5	5	6	7
58	7634	7642	7649	7657	7664	7672	7679	7686	7694	7701	1	1	2	3	4	4	5	6	7
59	7709	7716	7723	7731	7738	7745	7752	7760	7767	7774	1	1	2	3	4	4	5	6	7
60	7782	7789	7796	7803	7810	7818	7825	7832	7839	7846	1	1	2	3	4	4	5	6	6
61	7853	7860	7868	7875	7882	7889	7896	7903	7910	7917	1	1	2	3	4	4	5	6	6
62	7924	7931	7938	7945	7952	7959	7966	7973	7980	7987	1	1	2	3	3	4	5	6	6
63	7993	8000	8007	8014	8021	8028	8035	8041	8048	8055	1	1	2	3	3	4	5	5	6
64	8062	8069	8075	8082	8089	8096	8102	8109	8116	8122	1	1	2	3	3	4	5	5	6
65	8129	8136	8142	8149	8156	8162	8169	8176	8182	8189	1	1	2	3	3	4	5	5	6
66	8195	8202	8209	8215	8222	8228	8235	8241	8248	8254	1	1	2	3	3	4	5	5	6
67	8261	8267	8274	8280	8287	8293	8299	8306	8312	8319	1	1	2	3	3	4	5	5	6
68	8325	8331	8338	8344	8351	8357	8363	8370	8376	8382	1	1	2	3	3	4	4	5	6
69	8388	8395	8401	8407	8414	8420	8426	8432	8439	8445	1	1	2	2	3	4	4	5	6
70	8451	8457	8463	8470	8476	8482	8488	8494	8500	8506	1	1	2	2	3	4	4	5	6
71	8513	8519	8525	8531	8537	8543	8549	8555	8561	8567	1	1	2	2	3	4	4	5	5
72	8573	8579	8585	8591	8597	8603	8609	8615	8621	8627	1	1	2	2	3	4	4	5	5
73	8633	8639	8645	8651	8657	8663	8669	8675	8681	8686	1	1	2	2	3	4	4	5	5
74	8692	8698	8704	8710	8716	8722	8727	8733	8739	8745	1	1	2	2	3	4	4	5	5
75	8751	8756	8762	8768	8774	8779	8785	8791	8797	8802	1	1	2	2	3	3	4	5	5
76	8808	8814	8820	8825	8831	8837	8842	8848	8854	8859	1	1	2	2	3	3	4	5	5
77	8865	8871	8876	8882	8887	8893	8899	8904	8910	8915	1	1	2	2	3	3	4	4	5
78	8921	8927	8932	8938	8943	8949	8954	8960	8965	8971	1	1	2	2	3	3	4	4	5
79	8976	8982	8987	8993	8998	9004	9009	9015	9020	9025	1	1	2	2	3	3	4	4	5
80	9031	9036	9042	9047	9053	9058	9063	9069	9074	9079	1	1	2	2	3	3	4	4	5
81	9085	9090	9096	9101	9106	9112	9117	9122	9128	9133	1	1	2	2	3	3	4	4	5
82	9138	9143	9149	9154	9159	9165	9170	9175	9180	9186	1	1	2	2	3	3	4	4	5
83	9191	9196	9201	9206	9212	9217	9222	9227	9232	9238	1	1	2	2	3	3	4	4	5
84	9243	9248	9253	9258	9263	9269	9274	9279	9284	9289	1	1	2	2	3	3	4	4	5
85	9294	9299	9304	9309	9315	9320	9325	9330	9335	9340	1	1	2	2	3	3	4	4	5
86	9345	9350	9355	9360	9365	9370	9375	9380	9385	9390	1	1	2	2	3	3	4	4	5
87	9395	9400	9405	9410	9415	9420	9425	9430	9435	9440	0	1	1	2	2	3	3	4	4
88	9445	9450	9455	9460	9465	9469	9474	9479	9484	9489	0	1	1	2	2	3	3	4	4
89	9494	9499	9504	9509	9513	9518	9523	9528	9533	9538	0	1	1	2	2	3	3	4	4
90	9542	9547	9552	9557	9562	9566	9571	9576	9581	9586	0	1	1	2	2	3	3	4	4
91	9590	9595	9600	9605	9609	9614	9619	9624	9628	9633	0	1	1	2	2	3	3	4	4
92	9638	9643	9647	9652	9657	9661	9666	9671	9675	9680	0	1	1	2	2	3	3	4	4
93	9685	9689	9694	9699	9703	9708	9713	9717	9722	9727	0	1	1	2	2	3	3	4	4
94	9731	9736	9741	9745	9750	9754	9759	9763	9768	9773	0	1	1	2	2	3	3	4	4
95	9777	9782	9786	9791	9795	9800	9805	9809	9814	9818	0	1	1	2	2	3	3	4	4
96	9823	9827	9832	9836	9841	9845	9850	9854	9859	9863	0	1	1	2	2	3	3	4	4
97	9868	9872	9877	9881	9886	9890	9894	9899	9903	9908	0	1	1	2	2	3	3	4	4
98	9912	9917	9921	9926	9930	9934	9939	9943	9948	9952	0	1	1	2	2	3	3	4	4
99	9956	9961	9965	9969	9974	9978	9983	9987	9991	9996	0	1	1	2	2	3	3	3	4

INDEX